DISEASES OF SHEEP

DISEASES OF SHEEP

Edited by
W B MARTIN PhD, MRCVS, DVSM, FRSE
Director
Animal Diseases Research Association
The Moredun Institute
Edinburgh

BLACKWELL SCIENTIFIC PUBLICATIONS
OXFORD LONDON EDINBURGH BOSTON MELBOURNE

© 1983 by
Blackwell Scientific Publications
Editorial offices:
Osney Mead, Oxford, OX2 0EL
8 John Street, London, WC1N 2ES
9 Forrest Road, Edinburgh, EH1 2QH
52 Beacon Street, Boston, Massachusetts 02108, USA
99 Barry Street, Carlton, Victoria 3053, Australia

First published 1983

Typeset, printed and bound by
Richard Clay
The Chaucer Press
Bungay
Suffolk

Distributors
USA
   Blackwell Mosby Book Distributors
   11830 Westline Industrial Drive
   St. Louis, Missouri 63141

Canada
   Blackwell Mosby Book Distributors
   120 Melford Drive, Scarborough
   Ontario, M1B 2X4

Australia
   Blackwell Scientific Book Distributors
   31 Advantage Road, Highett
   Victoria 3190

British Library
Cataloguing in Publication Data

Diseases of Sheep
   1. Sheep—Diseases
   I. Martin, W. B.
   636.3'08'96   SF968

   ISBN 0–632–01008–8

CONTENTS

Contributors   vii
Foreword   ix
Preface   xi

SECTION 1   RESPIRATORY SYSTEM

1   Pasteurellosis   3
    N.J.L. GILMOUR AND K.W. ANGUS
2   Acute Respiratory Virus Infections   8
    J.M. SHARP
3   Chronic Respiratory Virus Infections   12
    J.M. SHARP AND W.B. MARTIN
4   Atypical Pneumonia   17
    G.E. JONES AND J.S. GILMOUR
5   Parasitic Bronchitis   23
    J. ARMOUR

SECTION 2   ALIMENTARY SYSTEM

6   Diseases of the Mouth and Teeth   29
    J. SPENCE
7   Clostridial Diseases   35
    D. BUXTON
8   Enteritis in Young Lambs   43
    D.R. SNODGRASS AND K.W. ANGUS
9   Coccidiosis in Lambs   49
    L.P. JOYNER
10 Johne's Disease   52
    N.J.L. GILMOUR AND K.W. ANGUS
11 Parasitic Gastroenteritis   56
    R.L. COOP AND M.G. CHRISTIE
12 Liver Fluke   62
    A. WHITELAW

SECTION 3   NERVOUS SYSTEM AND LOCOMOTION

13 Scrapie   71
    B. MITCHELL AND J.T. STAMP
14 Louping-ill   76
    H.W. REID
15 Listeriosis   79
    R.M. BARLOW
16 Swayback   82
    R.M. BARLOW
17 Polioencephalomalacia   85
    R.M. BARLOW
18 'Daft' Lamb Disease (DLD)   87
    R.M. BARLOW
19 Non-specific Bacterial Infections of
    the Central Nervous System (CNS)   89
    R.M. BARLOW
20 Tetanus   91
    D. BUXTON
21 Coenurosis   93
    B.M. WILLIAMS AND T. BOUNDY
22 Foot rot and Foot Conditions   98
    T. BOUNDY
23 Arthritis   104
    M.H. LAMONT
24 Osteodystrophic Diseases   111
    J.A. SPENCE

SECTION 4   REPRODUCTIVE SYSTEM

25 Enzootic (Chlamydial) Abortion   119
    I.D. AITKEN
26 Toxoplasmosis   124
    D. BUXTON
27 Border Disease   129
    R.M. BARLOW AND A.C. GARDINER
28 Vibriosis   133
    N.J.L. GILMOUR
29 Salmonellosis and Salmonella
    Abortion   135
    K.A. LINKLATER
30 Infertility   139
    A. GREIG AND D.W. DEAS

31 Ulcerative Balanitis and Ulcerative
    Vulvitis   144
    D.W. DEAS
32 Pregnancy Toxaemia   147
    E.J.H. FORD

33 Ventral Hernia and Vaginal Prolapse 151
    B. MITCHELL
34 Mastitis   153
    A.J. MADEL

SECTION 5   DISORDERS OF MINERAL METABOLISM

35 Deficiencies of Macro-elements in
    Mineral Metabolism   161
    A.J.F. RUSSEL

36 Disorders Related to Trace Element
    Deficiencies   168
    N.F. SUTTLE AND K.A. LINKLATER

SECTION 6   DISORDERS OF SKIN AND WOOL

37 Psoroptic Mange   181
    D.W. TARRY
38 Contagious Pustular Dermatitis   185
    J.A.A. WATT
39 Headfly and Blowfly Myiasis   189
    W.T. APPLEYARD

40 Pyodermas   193
    F.M.M. SCOTT, J. FRASER AND G.R.
    SCOTT
41 Photosensitization   197
    E.J.H. FORD
42 Dermatitis and Pruritic Conditions 200
    A.O. MATHIESON

SECTION 7   MISCELLANEOUS DISEASES

43 Tick Associated Infections   209
    G.R. SCOTT
44 Ovine Keratoconjunctivitis   214
    G.E. JONES

45 Tumours of Sheep   217
    K.W. ANGUS AND K.W. HEAD
46 International Importance of some
    Exotic Viral Infections of Sheep   220
    R.F. SELLERS

SECTION 8   POISONS AND POISONING

47 Plant and Inorganic Poisoning   231
    G.A.M. SHARMAN

48 Poisoning by Phenolic Compounds   239
    K.W. ANGUS

SECTION 9   FLOCK HEALTH

49 Flock Management for Health   245
    A.O. MATHIESON
50 Control of Gastrointestinal
    Helminthiasis   250
    J. ARMOUR
51 Hypothermia in Newborn Lambs   255
    F.A. EALES

52 Synchronization of Oestrus   261
    B. MITCHELL
53 Approach to the Investigation of a
    Disease Outbreak   263
    J.A.A. WATT
54 Samples and Sampling for the
    Diagnosis of Virus Diseases   266
    P.F. NETTLETON

Appendix   269
Index   271

CONTRIBUTORS

Animal Diseases Research Association,
Moredun Research Institute,
408 Gilmerton Road,
Edinburgh, Britain

I. D. AITKEN, PhD, BVMS, MRCVS,
Head of Department of Microbiology.
K. W. ANGUS, BVMS, FRCVS,
Department of Pathology.
W. T. APPLEYARD, BVM&S, MRCVS,
Department of Clinical Studies.
R. M. BARLOW, DSc, DVM&S, MRCVS,
Head of Department of Pathology.
D. BUXTON, PhD, BVM&S, MRCVS,
Department of Pathology.
M. G. CHRISTIE, B.A,
Head of Department of Parasitology.
R. L. COOP, PhD, BSc,
Department of Parasitology.
F. A. EALES, BSc, BVSc, MRCVS,
Department of Physiology.
J. FRASER, HNC,
Department of Microbiology.
A. C. GARDINER, BSc,
Department of Pathology.
J. S. GILMOUR, BVM&S, FRCVS,
Department of Pathology.

N. J. L. GILMOUR, DSc, PhD, BVM&S,
MRCVS,
Department of Microbiology.
G. E. JONES, PhD, BVSc, MRCVS, DTVM,
Department of Microbiology.
W. B. MARTIN, PhD, MRCVS, DVSM, FRSE,
Director.
B. MITCHELL, BVM&S, FRCVS,
Head of Department of Clinical Studies.
P. F. NETTLETON, MSc, BVMS, MRCVS,
Department of Microbiology.
H. W. REID, PhD, BVM&S, MRCVS, DTVM,
Department of Microbiology.
F. M. M. SCOTT, FIMLS,
Department of Microbiology.
J. M. SHARP, PhD, BVMS, MRCVS,
Department of Microbiology.
D. R. SNODGRASS, PhD, BVM&S, MRCVS,
Department of Microbiology.
J. SPENCE, BVM&S, MRCVS, DTVM,
Department of Pathology.
N. F. SUTTLE, PhD, BSc,
Department of Biochemistry.

Other Contributors

J. ARMOUR, PhD, MRCVS.
Titular Professor, Department of Veterinary
Parasitology, Bearsden Road, Glasgow.
T. BOUNDY, BVSc, MRCVS, FRAgS,
Kilaganoon, Montgomery, Powys, SY15 6HW.
D. W. DEAS, MRCVS,
Veterinary Investigation Officer, East of Scotland
College of Agriculture, Veterinary Investigation
Centre, Bush Estate, Penicuik, Midlothian,
EH26 0QE.
E. J. H. FORD, DVSc, MRCPath, FRCVS.
Professor of Veterinary Clinical Studies,

University of Liverpool, Veterinary Field Station,
Leahurst, Neston, South Wirral, Merseyside.
A. GREIG, BVM&S, FRCVS,
Assistant Veterinary Investigation Officer, East of
Scotland College of Agriculture, Veterinary
Investigation Centre, Greycrook, St. Boswells,
Roxburghshire.
K. W. HEAD, BSc, MRCVS,
Department of Veterinary Pathology, Royal
(Dick) School of Veterinary Studies, Summer-
hall, Edinburgh, EH9 1QH.
L. P. JOYNER, PhD, DSc, FI.Biol,

Department of Parasitology, Central Veterinary
Laboratory, New Haw, Weybridge, Surrey.

M. H. LAMONT, BVMS&S, PhD, MRCVS,
Ministry of Agriculture, Fisheries & Food,
Veterinary Investigation Centre, Madingley
Road, Cambridge.

K. A. LINKLATER, BVM&S, PhD, MRCVS,
Veterinary Investigation Officer, East of Scotland
College of Agriculture, Veterinary Investigation
Centre, Greycrook, St. Boswells, Roxburghshire.

A. J. MADEL, B.A, Vet.MB, Dip.AH, MRCVS,
Department of Animal Husbandry, Royal
Veterinary College, Boltons Park, Hawkshead
Road, Potters Bar, Herts.

A. O. MATHIESON, MSc, BVM&S, MRCVS,
Regional Veterinary Investigation Officer,
The East of Scotland College of Agriculture,
Bush Estate, Penicuik, Midlothian,
EH26 0QE.

A. J. F. RUSSEL, BSc, M.Agr.Sc, PhD,
Animal Production and Nutrition Department,
Hill Farming Research Organisation, Bush
Estate, Penicuik, Midlothian, EH26 0PH.

G. R. SCOTT, BSc, MS, PhD, MRCVS,

Centre for Tropical Veterinary Medicine, Easter
Bush, Roslin, Midlothian.

R. F. SELLERS, MA, ScD, PhD, BSc, MRCVS,
Director, Animal Virus Research Institute,
Pirbright, Woking, Surrey, GU24 0NF.

G. A. M. SHARMAN, BSc, MRCVS,
Veterinary Pathology Department, The Rowett
Research Institute, Greenburn Road, Bucksburn,
Aberdeen, AB2 9SB.

J. T. STAMP, CBE, DSc, FRSE, FRCVS,
Vale Bank, 48 Forth Street, North Berwick.

D. W. TARRY, PhD, BSc, MIBiol,
Department of Parasitology, Central Veterinary
Laboratory, New Haw, Weybridge, Surrey.

J. A. A. WATT, OBE, PhD, BSc, FRCVS,
Barrogill, Woodside Avenue, Grantown-on-Spey,
Morayshire.

A. WHITELAW, BSc, FRCVS,
Hill Farming Research Organisation, Bush
Estate, Penicuik, Midlothian, EH26 0PH.

B. M. WILLIAMS, DVSM, MRCVS,
Regional Veterinary Officer, Ministry of
Agriculture, Fisheries & Food, Hook Rise South,
Tolworth, Surbiton, Surrey, KT6 7NF.

FOREWORD

'A disease known is half cured'. Thomas Fuller, MD 'Gnomologia' 1732.

Two hundred and fifty years ago the significance of effective diagnostic skill was appreciated. The inability of animal patients to communicate renders acute observation particularly important to the veterinarian and few animals are more demanding of effective diagnosis than the sheep.

The Animal Diseases Research Association was founded in 1920 by a number of Scottish farmers who were concerned about the problems of disease in their livestock. Initially they funded two research workers and in 1925 they established the Moredun Institute in Gilmerton Road on which site the Institute as we know it today has been developed. Over the past sixty-two years great progress has been made by the research workers employed by the Association in the understanding of diseases. In this volume they have sought to bring much of that knowledge together for the first time in the hope that it will assist in accurate diagnosis by student, practitioner and researcher alike. Few establishments have more experience of ovine disease than the Moredun Institute, Edinburgh and the Director there, Dr W.B. Martin, is to be congratulated on the initiative which brought this book into being, having so skillfully elicited contributions from the most eminent workers in their respective fields and having edited the final work.

The Association thanks all who have contributed to this authoritative work, we recognise the need for this publication at this time and hope by publishing to further the sum of knowledge of sheep and the diseases which afflict them, because that, after all, has been the prime objective of the Association since its foundation.

J. Stobo
President, Animal Diseases Research Association

PREFACE

The current improvement in the economics of the sheep industry in Britain has provoked a modest expansion of the industry. This together with a rise in the value of sheep has stimulated a greater interest in sheep, their management and diseases. Veterinary surgeons, sheep specialists and other advisors to the sheep industry may have found the availability of published information on sheep and their diseases to be a disadvantage.

The staff at the Animal Diseases Research Association's Moredun Institute have been investigating the diseases of sheep since the Association was established in the early 1920s. The collected pool of knowledge and information among the members of staff at the Institute has provided an opportunity to present, in a collated form in this book, up-to-date information on the diseases which affect sheep. Naturally the information is directed particularly at the sheep industry in Britain and Europe but much of it should be applicable to the maintenance of sheep health in other parts of the world.

Most of the chapters have been written by members of staff at the Moredun Institute but valuable additional chapters have been contributed by experts from outwith the Institute.

As editor I earnestly hope that the information in this book will provide a useful source of information and reference for veterinary surgeons, veterinary students, sheep advisors and others concerned with the care, health and welfare of sheep. Livestock owners and shepherds may even find the information contained in this book of interest and help to them in caring for their stock.

Acknowledgements

In acknowledging the help I have received with this multi-author book there are inevitably many to be thanked.

I am most appreciative of the time and thought which the many authors have given to their contributions. If, as editor, I appeared at times to be unnecessarily precise I hope I shall be excused, but in a book of this kind some consistency of presentation must be maintained.

I wish also to thank the many others who have assisted in some way, in particular Miss Jane Goodier for her care and patience in typing the chapters, and Mrs Pam Kenworthy and Mrs Deirdre Holligan for their help in organising both my work and the collection and presentation of the chapters from authors.

The cooperation, help and advice I received from Mr Brian J. Easter C & G (Adv) regarding the photographs and from Mr Ian Swann, BVM&S, MRCVS, who prepared the index, I gratefully acknowledge.

Lastly I wish to record my appreciation of the encouragement I have received from members of staff at the Moredun Research Institute and the Directors of the Animal Diseases Research Association.

W.B. Martin, April 1983

ABBREVIATIONS

AGID	=	agar gel immunodiffusion
APD	=	average pore diameter
CF	=	complement fixation
CNS	=	central nervous system
CSF	=	cerebrospinal fluid
DM	=	dry matter
ELISA	=	enzyme-linked immunosorbent assay
EM	=	electron microscope (microscopy)
FAT	=	fluorescent antibody test
h	=	hours
HA	=	haemagglutination
ha	=	hectare
HAI	=	haemagglutination inhibition
HB	=	Heinz bodies
Hb	=	haemoglobin
iu	=	international units

IFAT	=	indirect fluorescent antibody test
kg	=	kilogram
pg	=	picogram
RIA	=	radioimmunoassay
SI units	=	standard international units
sp	=	species
UK	=	United Kingdom
USA	=	United States of America
LW	=	liveweight
SMCO	=	S-methylcysteine sulphoxide
Cu	=	copper
AAT	=	amino-aspartate transaminase
PCP	=	pentachlorophenol
PMSG	=	pregnant mares serum gonadotrophin
VTM	=	Virus transport medium

SECTION 1

RESPIRATORY SYSTEM

1 PASTEURELLOSIS

N.J.L. GILMOUR AND K.W. ANGUS

Synonym Enzootic pneumonia (for the pneumonic forms only)

There are two distinct clinical entities encompassed by the term pasteurellosis, pneumonic pasteurellosis (enzootic pneumonia) and systemic pasteurellosis. Both forms are caused by *Pasteurella haemolytica* of different biotype and serotype. *P. multocida* only occasionally causes pneumonia in sheep in temperate climates. It is likely that the pathogenesis and epidemiology of the pneumonia caused by *P. multocida* do not differ significantly from those of the pneumonia caused by *P. haemolytica* which are described in this chapter.

Pneumonic pasteurellosis was first described in Iceland[1] and subsequently has been shown to occur in many countries: Australia, Britain, Ethiopia, New Zealand, Norway, Sweden, East and South Africa, Somaliland and the USA among others. The systemic form of the disease was first described in Scotland in 1955.[2] Nothing is known of its general distribution.

It is impossible to give figures for the economic losses due to ovine pasteurellosis but there is no doubt that it is one of the most important infectious bacterial diseases of sheep. In the UK it is diagnosed more often than any other infectious sheep disease at Veterinary Investigation Centres.[3]

Cause

Pasteurella haemolytica is a small Gram-negative, aerobic coccobacillus. It is relatively easily identified and differentiated from *P. multocida*. (Table 1.1)

Table 1.1 The differentiation of *P. haemolytica* and *P. multocida*

	P. haemolytica	*P. multocida*
Haemolysis	+	−
Production of indole	−	+
Growth on MacConkey agar	+	−

Haemolysis is best appreciated by using plates prepared by overlaying nutrient agar with a thin layer of 7% sheep blood agar.

There are two biotypes of *P. haemolytica*[4] and it is important to differentiate between them since they cause two distinct syndromes. Biotype A causes pneumonic pasteurellosis and septicaemia in young lambs: biotype T causes systemic disease in older sheep. Biotype A was so designated because, in carbohydrate fermentation reactions, most strains of this biotype ferment arabinose but not trehalose. All biotype T strains ferment trehalose. Biotype A colonies are small and grey in colour after 24 hours incubation whereas colonies of biotype T strains are larger, 2 mm in diameter, and have brownish centres. *In vitro* studies have shown that biotype T strains are more resistant to penicillin, ampicillin, chloramphenicol, tetracycline, erythromycin and nitrofurantoin than biotype A strains. Basic fuchsin (0.2 μg/ml) in brain heart infusion broth inhibits A strains but not T strains. The two biotypes differ so markedly taxonomically and in their nucleic acid homologies[5] that it has been suggested that they are separate species. Differences in the epidemiology of the syndromes caused by A and T biotype tends to confirm this suggestion.

The two biotypes are further divided into 15 serotypes by an indirect haemagglutination (IHA) test.[6] Biotype A comprises serotypes 1, 2, 5, 6, 7, 8, 9, 11, 12, 13, 14 and biotype T 3, 4, 10 and 15. Recent discoveries of new serotypes 13, 14 in Ethiopia[7] and 15 in Scotland[8] suggest that further serotypes await discovery. Strains which cannot be typed by the existing typing sera exist and are isolated from both healthy and diseased sheep. The majority of untypable strains belong to biotype A.

Some serotypes are more common than others. Thompson and others[9] serotyped 406 strains which had been isolated from ovine pasteurellosis. Thirty-three per cent of the strains belonged to serotype A2 and serotypes T3, T4 and T10 made up 16, 19 and 12 per cent of the strains respectively. Serotypes A1 and A6 comprised 5 per cent each, the remaining A serotypes a total of 8 per cent and 6 per cent of isolates were untypeable. A5 was not known to occur in the UK at the time of the survey but has recently been isolated from an outbreak of pneumonia in sheep in southern England. Of the more recently discovered serotypes A13 and 14, the former has been found to be present in sheep in Scotland. Serotype T15, a new T serotype has been isolated from cases of systemic pasteurellosis in Scotland and England.

P. haemolytica occurs in the nasopharynges and tonsils of apparently normal sheep and again the two biotypes display different characteristics. Biotype A predominates in the nasopharynx and biotype T in the tonsils. In one survey[10] P. haemolytica was cultured from 95 per cent of tonsils and 64 per cent of nasopharyngeal swabs. Sixty-five per cent of tonsil isolates were biotype T whereas only 6 per cent of the nasopharyngeal isolates belonged to this biotype. Since the disease syndromes associated with the biotypes A and T are distinct these are described separately.

Disease associated with biotype A strains   The predominant disease caused by this biotype is pneumonia (enzootic pneumonia).

Clinical signs

In a proportion of cases of pneumonic pasteurellosis, clinical signs are not noticed and the animal is found dead. The clinical signs of acute, pneumonic pasteurellosis are dullness, anorexia, pyrexia of greater than 40.6°C (105°F) and varying degrees of hyperpnoea or dyspnoea. On auscultation, adventitious sounds are not a prominent feature and the respiratory sounds are loud and prolonged. There are often serous nasal and ocular discharges. A frothy fluid drooling from the mouth is often present in the terminal stages and is a prognostication of imminent death. In subacute or chronic cases of pneumonia the clinical signs may be transient and less obvious than in the acute disease.

Pathology

The initial finding at necropsy in most adult sheep with pneumonic pasteurellosis is extensive, subcutaneous ecchymotic haemorrhages in the throat region and over the ribs. On opening the thorax lesions of pneumonia, with pleurisy and pericarditis, are obvious. The pleural cavity often contains large amounts of straw-coloured fluid containing fibrin clots. The pneumonia tends to be in the cranial and ventral parts of the lungs, and the consolidated areas are often covered by a greenish, gelatinous, pleuritic exudate. In hyperacute cases which have died with no premonitory clinical illness, the lungs are swollen, heavy and cyanotic, with bright, purplish-red solid areas which exude a frothy, haemorrhagic fluid when incised (Plate 1). A feature of cases which have died after a brief illness is the presence of irregular, greenish-brown areas of necrosis in the consolidated portions of the lung, each area having a dark haemorrhagic margin. In cases of longer standing, a greyish-pink consolidation of cranial lobes predominates, and organizing pleural adhesions or pulmonary abscesses may be found.[11] Very rarely, cases occur in which the lungs are studded with swollen, greyish, tumour-like masses up to 5 cm in size; these are easily confused with the lesions of sheep pulmonary adenomatosis, unless histological examination is carried out.

In very young lambs, the effects of infection by A biotype P. haemolytica are those of a septicaemia rather than primary pulmonary disease, and the only visible changes may be petechiation of the liver, spleen, kidneys and epicardium, with swelling and hyperaemia of cervical and thoracic lymph nodes. In lambs 2–3 months old however, the predominant lesions are severe pleurisy and focal consolidation of the lungs. A common finding is that the entire thoracic cavity appears to be filled with coagulated fibrin, which coats the lungs and heart thickly. Closer inspection shows dark-red consolidation in the cranial lobes of the lungs, and a fibrinous pericarditis with sub-epicardial petechiation.

The microscopic changes are usually quite characteristic. The lung vasculature is intensely hyperaemic, and alveoli contain a pink-staining fluid rich in protein, often intermingled with masses of Gram-negative coccobacilli. There are usually extensive irregular areas of abnormality in which the central alveoli and vasculature have undergone necrosis, while more peripherally the alveoli are packed with intensely basophilic spindle-shaped leucocytes, or 'oat cells', which form whorls in alveoli, or appear to 'stream' from one alveolus to another (Plate 2). Interlobular septa are distended and oedematous, and lymph channels contain fibrin thrombi. The pleurisy is characterized by a copious fibrinous exudate, often incorporating masses of coccobacilli, presumably P. haemolytica. Air passages usually contain an exudate similar to that found in alveoli.

Diagnosis

Pasteurellosis is the commonest cause of acute pneumonia in sheep and a tentative diagnosis can be made on clinical signs and history. The presence of P. haemolytica in nasal swabs is of no diagnostic significance and there is as yet no evidence that serology is useful either on an individual or a flock basis. Many apparently healthy sheep have antibodies to a range of serotypes of *P. haemolytica*. A diagnosis is best made on necropsy when confirmation of the presence of pneumonic pasteurellosis is obtained by finding the acute inflammatory changes in the thorax and the lesions in the lungs. Gram-negative coccobacilli may be seen in smears from exudates and the cut surfaces of lung lesions but the isolation in culture of *P. haemolytica* in large numbers is required for confirmation of the presence of the disease. In untreated cases the recovery of small numbers of colonies of *P. haemolytica* is of little significance. Further confirmation of the diagnosis can be got from histopathology when the finding of lesions with 'oat cells' is pathognomonic. In acute cases *P. haemolytica* is usually isolated from livers, spleens and kidneys especially in lambs. In subacute pneumonia of up to 7 days' duration, numbers in the region of $10^7/g$ of lung are usually present in the lesions.

Routine serotyping of isolates of *P. haemolytica* is of little value but can be useful in epidemiological studies and especially when vaccine breakdowns are being investigated.

The presence of antibodies to *P. haemolytica* can be determined by IHA tests or mouse protection tests. As sheep may have titres to several serotypes these tests are of no value in establishing which serotype is the cause of an outbreak. No definite correlation exists between antibody titres and protection in sheep. Rising ELISA titres have been noted after experimental infection of sheep[12] and this test may prove to be of value in the diagnosis of subacute pneumonia.

Epidemiology

The majority of outbreaks of pneumonic pasteurellosis occur in May, June and July and many involve ewes and lambs. Flock outbreaks usually start suddenly with deaths, often in young lambs. In young lambs the disease is hyperacute and septicaemic rather than pneumonic. As lambs get older the disease becomes more confined to the thorax with prominent lesions of pleurisy and pericarditis. Beyond three months of age most cases are frankly pneumonic although sudden deaths with septicaemia rather than pneumonia may still occur. As the outbreak progresses over the next few days a number of sheep will be noticed with clinical signs of pneumonia. Observations of the flock show that some sheep have an occasional cough and slight oculo-nasal discharges. Morbidity and mortality vary from outbreak to outbreak, but rarely exceed more than 10 per cent of the flock. Pneumonic pasteurellosis also occurs in individual sheep sporadically rather than as part of a clearly defined flock outbreak.

It is generally assumed that environmental factors are important predisposing causes of pneumonic pasteurellosis but precise epidemiological studies are lacking. Some outbreaks can be linked to previous stressful situations as warm or cold, wet weather, and dipping, castration or dosing. There is evidence from experimental investigations[13] and epidemiological studies that infection with parainfluenza virus type 3 (PI3) and sheep pulmonary adenomatosis are factors predisposing to pneumonic pasteurellosis. Infection with PI3 virus produces a generally mild illness and cases of pneumonia in a flock can be attributed to superimposed infection of the lungs with *P. haemolytica* A biotype strains. In most cases flock outbreaks on individual farms are sporadic and do not occur every year although on some farms small numbers of sheep may succumb annually. There is a tendency for the prevalence of the disease to be higher overall in some years than in others. There are two possible explanations. Either environmental factors, e.g. climate, are particularly favourable for the disease over a wide area in some years or immunity to the predisposing virus infection is cyclic.

Control, prevention and treatment

The control of pneumonic pasteurellosis is difficult. Since the environmental predisposing factors are not well-defined or are part of normal sheep husbandry it is difficult to see how these can be altered or eliminated. The best hope for control lies in the stimulation of immunity to *P. haemolytica*. Vaccination with a PI3 virus vaccine did not prevent the disease in lambs subsequently challenged with PI3 virus and an aerosol of *P. haemolytica*[14] although it did reduce slightly the effects of the pasteurella infection. Vaccines which contain *P. haemolytica* antigen have been available for many years but there is no convincing evidence that they stimulate a satisfactory immunity. In the past testing of vaccines presented problems. There was no experimental model of pneumonic pasteurellosis and field trials are very difficult to conduct due to the sporadic nature of the disease on most farms.

Recent research at the Moredun Institute has been

concerned with the immunology of pneumonic pasteurellosis with the eventual aim of developing a vaccine of proven efficacy. At the time of writing an improved vaccine has been developed and is commercially available. It was found that vaccination against experimental challenge was serotype specific, i.e. a vaccine must be multivalent and contain at least the most important serotypes with which it is likely to be challenged under natural conditions. Thus at least five serotypes A1, A2, A6, A7 and A9 would have to be included. To ensure that sufficient antigen of each serotype was presented to the animal, extracts of these strains were prepared, incorporated into oil adjuvant vaccine and tested for the ability to protect against experimental challenge.[15] Extracts of some serotypes were more immunogenic than others. Good protection has been achieved against serotypes 1, 6 and 9 on a number of occasions using monovalent and multivalent vaccines. It has proved more difficult to get the same protection against the important serotype A2 and further work is necessary to get the optimum immunogens for this serotype. There has been no challenge of an A7 serotype vaccine so far.

Pneumonic pasteurellosis can be treated by the administration of penicillin, ampicillin or terramycin. However, we have found some biotype A strains to be resistant to penicillin. In the USA Chang and Carter[16] found half of their pig and cattle strains were resistant to penicillin, streptomycin and terramycin.

Diseases Associated with Biotype T Strains P. haemolytica biotype T serotypes 3, 4, 10 and 15 (T3, T4, T10 and T15) are the cause of systemic pasteurellosis. This is epidemiologically and pathogenetically distinct from the pneumonic form of pasteurellosis.

Clinical signs

The main feature of this condition is sudden death, so affected sheep are seldom seen alive. Those which are seen are usually recumbent, extremely depressed, dyspnoeic and frothing at the mouth.

Pathology

The carcass is usually that of a young sheep in good condition, which has died suddenly. Subcutaneous haemorrhages are found over the neck and thorax, and ecchymoses are also frequently seen on the pleura and diaphragm, or under the epicardium. The lungs are swollen and oedematous or haemorrhagic, and frothy blood-tinged fluid exudes from the bronchioles. Consolidation is not a feature, however. Lesions also occur in the pharynx and upper alimentary tract. In the former site, they take the form of necrotic erosions which are especially prominent around the tonsillar crypts (Plate 3). Similar erosions may also be found in the nasal mucosa, tongue or soft palate. Necrosis of the oesophageal lining, with extensive sloughing of mucosa, may be present and similar necrotic lesions are variably found in the omasum and rumen. The abomasum may contain considerable areas of haemorrhagic inflammation, or shallow haemorrhagic ulcers, most numerous at the pyloric end. Rarely, similar lesions may be seen in the duodenum.[2]

The liver is usually swollen and congested, and may contain numerous small (0.5–5 mm) grey, necrotic foci scattered throughout its substance. Small infarcts may be seen in the spleen, and the kidneys often resemble those of sheep dying of pulpy kidney disease. The tonsils and retropharyngeal lymph nodes are usually enlarged and oedematous.

Microscopically, the necrotic lesions in the pharynx and alimentary tract show necrosis of the mucosa with extensive sloughing. Underlying tissues are hyperaemic but exhibit surprisingly little cellular reaction. Masses of Gram-negative coccobacilli and Gram-positive cocci can be seen adhering to the luminal surface of many ulcers or eroded areas and similar masses occlude local vessels and lymphatics. Lesions in the lungs, liver, spleen, adrenals and less frequently kidneys can be attributed to the dissemination of bacterial emboli in the terminal arterial system. The lesions consist of masses of Gram-negative coccobacilli, usually surrounded by zones of necrosis enclosed by basophilic spindle-shaped leucocytes (oat cells). In the brain, serum protein leakage in the cerebro-cortical leptomeninges, with mononuclear cell infiltrates in the choroid plexuses of the lateral and fourth ventricles have been reported recently.[17] From the evidence of these findings and the results of recent experimental work with T biotypes in sheep, possible pathogenetic mechanisms can be postulated (Fig. 1.1). It is presumed that multiplication of T biotypes resident in the tonsils occurs under the influence of poorly-understood environmental factors, e.g. change in pasture, with the development of necrotizing lesions in the pharynx and upper alimentary tract. Bacterial emboli from these sites pass by way of the general circulation to the lungs and other organs, where further multiplication and toxin release cause the death of the animal.

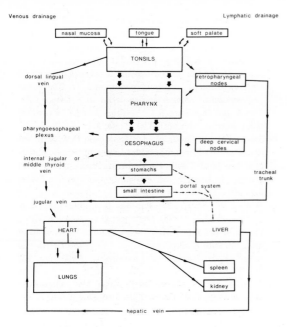

Fig. 1.1 Schema of pathogenesis of infection with T biotype strains of *P. haemolytica.*

An alternative hypothesis sites the primary lesions in the forestomachs and intestines, emboli passing to the lungs via the liver and portal system.

Diagnosis

The diagnosis of systemic pasteurellosis depends on the isolation in culture of large numbers ($> 10^7$/g of tissue) of *P. haemolytica* from the lungs, livers and spleens of sheep with the gross pathological changes described. Also biotype T strains can be isolated from the nasopharynges and tonsils of apparently normal sheep and in small numbers from other sites including the lungs but their presence in these sites can be ignored. However, it should be remembered that antibiotic therapy may reduce the numbers of bacteria in the lesions.

Epidemiology

Most outbreaks of systemic pasteurellosis conform to a typical pattern. The disease affects 6–9 month old sheep during October, November and December. The onset of the disease frequently coincides with folding on rape or turnips, or the change to improved pastures and these have been postulated as predisposing causes. However, there is as yet no proof that they are so. A change to wet, cold weather has also been noted as a possible contributing factor.

A typical episode usually starts with a number of sudden deaths, but the number of deaths then quickly drops over the next few days. Mortality is quite variable but seldom exceeds 10 per cent of the flock. Sporadic deaths due to systemic pasteurellosis do occur at other times of the year and in all ages of sheep but even less is known about the factors which predispose to these.

Control, prevention and treatment

Because of the epidemiology of the disease, control would be best achieved by vaccination. However, there is no evidence that, at the moment, vaccination is effective in preventing the disease. As with pneumonic pasteurellosis, field trials are difficult due to the sporadic nature of the disease on individual farms and from year to year. Experimental reproduction of the disease with small challenges has not been successful until recently.

Most isolates of *P. haemolytica* biotype T are sensitive to oxytetracycline (25 µg discs) but as sheep are seldom seen in the early stages of the disease therapy is seldom possible and there are no reports on the value of prophylactic therapy.

Since stress may play some part in predisposing to this disease, flock management should be designed to minimize the stress involved in changes of environment and nutrition during the period from October to December.

Other P. haemolytica infections—P. haemolytica is the cause of other pathological conditions but these occur less frequently than pneumonic and systemic pasteurellosis. Strains of *P. haemolytica*, usually of biotype A, can be isolated from mastitis in ewes and this condition is occasionally fatal. Arthritis is a common sequel to experimental intravenous inoculation of T biotypes of *P. haemolytica* and is also occasionally diagnosed. Sporadic pasteurella meningitis affects ewes and lambs. Diagnosis of all the above mentioned conditions depends on the isolation of large numbers of *P. haemolytica* from the lesions.

REFERENCES

1 Dungal N. (1931) Contagious pneumonia in sheep. *Journal of Comparative Pathology & Therapeutics,* **44**, 126–43.

2 Stamp J.T., Watt J.A.A. & Thomlinson J.R. (1955) *Pasteurella haemolytica* septicaemia of lambs. *Journal of Comparative Pathology and Therapeutics,* **65**, 183–96.

3 Ministry of Agriculture Fisheries and Food: (1979) Veterinary Investigation Diagnosis Analysis II. 1978.

4 Smith G.R. (1961) The characteristics of two types of *Pasteurella haemolytica* associated with different pathological conditions in sheep. *Journal of Pathology & Bacteriology,* **81**, 431–40.

5 Biberstein E.L. & Francis C.K. (1968) Nucleic acid homologies between the A & T types of *Pasteurella haemolytica. Journal of Medical Microbiology,* **1**, 105–8.

6 Biberstein E.L., Gills M. & Knight H. (1960) Serological types of *Pasteurella haemolytica. Cornell Veterinarian,* **50**, 283–300.

7 Pegram R.G., Roeder P.L. & Scott J.M. (1979) Two new serotypes of *Pasteurella haemolytica* from sheep in Ethiopia. *Tropical Animal Health and Production,* **11**, 29–30.

8 Fraser J., Laird S. & Gilmour N.J.L. (1982) A new T biotype serotype of *Pasteurella haemolytica. Research in Veterinary Science,* **32**, 127.

9 Thompson D.A., Fraser J. & Gilmour N.J.L. (1977). Serotypes of *Pasteurella haemolytica. Research in Veterinary Science,* **22**, 130– 31.

10 Gilmour N.J.L., Thompson D.A. & Fraser J. (1974) The recovery of *Pasteurella haemolytica* from the tonsils of adult sheep. *Research in Veterinary Science,* **17**, 413–14.

11 Gilmour N.J.L. (1980) *Pasteurella haemolytica* infections in sheep. *Veterinary Quarterly.* **2**, 191–98.

12 Donachie W. & Jones G.E. (1982) The use of ELISA to detect antibodies to *Pasteurella haemolytica* in sheep with experimental chronic pneumonia. in *The Elisa: Enzyme-linked Immunosorbent assay in Veterinary Research and Diagnosis.* R.C. Wardley and J.R. Crowther (eds). Martinus Nijhoff, The Hague 102.

13 Sharp J.M., Gilmour N.J.L., Thompson D.A. & Rushton B. (1978) Experimental infections of specific pathogen free lambs with Parainfluenza virus type 3 and *Pasteurella haemolytica. Journal of Comparative Pathology,* **88**, 237–43.

14 Wells P.W., Sharp J.M., Rushton B., Gilmour N.J.L. & Thompson D.A. (1978) The effect of vaccination with a Parainfluenza type 3 virus on pneumonia resulting from infection with Parainfluenza type 3 virus and *Pasteurella haemolytica. Journal of Comparative Pathology,* **88**, 253–9.

15 Gilmour N.J.L., Martin W.B., Sharp J.M., Thompson D.A. & Wells P.W. (1979) The development of vaccines against pneumonic pasteurellosis in sheep. *Veterinary Record.* **104**, 15.

16 Chang W.H. & Carter G.R. (1976) Multiple drug resistance in *Pasteurella haemolytica multocida* and *pasteurella* from cattle and swine. *Journal of the American Veterinary Medical Association,* **169**, 710–12.

17 Dyson D.A., Gilmour N.J.L. & Angus K.W. (1981) Ovine Systemic Pasteurellosis caused by *Pasteurella haemolytica* Biotype T. *Journal of Medical Microbiology,* **14**, 89–95.

ACUTE RESPIRATORY VIRUS INFECTIONS

2

J.M. SHARP

The acute respiratory disease complex of sheep is dominated by pneumonic pasteurellosis caused by *Pasteurella haemolytica* biotype A. Therefore, although several viruses have been isolated from sheep with acute respiratory illness, it seems unlikely that a syndrome exists which is attributable to virus infection alone. Nevertheless, evidence is accumulating which indicates that some of these viruses may be involved in the aetiology of ovine respiratory disease.

For example, viruses have been isolated from a high proportion of outbreaks of acute illness and also have been closely related to high levels of pneumonia in slaughtered lambs. Several studies have suggested that parainfluenza virus type 3 and adenoviruses are particularly important, whereas the roles of other viruses such as reovirus and respiratory syncytial virus remain to be established.

PARAINFLUENZA VIRUS TYPE 3 (PI3)

PI3 is an enveloped RNA virus that matures by budding from the surface of infected cells. The envelope of the virion contains at least two major glycoproteins and these appear to be necessary for stimulation of immunity and haemagglutination of guinea pig erythrocytes. There is only one serotype of ovine PI3, which is antigenically related to, but distinct from, bovine and human strains of PI3.

Clinical signs

PI3 virus is associated with a spectrum of illness in sheep. The majority of infections are inapparent or of a mild nature. However, outbreaks of acute illness associated with PI3 virus have been recorded which are characterized by high morbidity. Affected animals usually are afebrile, may cough frequently and have a copious serous nasal discharge, sometimes accompanied by a conjunctival discharge. The illness following experimental inoculation largely depends upon the route of inoculation. Intranasal instillation or aerosol exposure of lambs with PI3 virus results in viral replication in the upper respiratory tract, but with no clinical signs whereas combined intranasal and intratracheal inoculation causes a severe respiratory illness characterized by pyrexia, tachypnoea, dyspnoea and dullness lasting 3–5 days between 3 and 7 days after inoculation .[1]

Pathology

The lungs of lambs inoculated experimentally with PI3 virus contain linear or patchy areas of dull red consolidation in the apical lobes (Fig. 2.1). Lesions are most extensive 6–8 days after inoculation. The essential histological features of these lesions are hyperplasia of the bronchiolar epithelium, infiltration of interalveolar septa by mononuclear cells and cellular exudate in the bronchiolar lumen. Acidophilic intracytoplasmic inclusion bodies may be detected in the bronchiolar epithelium up to 6 days after infection.[2] These lesions resolve fairly quickly although a residual interstitial pneumonia and focal alveolar epithelialization persists for at least 28 days after inoculation.

Diagnosis

Infection by PI3 virus can be confirmed by isolation of the virus from swabs or aspirates from the upper respiratory tract during the first 6 days of infection, which usually coincides with the presence of a copious serous nasal discharge, or from pieces of tissue from the respiratory tract. More usually, a rise in serum antibody titre may be detected in paired sera by either the haemagglutination inhibition (HAI) test or the more sensitive ELISA. However, there are limitations in this approach as it is not uncommon for infections by PI3 virus to occur in the absence of an apparent rise in serum antibody titre.

Epidemiology

PI3 infections of sheep occur worldwide and workers in many countries have reported the presence of antibodies in the sera of both pneumonic and healthy sheep. It appears that such infections are common as

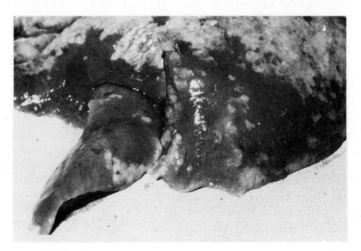

Fig. 2.1 Lungs of specific pathogen free lamb infected with ovine PI3 virus and killed 5 days later. Extensive consolidation is evident in apical, cardiac and diaphragmatic lobes.

the proportion of sheep with antibody is rarely less than 70–80 per cent. Most lambs acquire antibodies to PI3 virus via colostrum and infections are rarely detected while these are present. However, these antibodies wane quickly and the lambs become susceptible to infection with PI3 virus, so that most appear to become infected within the first 12 months, although outbreaks have been detected in adult sheep up to 5 years of age. The method by which the virus is maintained in the flock is not known although it seems likely that some animals must become persistently infected, even in the presence of an immune response.

Although most infections with PI3 virus pass unnoticed, field observations indicate that the virus may predispose sheep to infection by bacteria, notably *P. haemolytica*. These observations have been borne out by experimental findings which have shown that prior infection of lambs with PI3 virus exacerbates that induced by *P. haemolytica*. The clinical illness and lesions produced by combined infection with PI3 virus and *P. haemolytica* are identical with those observed in the naturally-occurring pneumonic pasteurellosis and lesions in surviving lambs persist for at least 8 weeks after infection.[3]

Control

The available evidence suggests that PI3 virus is involved in a proportion of outbreaks of acute respiratory disease, many of which also involve *P. haemolytica*. Some means of control would be desirable and this might be best achieved by incorporating PI3 virus in any vaccines which are designed for the prophylaxis of pneumonic pasturellosis.

Immunity to PI3 virus can be stimulated by local or parenteral administration of PI3 antigens, which will prevent virus replication, clinical illness and pneumonic lesions.[1] Such vaccines also have the advantage of being able to reduce the extent of the pneumonic lesions produced by combined infection with PI3 virus and *P. haemolytica*. This effect could supplement the effect of other components in any vaccine directed against *P. haemolytica*.

ADENOVIRUSES

Adenoviruses are unenveloped icosahedral viruses comprised of 252 capsomeres. Those at the 12 vertices have a filamentous projection which is involved with the haemagglutination exhibited by some serotypes. Each of the isolates of ovine adenovirus has been shown to possess the group-specific antigen which is common to all mammalian adenoviruses. In addition, 5 distinct serotypes of ovine adenovirus (OA) have been determined and there are many untyped isolates, at least one of which is not neutralized by antisera to bovine, ovine or porcine adenoviruses. An adenovirus related to bovine adenovirus type 2 (BA2) has been isolated from lambs in Hungary.

Clinical signs

Adenoviruses have been isolated from sheep with a variety of clinical conditions ranging from apparently healthy to severe pneumoenteritis. Most isolates have come from apparently healthy lambs although in Hungary OA1, OA5 and BA2 often are associated with outbreaks of respiratory illness in large groups of lambs housed for fattening.[4]

The clinical illness following experimental inoculation of lambs with adenoviruses appears to be related to the individual serotypes and in some cases, strains within serotypes. OA types 1–4 appear to cause little illness, although the Hungarian isolate of OA1 induces anorexia, sneezing and pneumonia. OA5 and the New Zealand isolate, WV757/75, which may be a prototype OA6, induce a serous or mucoid nasal discharge and pyrexia. One untyped adenovirus, WV419/75, causes mild upper respiratory tract illness with possible central nervous system involvement.

Pathology

Lesions tend to be found only in lambs which have been inoculated with serotypes that cause clinical illness, that is OA1, OA5, BA2 and OA strains WV419/75 and WV757/75. The apical and cardiac lobes of the lungs from such lambs contain varying extents of atelectasis and dull red consolidation which may last for more than 14 days. The bronchial and mediastinal lymph nodes can be enlarged. The principal lesion appears to be proliferative bronchiolitis with an associated exudate of desquamated epithelial cells, macrophages and lymphoid cells. The bronchiolitis induced by some serotypes may extend into alveoli to form a bronchopneumonia. These lesions appear to be most extensive around 7 days after inoculation, but have largely regressed by day

14–21.[5] A particular feature of experimental infection by OA strain WV419/75 was cytomegaly and karyomegaly of bronchiolar epithelial cells, which has been reported in a naturally-occurring pneumonia of lambs attributed to adenovirus.

Some serotypes induce lesions in organs other than the respiratory tract such as focal hepatic necrosis and lymphangitis by OA4 and nephritis by OA5.

Diagnosis

Because ovine adenoviruses possess the mammalian group specific antigen it is common to rely on serological tests which detect antibodies to this antigen. These tests, usually complement fixation (CF) or agar gel immunodiffusion (AGID), are convenient and rapid to perform but have disadvantages. They are less sensitive and less specific than microneutralization or HAI tests, and the immune response to the group antigen is slow, requiring at least 4–5 weeks to become apparent. Microneutralization or HAI tests are much more sensitive but because they are specific for each serotype they are not employed routinely.

Virus isolation remains an important adjunct to serology as a means of diagnosis for ovine adenoviruses. It is unlikely that the list of ovine adenoviruses is complete and in any unisolated serotype, as with many bovine serotypes, the common group antigen may be absent or a minor component. Infection by such serotypes would therefore go undetected if serology was the sole means of diagnosis.

Epidemiology

Antibodies to the adenovirus group antigen have been detected in sheep sera in many countries, indicating the wide distribution of this virus group. Detailed serological and virological studies in a few countries have indicated that such infections are common, particularly in young lambs. The prevalence of antibodies to individual serotypes varies between 20 per cent and 70 per cent and most of these infections occur before the lambs are 12 months old. Longitudinal surveys of selected flocks have shown that few viruses could be isolated whilst colostrum-derived antibodies were high, but when the majority of the flock became susceptible, adenoviruses could be isolated frequently from the faeces and sometimes the upper respiratory tract. The widespread occurrence of adenovirus infections at an age when respiratory illnesses are commonplace in lambs has compounded the difficulty in determining the role of such viruses and there is a need for carefully controlled epidemiological studies.

Ovine adenoviruses have been shown experimentally to form persistent infections and virus may be excreted for at least 80 days after infection. Such persistently-infected sheep must form an important role in the maintenance of infection within a flock. The interaction of ovine adenoviruses with *P. haemolytica* has been studied experimentally and at least one untyped strain, WV419/75, can enhance the pneumonia caused by *P. haemolytica.*

Control

Because adenoviruses commonly cause inapparent infections of young sheep, the need for their control in most countries is unclear. This general observation does not seem to apply in Hungary where the collection of young lambs from many sources on to large fattening units is believed to tranform this picture and adenoviruses, particularly OA1 and BA2, are major pathogens. Consequently, both monovalent and bivalent vaccines have been developed which protect lambs both by ingestion of colostrum from vaccinated ewes and by active immunization.

MISCELLANEOUS VIRUSES

Reovirus types 1, 2 and 3 have been isolated from sheep but, as in other species, their role in disease appears equivocal. One exception to this may be reovirus type 1 which is regarded in Hungary as one part of the pneumoenteritis complex.

Respiratory syncytial virus (RSV) is regarded as a cause of serious respiratory illness in human infants and cattle. Antibodies to RSV have been detected in the sera of sheep in North America, but the position in other countries is unknown. Experimental inoculation of lambs with bovine RSV has induced a mild clinical response consisting of a transient pyrexia, hyperpnoea and dullness. The lungs of these lambs contained small areas of consolidation consisting of interstitial pneumonia and bronchiolitis. The pattern of virus excretion suggested that the principal site of replication was the lower respiratory tract.

The importance of RSV infection of sheep remains unclear and may be difficult to resolve. The mild clinical illness may allow natural infections to go unrecognized and chance isolation of the virus may be frustrated by its fragility and its principal site of replication in the lower respiratory tract.

REFERENCES

1 Wells P.W., Sharp J.M. & Smith, W.D. (1977) Experimental assessment of parainfluenza type 3 virus vaccines. In *Respiratory Diseases in Cattle.* (Ed. Martin W.B.) Current Topics in Veterinary Medicine, Vol.3. pp. 515–20. Martinus Nijhoff, The Hague.
2 Hore D.E. & Stevenson R.G. (1969) Respiratory infection of lambs with an ovine strain of parainfluenza virus type 3. *Research in Veterinary Science,* **10,** 342–50.
3 Sharp J.M., Gilmour N.J.L., Thompson D.A. & Rushton B. (1978) Experimental infection of specific pathogen-free lambs with parainfluenza virus type 3 and *Pasteurella haemolytica. Journal of Comparative Pathology,* **88,** 237– 43.
4 Belak S. (1980) Properties of ovine adenoviruses. *Acta Veterinaria Academiae Scientiarum Hungaricae,* **28,** 47–55.
5 Davies D.H., Dungworth D.L. & Mariassy A.T. (1981) Experimental adenovirus infection of lambs. *Veterinary Microbiology,* **6,** 113– 28.

CHRONIC RESPIRATORY VIRUS INFECTIONS   3

J.M. SHARP AND W.B. MARTIN

Two virus infections produce chronic progressive pulmonary disease in sheep in several countries throughout the world. These diseases are pulmonary adenomatosis and maedi-visna. A retrovirus has been isolated from sheep with each disease, but each virus is distinct antigenically and in other ways.

In countries where these diseases are recognized, both may be present in the same flock and can occur concurrently in individual sheep. A few countries however recognize only one of these diseases at present.

As both infections result in chronic progressive respiratory tract disease in sheep and may on occasion be confused, each is described in this chapter.

SHEEP PULMONARY ADENOMATOSIS

Synonym Jaagsiekte (Afrikaans); pulmonary carcinoma

Sheep pulmonary adenomatosis (SPA) is a contagious disease produced by a tumour in the lungs of sheep which is experimentally transmissible. SPA does not affect cattle or other animals although it has been suggested, without firm evidence, that goats can be infected. The disease has been recognized in over 20 countries on the continents of Europe, Africa, America and Asia, and in a wide variety of breeds.[1]

Cause

The precise cause of SPA is not known. Two viruses have been recovered from affected sheep. One, which is most likely to be the causal virus, is a retrovirus and the other is a herpesvirus. It may be that both act synergistically but more probably the retrovirus induces the characteristic cellular proliferation which results in the development of the tumour in the lungs of sheep. The herpesvirus is likely to have no aetiological role but may replicate in the alveolar macrophages either because of their abundance in the tumour or possibly as a result of immunosuppression in sheep affected with the tumour.

The presence of a retrovirus in tumour tissue was first recorded in 1971[2] which was subsequently shown to have the physical and biochemical characteristics of an RNA tumour virus of the family Retroviridae.[3] A feature of the virus like others of this group, is that it contains an enzyme, RNA-directed DNA polymerase or reverse transcriptase, which is essential for its replication. Reverse transcriptase can, at present, be detected more readily than the virus particle in SPA tumour tissue or fluid. No means of culturing the virus has been found, and in natural cases particles are not readily observed with the electron microscope.

The herpesvirus alone does not induce tumour formation nor does it produce overt clinical illness when injected into specific-pathogen-free lambs.[4] Recently however, evidence that it can cause latent infection in such lambs has been obtained at the Moredun Institute.[5]

Clinical signs

The incubation period in naturally-infected sheep appears to be long so that the disease is not usually seen till sheep are about 2–4 years old. Tumours have, however, been seen exceptionally in lambs 5–6 months old and sheep 11 years of age.

Experimentally clinical illness and tumour formation occur within 5–12 months after the intratracheal inoculation of tumour tissue or fluid into lambs several months old. In very young lambs tumours can occur more rapidly especially when they are inoculated with concentrated material.

Although small areas of adenomatous tissue may be present without producing obvious clinical signs, tumours which are sufficiently large to interfere with normal lung function result in respiratory embarrassment which is most obvious following exercise. The degree of rapid or exaggerated breathing, often associated with noticeable movement of the abdominal wall (abdominal lift), depends on the extent of the tumour and the loss of normal functional lung. In advanced cases high-pitched and moist sounds may be heard on auscultation or even by the unaided ear.

Appetite is maintained though loss of weight is obvious. Death inevitably occurs, often suddenly and from a complicating *Pasteurella haemolytica* pneumonia.

Pathology

SPA lesions are confined to the lungs but occasionally the associated lymph nodes also show changes. Affected lungs are larger than normal and infiltrated with areas of tumour which may vary from small discrete nodules, measuring only 0.5–2 cm, to extensive tumours involving the entire ventral half of the diaphragmatic and other lobes. Usually tumour tissue is present in both lungs although the extent on either side does vary. Tumours are solid, grey or light purple with a shiny, translucent sheen. Around the tumour the lung generally looks normal. The respiratory passages are filled with frothy, white fluid which flows out of the trachea when it is cut or pendant (Fig. 3.1).

Pleurisy may be evident over the surface of the tumour and in some cases abscesses are present in the adenomatous tissue.

Adult sheep, which on post-mortem examination appear to have died from acute pasteurellosis should have their lungs examined carefully, as lesions of SPA may be masked by the acute reaction to the pasteurellae.

Histological examination of SPA lesions shows

Fig. 3.1 Tumour occupying the lower portions of all lobes in the lungs of a sheep affected with SPA. There is a distinct margin between the tumour and the rest of the lung (black arrows). The frothy fluid which accumulates in the respiratory tract of affected sheep can be seen exuding from the trachea (open arrow).

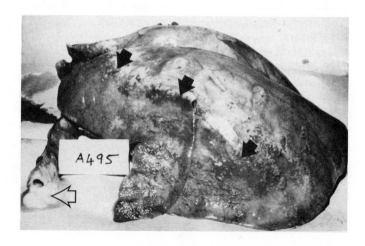

areas of lungs where cuboidal or columnar cells replace the normal thin alveolar cells. Sometimes these abnormal cells form papilliform growths which project into the alveoli. Intrabronchiolar proliferation may also be present.[6] A third type of change consists of nodules of loose connective tissue in a mucopolysaccharide substance. Accumulations of large macrophage cells occur in the lung tissue around the tumours.

Ultrastructural studies have shown that the cells forming the alveolar lesion are type II alveolar cells and those in the bronchiolar tumours, the cells of Clara.[6]

Diagnosis

No serological test which can unequivocally diagnose SPA has been described.

Antibodies to the herpesvirus, which can be detected by neutralization and ELISA tests, are present in a percentage of sheep with and without SPA. Isolation of the herpesvirus from tumour tissue using special techniques is possible but this would not be confirmatory or useful in diagnosis.

It has been suggested that serum protein concentrations, especially IgG, are altered in sheep with SPA, and that such alterations may be of value in identifying animals which are incubating the disease. However, fluctuations in the level of serum proteins occur for many reasons unrelated to SPA and therefore their measurement does not assist a diagnosis during the preclinical stages of the disease.

Clinical detection of a case of SPA should be based on the presence, in a single mature sheep of an afebrile, wasting disease, with marked respiratory signs. A useful aid to diagnosis is to raise the hindquarters and lower the head of the sheep which causes mucoid fluid, sometimes copious in amount, to run from the nostrils (Fig. 3.2). Large tumours are unmistakable in appearance at necropsy but confirmatory histology may be necessary where the amount of adenomatous tissue is small or secondary bacterial pneumonia has developed. Where tumours are not obvious at least 1–2 samples, for histological examination, should be taken from each lobe.

Epidemiology

The precise roles of the herpesvirus and the retrovirus have not been fully elucidated as yet.

SPA has been recognized for many years to be

Fig. 3.2 Sheep affected with SPA. When the head is lowered fluid flows from the nostrils into the container (arrows). As much as 300 ml may be collected although 30–50 ml is more usual.

naturally and experimentally transmissible by the respiratory route.[7] It is probable that an infected sheep, even before it develops obvious respiratory signs, excretes virus-containing droplets as it breathes. Later as the disease progresses, it will discharge quantities of infective respiratory fluid, especially during feeding when the head is lowered. This causes the sheep to snuffle and doubtless creates an aerosol of infected droplets. Close confinement of a flock obviously increases the probability of transmission so that housing, with trough feeding and watering may allow the infection to spread.

SPA is generally introduced into flocks and even countries by the acquisition of infected sheep. Losses from SPA can be high when the disease first appears in flocks, as occurred in Iceland where the annual loss for SPA was recorded as reaching 50–80 per cent. In countries where the disease is endemic it may be that SPA is more common in certain areas or breeds. This appears to be the case in Scotland where losses in infected flocks are generally between 2 and 10 per cent annually.

Although an association of the disease with cold conditions has been suggested this has not been proven and close contact between sheep is probably the vital factor in transmission.

Control and treatment

No method of treatment is recognized or advised.

Control should be based on regular inspection of adult sheep in affected flocks. Any showing loss of weight or signs of respiratory disease should be isolated and veterinary advice sought. Prompt culling of any suspicious animal is advisable. While these methods are unlikely to eradicate SPA from a flock in which the disease is endemic, reduction in the prevalence of infection may be obtained.

MAEDI-VISNA

Synonyms zwoegerziekte (Dutch), 'Montana' progressive pneumonia, la bouhite (Fr).

Maedi and visna are two Icelandic words, meaning 'dyspnoea' or difficult breathing and 'wasting' respectively, which describe the respiratory and nervous forms of the disease following infection by the same virus. Maedi has been recognized now as occurring in most countries in Europe and several in Africa and Asia. It is also present in the USA and Canada. In 1979 it was described in Britain where it occurs in imported breeds of sheep.[8]

Cause

The causal virus belongs to the sub-family Lentivirinae within the Retroviridae family, and is related to the RNA tumour viruses. The virion, which matures by budding from the cell membrane, contains two major proteins, an internal core protein of 25000 mol wt (p 25) and a membrane glycoprotein of approximately 135 000 mol wt (gp 135). The internal p 25 is a stable antigen containing group-specific determinants, whereas the gp 135 appears to contain additional type-specific antigens, detectable by neutralization tests, which are highly variable. Thus marked antigenic variation between isolates of maedi-visna virus is recognized. Maedi-visna virus contains reverse transcriptase.

A retrovirus which is related antigenically to maedi-visna virus, causes arthritis, leucoencephalitis and interstitial pneumonia in goats. Present knowledge does not allow the two viruses to be distinguished by laboratory tests and the possibility of transmission from goats to sheep is unknown.

Clinical signs

Following infection, there is a long incubation period of 3–4 years and as the onset of disease is insidious clinical signs are seen mainly in adult sheep. Infection with maedi-visna virus can result in two main forms of the disease; a progressive pneumonia (maedi),which is the more common syndrome or a nervous form (visna). Sometimes both forms are present in one animal. A third manifestation of the infection in affected flocks has been described and consists of chronic arthritic changes, principally of the carpal joints.[9]

One of the first indications of illness in maedi is a deterioration in body condition and affected sheep lag behind the rest of the flock. As the disease progresses, respiratory embarrassment increases until the sheep is obviously dyspnoeic. Such a sheep may show clinical signs for several months before it eventually dies.

Sheep affected with visna show a variety of signs associated with CNS disease. During the early stages these signs are often subtle, such as a change in behaviour, and it is only in the later stages that definite signs such as circling, ataxia and posterior paresis are seen.

Pathology

The lungs of sheep affected by maedi do not collapse when removed from the thorax and often retain the impressions of the ribs. As the disease progresses, both the lungs and their associated lymph nodes increase in weight. In advanced cases the weight of the lungs may be 2–3 times the normal weight for that breed, e.g. 1 kilo or more.[10] The lesions are not focal but distributed throughout the lungs so that affected lungs are uniformly discoloured or mottled grey-brown and of a firm texture. The prime histological feature is infiltration of the interalveolar spaces by mononuclear cells often associated with peribronchiolar lymphoid hyperplasia and smooth muscle hypertrophy.

There are no gross alterations observable in the

CNS of sheep affected by visna but, histologically, there are chronic inflammatory changes in the brain, particularly in the periventricular areas and choroid plexus. There is also prominent lymphocytic infiltration, perivascular cuffing and variable amounts of demyelination.

Diagnosis

The clinical signs induced by infection with maedi-visna virus are not specific enough to enable a firm diagnosis to be made and confirmation must be attempted by laboratory procedures. Those used in the past have been histopathology and virus isolation, both of which are slow and have major drawbacks so that confirmation now is derived mainly from serological tests.

Several techniques are available to detect antibodies in the serum of infected sheep and these differ in specificity and sensitivity. Virus neutralization appears to detect antibodies to type-specific antigens and can be used to differentiate isolates of virus. These antibodies usually appear several months to years after infection. CF, AGID and ELISA tests detect antibodies to group-specific antigens, both p 25 and gp 135, and therefore are more useful as diagnostic tests, particularly the ELISA and AGID tests.

The results obtained with these tests are influenced by the composition of the test antigen and the unpredictable nature of the humoral antibody response in individual sheep. Despite these qualifications, it appears that antibody to the internal core protein is slow to appear (between 3 and 24 months) and develops in only a proportion of infected sheep, whereas those to the glycoprotein group-specific antigens appear more quickly (between 1 and 6 months) and are present in the majority of infected sheep. It is also known that a small proportion of sheep, from which virus can be isolated, do not develop antibodies which are detectable by any of the presently available tests and that these sheep are capable of infecting others.

Epidemiology and pathogenesis

Maedi and visna have been recognized in several countries throughout the world and, in recent years, they have appeared in northern Europe and Britain following the importation of breeding stock. In countries where these diseases are endemic, 40–60 per cent of adult sheep sera contain antibodies to the virus.

The routes by which infection is transferred between sheep are not known, although there must be at least two forms of horizontal transmission. The main form of transmission, at least where the infection is endemic, appears to be from dam to offspring and virus has been isolated from both colostrum and milk. Infection in utero is considered to occur infrequently if at all. A second form of horizontal transfer between adults has been well documented but the routes of virus secretion and transmission have not been determined. Transmission by the respiratory route is thought most likely.

During the course of the disease, the virus establishes a persistent infection of the lungs, CNS and haemopoietic organs, despite a humoral and cellular immune response. The important mechanism whereby maedi-visna virus is able to do this appears to be by integration of the viral genome within the chromosome of the host cell. If the genetic information of the virus is not expressed at the surface of the cell as new antigens, there will be no target for the immune response and the virus infection will be able to persist. This appears to be the case, as only a small proportion of the potential target cells become infected and within these the replication of the virus is restricted so that little viral protein or few mature virions are produced. While integration can explain the persistence of the virus, it is inadequate to account for the circumvention of the immune responses, which must occur to allow horizontal transmission, or the pronounced lymphocytic infiltration that is a feature of the histopathology. It is known that during the long course of the infection, mutation of the virus occurs so that new antigenic types appear. The epidemiological significance of these mutants is not yet clear because although they seem to appear in response to immune pressure, the old antigenic types may continue to be isolated. However, because the mutants are antigenically distinct, it may be that they are important in the transmission of infection between sheep and act as a recurring stimulation of the immune response leading to the progressive lymphocytic infiltration.

Control

It seems unlikely that vaccination could ever be used to control maedi-visna because of the ability of the virus to mutate around the immune response, and the implication of this response in the development of the lesion. Means of control, therefore, are limited to restricting the spread of infection to susceptible animals and eliminating infection from the flock.

Two approaches are currently under evaluation.[11] In the first, the flock is closed and sheep are tested at intervals for antibodies to maedi-visna. All sheep with antibodies are killed. The results so far suggest that all positive reactors can be eliminated in a few years. An Accreditation Flock Scheme was introduced in Britain in 1982 by the Ministry of Agriculture Veterinary Services which is aimed at applying these control methods.

The second approach is based on the contentious observation that intrauterine transmission does not occur, or only infrequently, and that the major means of transmission is in the dam's colostrum to her offspring. If the lambs are removed from their dams at birth before they have taken colostrum, and reared in isolation, it is possible in a short time to establish a maedi-free nucleus flock. This scheme can be difficult to implement because the lambs are deprived of colostrum and adherence to strict hygiene is required. Alternatively the lambs can be fed colostrum from a maedi-free source or bovine colostrum may be used.

REFERENCES

1 Wandera J.G. (1971) Sheep pulmonary adenomatosis (jaagsiekte). *Advances in Veterinary Science and Comparative Medicine*, **15**, 251– 83.

2 Perk K., Hod I. & Zimber A. (1971) Pulmonary adenomatosis of sheep (jaagsiekte) I. ultrastructure of the tumour. *Journal of National Cancer Institute*, **46**, 525–30.

3 Perk K., Michalides R., Speigelman S. & Schlom J. (1974) Biochemical and morphological evidence for the presence of an RNA tumor virus in pulmonary carcinoma of sheep (jaagsiekte). *Journal of the National Cancer Institute*, **53**, 131–35.

4 Martin W.B., Angus K.W., Robinson G.W. & Scott F.M.M. (1979) The herpesvirus of sheep pulmonary adenomatosis. *Comparative Immunology, Microbiology and Infectious Diseases*, **2**, 313–25.

5 Scott F.M.M., Sharp J.M. & Angus K.W. (1982) Proceedings of EEC. Seminar on Latent, Persistent Herpesvirus Infections in Veterinary Medicine. September 1982, Tubingen, FDR CEC Publication, in press.

6. Nisbet D.I., Mackay J.M.K., Smith W. & Gray E.W. (1971) Ultrastructure of sheep pulmonary adenomatosis (jaagsiekte). *Journal of Pathology*, **10**, 157–62.

7 Dungal N. (1946) Experiments with jaagsiekte. *American Journal of Pathology*, **22**, 737–59.

8 Dawson M., Chasey D., King A.A., Flowers M.J., Day R.H., Lucas M.H. & Roberts D.H. (1979) The demonstration of maedi/visna virus in sheep in Great Britain. *Veterinary Record*, **105**, 220–3.

9 Oliver R.E., Gorham J.R., Parish S.F., Hadlow W.J. & Narayan O. (1981) Ovine progressive pneumonia. Pathologic and virologic studies on the naturally occurring disease. *American Journal of Veterinary Research*, **42**, 1554–9.

10 Ressang A.A., de Boer G.F. & de Wijn G.C. (1968) The lung in zwoegerziekte. *Pathologia Veterinaria*, **5**, 353–69.

11 De Boer G.F., Terpstra C. & Houwers, D.J. (1979) Studies in epidemiology of maedi/visna in sheep. *Research in Veterinary Science.* **26**, 202–8.

4 ATYPICAL PNEUMONIA

G.E. JONES AND J.S. GILMOUR

Synonyms Apical pneumonia, lobar or enzootic pneumonia (Australia and New Zealand), chronic non-progressive pneumonia,[1] proliferative exudative pneumonia (experimentally reproduced pneumonia indistinguishable from atypical pneumonia).[2]

'Atypical pneumonia' (AP) of sheep, so-called to distinguish it from the classic, generally fatal 'enzootic pneumonia' or pasteurellosis, was first described and defined on pathological grounds in 1963.[3] The definition of the disease has since been narrowed but still depends on pathological criteria, due to the complex nature of the causal factors involved, although knowledge of the micro-organisms concerned is much improved. AP is a proliferative exudative pneumonia in which all or part of the anterior lobes of the lungs become consolidated. Pleurisy may or may not be present. The disease, which affects sheep between 2 and 12 months old, is chronic, non-progressive, often sub-clinical and infrequently fatal. Its occurrence and morbidity in flocks appears to be largely dependent on the type of management practised. In this chapter, AP will be used to denote the naturally-occurring disease, and proliferative exudative pneumonia (PEP) its experimentally reproduced form.

Cause

The principal agents implicated in AP in the UK are *Mycoplasma ovipneumoniae* and *Pasteurella haemolytica* biotype A serotypes (hereafter termed '*P. haemolytica A*'), but a number of other organisms are occasionally isolated from AP lesions which possibly modify the disease to a greater or lesser extent. The isolation rates of these organisms from the lungs of sheep observed in two surveys performed in Scotland, one of housed lambs[4] and the other of abattoir slaughtered animals, are shown in Table 4.1.

M. ovipneumoniae

This typical member of the Order *Mycoplasmatales* ferments sugars and produces centreless colonies on agar media. Table 4.1 shows that a high degree of association was found between the presence of *M. ovipneumoniae* and lesions of AP; lower isolation

rates were obtained from lungs in which bronchopneumonia, interstitial or cuffing pneumonia were the only observable changes. *M. ovipneumoniae* was also isolated, sometimes at high titre, from about 50 per cent of lungs in which no changes were apparent. These observations from the field are reflected by findings from experimental studies. Inoculation into the lungs of sheep of *M. ovipneumoniae* cultures or, more successfully, lung homogenates containing the organism reproduces PEP, interstitial or cuffing pneumonia in variable proportions of animals, and no observable changes in the remainder despite pulmonary colonization.[1, 2] It is concluded that *M. ovipneumoniae* is the primary agent of AP, but that the organism is a facultative pathogen which probably elicits pathological change under natural conditions only where defence mechanisms of the host are compromised. Wide inter-strain variations in antigenicity occur in *M. ovipneumoniae*, but the relevance of this to pathogenicity is unknown.

P. haemolytica A

This Gram-negative bacterium is considered to be the major (possibly sole) secondary and exacerbating factor of AP in the UK. The A2 serotype accounts for about half the isolations of *P. haemolytica*, although almost all 11 A serotypes have been cultured from field cases of AP. Table 4.1 shows that all cases of AP seen in young housed lambs yielded *P. haemolytica* A from the lungs, but only a quarter of the abattoir slaughtered sheep did. This discrepancy probably reflects both the capacity of *M. ovipneumoniae* per se to produce the disease and, of particular relevance to

Table 4.1 Microbiological findings in the lungs of lambs examined in two surveys in Scotland. Percentage of animals with the indicated type of histopathological change from which each organism was isolated

Organism	Housed lambs*			Abattoir-slaughtered lambs†			
	AP (15)‡	Interstitial or cuffing pneumonia (11)	No visible lesions (8)	AP (197)	Broncho-pneumonia (65)	Interstitial or cuffing pneumonia (711)	No visible lesions (56)
M. ovipneumoniae	100	91	50	95.4	78.5	50.4	48.0
P. haemolytica A	100	9	0	25.4	10.8	3.1	1.8
M. arginini	40	42	25	16.2	6.2	5.1	5.4
Chlamydia	ND	ND	ND	2.4	8.9	2.5	2.1

AP—Atypical pneumonia
ND—Not done
*Observations in one flock
†Observations involving 73 flocks or sources
‡Figures in parentheses indicate number of lungs within each category sampled in the survey

the older animals of the abattoir survey, that *P. haemolytica* A is eliminated from PEP lesions by 8–11 weeks post-inoculation.[5]

Except at very high titres *P. haemolytica* A neither produces lesions nor even establishes when inoculated into the lungs of conventional sheep over 2–3 months old.[2] However, if cultures or lung homogenates containing *M. ovipneumoniae* are inoculated simultaneously or up to 7 days previous to the *P. haemolytica* A (administered intranasally or intratracheally at numbers as low as 10000 organisms) the bacterium will establish in the lungs and subsequently provoke signs of acute pneumonia.[6, 7] The evidence suggests that *P. haemolytica* A establishes not only in consolidated lesions of PEP, but also in the less severe forms of pneumonia which *M. ovipneumoniae* can otherwise induce, and that these milder lesions are transformed and extended by the pasteurella into PEP. The paradox of this mycoplasma–pasteurella relationship is that *M. ovipneumoniae*, presumably through the inflammatory cell exudate which it induces, appears to restrict the invasiveness of the bacterium and thereby reduce its lethality.

Other micro-organisms associated with AP

Mycoplasma arginini

This mycoplasma is a common inhabitant of ovine tonsillar tissue. Table 4.1 shows that in the survey of housed lambs, *M. arginini* was isolated at similar frequencies from lungs with AP and with interstitial or cuffing pneumonia, and at a lower frequency from apparently normal lungs. Isolation rates from abattoir slaughtered sheep were much lower overall, but showed a threefold difference between lungs with AP and those without, due largely to the preponderance of *M. arginini* in some flocks in which AP was rife and severe. These field findings suggest that *M. arginini* neither causes nor modifies chronic ovine pneumonia, but that it can establish and flourish in the lungs of sheep kept under conditions of poor management. Experimental inoculation of *M. arginini* alone into the lungs of sheep produces neither pulmonary establishment nor disease.

Chlamydia psittaci ovis

Strains of chlamydia isolated from a variety of sources have been shown experimentally to reproduce a pneumonia in sheep that has some of the pathological features of AP. However, Table 4.1 shows that very few isolations of chlamydia were obtained from the lungs of slaughtered sheep, and that no association was apparent between presence of the organism and any form of histopathological change. Thus chlamydia are probably of no importance in chronic ovine pneumonias.

Pasteurella multocida and Acholeplasma oculi

These organisms have both been isolated on rare occasions from pneumonic sheep lungs in the UK; their relevance to disease production is unknown.

Other bacteria and mycoplasmas which have been isolated from AP cases, including *P. haemolytica* biotype T serotypes, *Branhamella ovis*, haemolytic staphylococci, streptococci and coliforms, *Acholeplasma laidlawii*, *Mycoplasma conjunctivae* and ureaplasmas, very probably have no role in AP.

No isolations of any virus from cases of AP have been reported, but in a survey of housed lambs serological evidence indicated that parainfluenza 3 (PI3) virus infection preceded the pulmonary establishment of mycoplasmas and bacteria and the development of AP.[4] This suggests that under natural circumstances PI3 virus could sometimes have a role in initiation of the disease. Natural, inadvertent superinfection of PEP with PI3 virus was observed to produce no apparent alteration in lung pathology and no additional clinical signs other than a fall in appetite and growth rate.[7]

Clinical signs

AP can affect over 40 per cent of lambs in a flock, but signs are frequently mild and may be overlooked by the flockmaster. The cardinal signs are chronic coughing (over weeks or months), often accompanied by dyspnoea or hyperpnoea, and best demonstrated by exercise; a mucopurulent nasal discharge; and depression. Sporadic deaths may occur. Observations of PEP indicate that the severity of signs exhibited is largely influenced by *P. haemolytica* A involvement. Where the bacterium is absent sheep generally remain bright and do not develop pyrexia. When *P. haemolytica* A is included in the challenge infection pyrexia usually follows 2–5 days after its administration irrespective of whether this is simultaneous with, or 7 days following, the *M. ovipneumoniae* component of the system.[6, 7] The pyrexia may persist for some days or recur intermittently on several occasions in the weeks following inoculation. Depression accompanies the pyrexia, and death occurs in 10–25

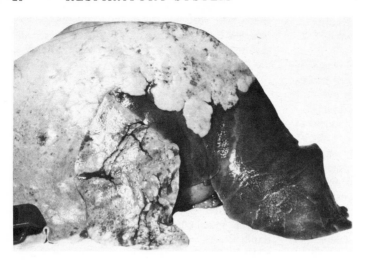

Fig. 4.1 *Atypical pneumonia.* Reddish consolidation involving a large area of anterior lung.

per cent of animals from 2–90 days post-inoculation.

Experimental studies have shown that chronic pneumonia can markedly reduce the appetite and growth rate of lambs, particularly in the first few weeks following inoculation.[7] In one experiment infected sheep ate 25 per cent more feed, required 9 weeks longer to reach similar mean live and carcass weights, and appeared to have lower feed conversion efficiencies than controls. Individual growth rates were significantly correlated with the degree of lung consolidation observed at slaughter. However, for reasons which are not yet understood, PEP has not consistently been found to reduce appetite and growth rates and its effects on the performance of experimental animals may on occasion be limited to a diminished efficiency of food utilization.

The sole study undertaken on the effect of naturally-occurring chronic pneumonia on the performance of lambs was in an extensively maintained flock in New Zealand.[8] This survey found that only moderate and severe pneumonia, which affected 6.5 per cent of lambs, significantly reduced carcass weights by a mean of 0.45 kg per animal. It should be noted, however, that even normal lambs of this survey showed only moderate growth rates, and that slaughter was at 4–5 months old, only shortly after the age at which AP generally develops.[4]

Pathology

Lesions of AP are confined usually to the apical, cardiac and, less frequently, anterior borders of the diaphragmatic lobes of the lung (Fig. 4.1). The clearly-demarcated areas of consolidation vary in colour from grey to red-brown, and may be accompanied by dull red bands of collapsed tissue. Slices of consolidated tissue reveal the lesions to be firm, with grey-white nodules often prominent on the cut surface. A pleurisy, characterized by the production of fibrinous 'tags' or adhesions, is occasionally present.

The original description of AP[3] divided the disease into two histological forms, interstitial pneumonia and lymphoid hyperplasia. Interstitial pneumonia was the more common, accounting for about 90 per cent of cases seen, and its macroscopic and microscopic features are those of the narrower definition of AP adopted in this chapter. This pneumonia is thought to constitute the commonest and most severe form of disease produced by the combination of *M. ovipneumoniae* and *P. haemolytica* A. The range of tissue responses seen is wider than the term 'interstitial' suggests, and includes nodular lymphoid hyperplasia and bronchiolar epithelial hyperplasia as well as an increased number of mononuclear cells in the alveolar septa (Fig. 4.2). Many of the latter appear to be lymphocytes and these may also form cuffs around bronchioles and blood vessels. Some alveoli may be collapsed but most contain an exudate in which macrophages are a constant feature and neutrophil clusters are a variable finding (Fig. 4.3). Infrequently a pseudoepithelialization of alveoli is seen; this appears to be the result of hyperplasia of Type 2 cells. A final feature of this lesion, and one which characterizes AP, is the presence of nodular 'scars' situated within or close to bronchiolar walls. These have a hyaline appearance and may project into the lumen of the bronchiole with only a thin epithelial covering. Sequential examinations of PEP,

Fig. 4.2 *Atypical pneumonia.* Marked hyperplasia of lymphoid nodules and bronchiolar epithelium. (H & E x 32)

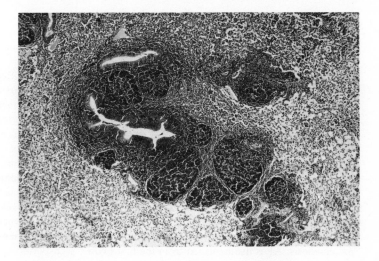

which is pathologically indistinguishable from AP, demonstrated that hyaline scars may be present at all stages, but with greatest frequency around 2 months post-infection.[5] The suppurative exudate disappeared at about 3 months post-infection, and this coincided with a failure to recover *P. haemolytica* A from the lungs. Lymphoid nodular hyperplasia, however, appeared to develop slowly and remained the prominent feature at the end of the observation period of 7 months, by which time the macrophage exudate and bronchiolar epithelial hyperplasia were apparently receding. A significant feature was that *M. ovipneumoniae* was still recovered from the lungs at these late stages.

Lymphoid hyperplasia, the second form of AP according to the original description,[3] has extensive cuffs of lymphoid tissue surrounding airways and blood vessels. Compression of airways and a mild macrophage exudate into surrounding alveoli may also be present. This form of lesion has been observed in young lambs before the age at which AP commonly develops,[4] and in the late, resolving stages of PEP.[5] In both situations *M. ovipneumoniae* was the only organism cultivated, and experimental pathogenicity studies support the conclusion that lymphoid hyperplasia is a common, generally late manifestation of pulmonary infection with the mycoplasma.

Fig. 4.3 *Atypical pneumonia.* Consolidated lung demonstrating bronchiolar epithelial hyperplasia, macrophage and neutrophil exudates and a hyaline scar (arrow) (H & E x 80)

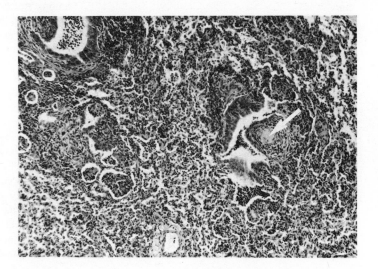

Diagnosis

The exhibition of signs of respiratory disease by sheep under one year old is indicative, but confirmation depends on observation of the characteristic lesions of AP at slaughter. Nasal swabs generally reveal a high proportion of sheep in the flock to be carrying *M. ovipneumoniae* and, less consistently, *P. haemolytica* A, but this finding does not invariably imply the presence of AP. Recently developed ELISA tests for antibodies to both organisms have proved promising as diagnostic aids, although high antibody titres to *M. ovipneumoniae* are also often found in sheep suffering from pulmonary adenomatosis (SPA).

Differential diagnoses include SPA, maedi and pasteurellosis. AP is distinguished from the first two on the basis of age distribution and pathology, and from pasteurellosis on clinical symptoms and pathology. *M. ovipneumoniae* may be isolated from the respiratory tract of animals affected with these diseases. Although rarely associated with clinical disease, dictyocauliasis (*Dictyocaulus filaria*) may result in lesions grossly resembling those of AP, but usually restricted to the ventral borders of the diaphragmatic lobes.

Epidemiology and transmission

Transmission of AP is probably by the respiratory route, a presumption supported by the ability to reproduce the disease by administration of pneumonic lung homogenates either as suspensions inoculated intratracheally,[2] or as aerosols applied to the nares.[1]

M. ovipneumoniae and *P. haemolytica* A are both found commonly in lowland flocks, and with diminishing prevalence as the stocking density decreases and the altitude of maintenance rises. Correspondingly, AP is largely a disease of intensive and semi-intensive systems of husbandry, and rarely occurs in extensively maintained animals. AP may affect over 40 per cent of lambs kept in poorly ventilated houses, and up to 30 per cent of animals kept outdoors. Endemic and epidemic forms of AP can be distinguished. The endemic form occurs in home-bred lambs and involves an escalating level of infection with the micro-organisms concerned (indicated by isolation rates from both nasal passage and lungs) in the lambs and their dams. Pulmonary infection with *M. ovipneumoniae* precedes that with *P. haemolytica* A, but PI3 virus infection may antecede both.[4] The

increasing incidence of AP after 2 months of age probably reflects both the suppressive effect of colostrally-acquired antibody, which wanes to minimal levels by 7 weeks of age, and the importance of weight of infection on development of the disease.[4] Other factors clearly operate in the endemic situation, since farms in the same locale and employing similar systems of husbandry often have very different prevalences of AP, but the exact nature of these influences, whether microbiological, environmental, genetic or nutritional, is unknown.

The epidemic form occurs most commonly where batches of lambs from different sources bought in for fattening over the autumn and winter months are mixed together. The subsequent development of AP probably stems both from the adverse effects of transportation and change of feeding regime and from the mutual exchange between sheep of different strains of mycoplasma and pasteurella.

Control, prevention and treatment

Control requires that those factors which debilitate the host and favour the transmission of airborne pathogens be reduced or eliminated. The simplest means to achieve both aims is by reduction of the stocking density. Good ventilation is essential in housed systems to prevent the build-up of respiratory pathogens and noxious gases such as ammonia, and to reduce dust. Batches of bought-in lambs should be reared separately, at least until the effects of transportation and change have been overcome.

No vaccine against *M. ovipneumoniae* has been developed, but due to the wide antigenic variation among strains of the mycoplasma and the difficulty experienced in producing efficacious vaccines against respiratory mycoplasmas of other animal species, the prospect does not seem to offer much hope. An alternative is vaccination against *P. haemolytica* (considered in detail in the Chapter on Pasteurellosis), since the most important economic aspects of AP appear to be attributable to the bacterium.

The effectiveness of three parenteral and one oral therapeutic agent against experimentally reproduced AP have been evaluated.[9] Oxytetracycline, tylosin and penicillin were found to prevent almost entirely the development of pneumonia when administered parenterally on the day of challenge and for 12 successive days thereafter: penicillin and tylosin did not, however, entirely eliminate bacteria and mycoplasmas from the lungs. Ronidazole *per os* was found to be as effective as the parenteral drugs only when given at levels (100 mg/kg) which produced toxic side effects.

REFERENCES

1 Alley M.R. & Clarke J.K. (1979) The experimental transmission of ovine chronic non-progressive pneumonia. *New Zealand Veterinary Journal,* **27**, 217–20.

2 Gilmour J.S., Jones G.E. & Rae A.G. (1979) Experimental studies of chronic pneumonia of sheep. *Comparative Immunology, Microbiology and Infectious Diseases,* **1**, 285–93.

3 Stamp J.T. & Nisbet D.I. (1963) Pneumonia of sheep. *Journal of Comparative Pathology and Therapeutics,* **73**, 319–28.

4 Jones G.E., Buxton D. & Harker D.B. (1979) Respiratory infections in housed sheep, with particular reference to mycoplasmas. *Veterinary Microbiology,* **4**, 47–59.

5 Gilmour J.S., Jones G.E., Keir W.A. & Rae A.G. (1982) Long-term pathological and microbiological progress in sheep of experimental disease resembling atypical pneumonia. *Journal of Comparative Pathology* **92** 229–38

6 Jones G.E., Gilmour J.S. & Rae A. (1978) Endobronchial inoculation of sheep with pneumonic lung-tissue suspensions and with the bacteria and mycoplasmas isolated from them. *Journal of Comparative Pathology,* **88**, 85–96.

7 Jones G.E., Field A.C., Gilmour J.S., Rae A.G., Nettleton P.F. & McLauchlan M. (1982) The effects of experimental chronic pneumonia on body weight, feed intake and carcass composition of lambs. *Veterinary Record,* **110** 168–73

8 Kirton A.H., O'Hara P.J., Shortridge E.H. & Cordes D.O. (1976) Seasonal incidence of enzootic pneumonia and its effect on the growth of lambs. *New Zealand Veterinary Journal,* **24**, 59–64.

9 Alley M.R. & Clarke J.K. (1980) The effect of chemotherapeutic agents on the transmission of ovine chronic non-progressive pneumonia. *New Zealand Veterinary Journal,* **28**, 77–80.

5   PARASITIC BRONCHITIS

J. ARMOUR

Synonyms Hoose or Husk

Parasitic bronchitis in sheep in Britain is a minor problem compared to the similar condition occurring in cattle. However, in other parts of the world such as the south east of Europe, the Mediterranean area and the Middle East the clinical disease is seen much more frequently. Husk is primarily a condition of young sheep and is characterized by coughing and weight loss.

Cause

The condition is caused by *Dictyocaulus filaria*; the adult parasites being located in the trachea and bronchi.

Clinical signs

The principle clinical signs are coughing, tachypnoea and weight loss. Appetite is often reduced and in chronic cases the weight loss is severe. Pyrexia is uncommon except where there is a complicating secondary infection, in such cases there is a discharge from eyes and nose and dyspnoea may occur.

Life-cycle and pathogenesis

The worms are ovo-viviparous and the eggs, containing first-stage larva L1, hatch in the air passages or the gut after being coughed up from the trachea and being swallowed. In the faeces the L1 moult and develop through the L2 to the L3 which is the infective stage. The L3 retain the sheaths of the previous stages and therefore never feed but possess an abundance of nutrient material easily distinguishable as dense dark-stained granules in the intestinal cells of the larvae. The sluggish larvae migrate onto the pasture and if ingested by a susceptible sheep they penetrate the wall of the small intestine and reach the lungs via the mesenteric lymph glands, the thoracic duct, anterior vena cava and heart. Moulting to the L4 stage occurs by the seventh day after infection by which time they have reached the alveoli. Thereafter the larvae move into the bronchioles, the bronchi and trachea reaching the mature adult stage about 4 weeks post-infection.

The presence of the larvae in the alveoli stimulates a cellular infiltration of neutrophils, eosinophils, macrophages and multinucleated giant cells. As the larvae move up the bronchioles a similar infiltrate results and the bronchiolar walls become infiltrated with cells, mainly eosinophils. These changes lead to alveolar collapse, characterized clinically by tachypnoea and mild coughing. In heavy infections emphysema occurs.

Once in the main bronchi, the L5 and adult worms cause a severe reaction in the bronchial epithelium which becomes hyperplastic and infiltrated with neutrophils and eosinophils. Some of the eggs laid by the adult worms and L1 which have hatched are aspirated into the bronchioles and alveoli and pneumonia develops due to the inflammatory cellular reaction to these eggs and larvae. Clinically, these changes are characterized by tachypnoea and more severe coughing. The pneumonia is often complicated by bacterial infection and where this is severe, dyspnoea may be present. Epithelialization of the alveoli occurs during the patent infection and can

sometimes be widespread even in sheep harbouring comparatively few lungworms.

At post-mortem examination, the pneumonic areas are widespread in the lungs and the surface of the lungs is often studded with purulent areas caused by secondary infection. Emphysema and oedema may be complicating features.

Diagnosis

Diagnosis is based primarily on the clinical signs and seasonal occurrence of the disease. The demonstration of the characteristic L1 in the faeces, using either conventional flotation techniques or the Baermann method, is a useful aid to diagnosis. This larva which is shown in Fig. 5.1 is 560 μm in length and has a characteristic button-like structure on its anterior end.

At post-mortem examination the adult parasites, which range from 4 cm (males) to 10 cm (females) in length, are easily recognized on opening the main bronchi. Where heavy larval infections predominate these may be recovered by opening the smaller air passages and placing the opened lungs face down on a sieve in a funnel of warm water; the larvae will migrate out of the lungs into the warm water and can be recovered from the neck of the funnel within a few hours. Alternatively, the Inderbitzen technique[1] may be employed. In this method the heart is removed with the lungs and opened to permit the introduction of a broad catheter into the right ventricle. The other end of the catheter is attached to a pressure tap and the flow of water through the pulmonary artery flushes the larvae (or adults) through the lungs and up the trachea which is directed onto a sieve to collect

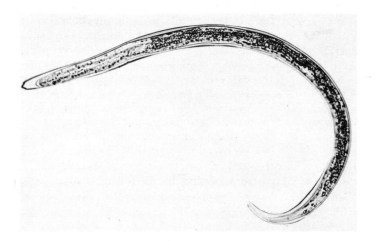

Fig. 5.1 First stage larva of *D. filaria.*

the lungworms. A clamp is applied to the other pulmonary vessels to prevent rupture of these.

Epidemiology and transmission

In the faeces, development of the L1 to the infective L3 stage only occurs during the spring, summer and autumn months. It is not definitely known how many cycles of infection occur each year; as with other trichostrongyles a maximum of two cycles per annum is the probable situation in western Europe. Development of the free-living stages requires adequate moisture and the optimal temperature for development and survival is 19°C. Larvae therefore accumulate on pastures in the second half of the summer [2] and outbreaks are seen mainly in the early autumn i.e. September and October, but sometimes occur earlier in the summer. Clinical husk is seen mainly in lambs but adult sheep not previously exposed can also be severely affected.

Infection is carried on from year to year by:

1. larvae overwintering on the pasture
2. carrier sheep

It is possible that the infections carried over the winter by carrier sheep are hypobiotic larvae which mature in the spring. This has been reported from Canada [3] and the present author has also observed an increase in *D. filaria* larval numbers in the faeces of ewes in the spring.

Sheep are known to develop an acquired immunity to *D. filaria* infection but during the peri-parturient period the relaxation of immunity known to occur with gastrointestinal nematodes may also affect immunity to lungworms and allow either reinfection or maturation of hypobiotic larvae.

Treatment and control

The following drugs are effective in the treatment of ovine parasitic bronchitis:

    levamisole (Nemicide, ICI Ltd, Macclesfield, Cheshire, England) at 7.5 mg/kg, and the benzimidazoles:

    fenbendazole (Panacur, Hoechst UK Ltd., Milton Keynes, Bucks, England) at 7.5 mg/kg

    oxfendazole (Systamex, Wellcome Foundation Ltd., Berkhamsted, Herts, England) 5 mg/kg

    albendazole (Valbazen, Smith, Kline and French Labs. Ltd, Welwyn Garden City, England) at 5 mg/kg.

Specific control measures are not adopted in Britain for the control of husk. However, the methods and drugs recommended for the control of parasitic gastroenteritis (see page 56) should also be effective in the prevention of ovine parasitic bronchitis, i.e. prophylaxis by anthelmintics, alternate grazing of sheep and cattle or a combination of these strategies.

An irradiated larval vaccine has been developed for the control of *D. filaria* in Yugoslavia and some Middle Eastern countries. This vaccine has given good results but since the disease occurs only sporadically in Britain, this method of control has not been pursued.

OTHER PARASITIC PNEUMONIAS OF SHEEP

Other lungworms which occur in British sheep include *Muellerius capillaris, Protostrongylus rufescens, Cystocaulus ocreatus* and *Neostrongylus linearis.* Of these *M. capillaris* and *P. rufescens* are by far the most common.

Muellerius capillaris

This is the commonest lungworm of sheep and occurs in more than 90 per cent of British sheep. The parasite has an indirect life-cycle, the intermediate host being several species of land molluscs. Infection is by ingestion of the snail and migration like *D. filaria* is via the lymphatic route.

The lesions produced are usually around terminal bronchioles and alveoli of the caudal lobes and are usually sub-pleural in location. Macroscopically they appear as multiple nodules, grey or greenish in colour and about 1 mm in size. Histologically, the adult parasites, eggs and first stage larvae are found in the alveoli or alveolar ducts often surrounded by eosinophils, lymphocytes, macrophages and giant cells. Although the parasite produces readily identifiable lung damage the significance of the infection in terms of clinical signs and reduced productivity is not known and infection is usually discovered at necropsy or during routine faecal examination when the first stage larvae are seen.

Control is difficult in view of the indirect life-cycle of the parasite. The modern benzimidazoles, i.e. fenbendazole, oxfendazole and albendazole used in treating *D. filaria* infections are reported as being effective in treatment.

Protostrongylus rufescens

This parasite also has an indirect life-cycle utilizing a wide range of land snails and slugs as intermediate hosts. Migration to the lungs is via the lymphatic system and as the adult parasites are found in the small bronchioles the lesions are therefore more widespread than with *M. capillaris*.

Like *M. capillaris*, diagnosis is usually made post-mortem or by identifying the L1 during routine faecal examinations. The modern benzimidazoles are recommended for treatment.

REFERENCES

1 Inderbitzin F. & Eckert J. (1976) Experimentell erzeugte Entwicklungschemmung bei *Dictyocaulus viviparus* des Rindes. *Zeitschrift fur Parasitenkunde*, **50**, 218.

2 Al-Sammarrae S.A. & Sewell M.M.H. (1977) Studies on the epidemiology of *Dictyocaulus filaria* infection in Blackface sheep on a low-ground Scottish farm. *Research in Veterinary Science*, **23**, 336–9.

3 Ayalew L., Frechette J.L., Malo R. & Beauregard C. (1974) Seasonal fluctuations and inhibited development of populations of *Dictyocaulus filaria* in ewes and lambs. *Canadian Journal of Comparative Medicine*, **38**, 448–56.

SECTION 2

ALIMENTARY SYSTEM

DISEASES OF THE MOUTH AND TEETH

6

J.A. SPENCE

Many significant diseases show, as part of their symptomatology, lesions of the mouth, e.g. foot-and-mouth disease and bluetongue and thus the site is of great diagnostic significance. However, these systemic conditions are covered in full elsewhere and therefore this section includes only those conditions in which lesions are limited to the oral cavity which may be significant economically or those that should be considered in the differential diagnosis of systemic diseases.

The sheep is very prone to dental disease. In many parts of the world where sheep rearing is a major industry the results of dental disease—loss of weight and a reduced ability to thrive under suboptimal conditions—and its high prevalence reduce flock profitability. Culling of breeding ewes is necessary before the end of their useful reproductive life with consequently increased flock replacement costs.

Dental conditions take one of a number of forms: early incisor or cheek tooth loss, excessive tooth wear, irregular wear, deformities of the mandible and problems of occlusion are examples. A number of syndromes are recognized that incorporate one or more of these forms of dental disease, e.g. broken mouth and caries in Britain, paradontal or periodontal disease, acute ulcerative gingivitis and excess incisor wear in New Zealand and Australia and mandibular dentigerous cysts, 'wavy mouth' and occlusal defects in many countries. The delineation of these specific syndromes in the literature is imprecise and it is probable that many of them are synonymous. In this chapter, therefore, dental conditions are discussed according to their presenting clinical signs rather than by specific syndrome, although relationships between syndromes are discussed where necessary.

INCISOR LOSS

Broken mouth is the common term applied to incisor loss in sheep in Britain where it is the most significant dental problem recognized (Fig. 6.1). Early incisor loss has been reported world-wide but is most often seen in certain areas of New Zealand where it is called paradontal or periodontal disease of sheep. In

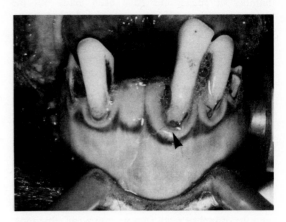

Fig. 6.1 A typical case of terminal broken mouth. A central incisor has been lost and the remaining teeth have moved within the jaw giving gaps between the teeth. The exposed crown appear larger than normal and the gingiva surrounding the teeth looks swollen and "rounded" (arrow).

Britain 70 per cent of ewes coming to slaughter have evidence of incisor loosening or loss, and abattoir surveys show that it occurs country-wide .[1] However, broken mouth is not evident in all flocks; some are completely free of the condition whilst in others it may appear in animals at any age between 3 and 8 years of age, the prevalence within the flock varying between 5 and 70 per cent.

Its economic significance varies with incidence, flock management and farm type.[2] On hill and upland farms where grazing is poor, farmers cull ewes before the end of their useful reproductive life since they believe that animals developing broken mouth are then unable to maintain themselves on such grazing. However, the sale price of the broken mouth ewe is now so much less than sound-mouthed contemporaries (33 per cent difference in 1981) that some farmers now retain such animals though they require extra feeding through the winter and spring in an attempt to maintain body condition and produce a good lamb. There is some evidence that, despite this treatment, such ewes produce lighter weaned twins than sound-mouthed ewes of the same age.

On lowground farms the economic impact of broken mouth is less since the better pasture conditions maintain body weight and production despite incisor loss. Culling for dental disease under such conditions is less common.

The cause of early incisor loss is unknown. A number of suggestions have been made but the evidence for any one is inconclusive. An imbalance in the nutritional calcium to phosphorus ratio, or lack of calcium and an effect on the bony tooth supports has been unable to substantiate this. Improved pasture management, e.g. reseeding and liming, tough diets, e.g. root crops, excess wear on cheek teeth, faulty occlusion, high levels of oestrogen in the diet and problems in hogg nutrition, e.g. root feeding, are further unsubstantiated proposals.
further unsubstantiated proposals.

Clinically, broken mouth is usually noticed in its terminal stages of tooth loosening and loss, but careful examination of incisors and gingivae can show dental changes long before this stage. Within months of incisor teeth coming into wear, gingivitis, characterized by patchy gingival oedema, reddening of the gingival margin and capillary fragility, is evident in all sheep. In animals that remain free of broken mouth this may persist throughout life but remain slight and difficult to see. On farms where broken mouth develops, the gingivitis worsens within months of eruption and symptoms are more readily seen; pus may be expressed from the sulcus and the gingival margin becomes uneven and wrinkled. There is some suggestion that the intensity of the gingivitis fluctuates and that it is worst in the summer, but this has yet to be substantiated.

As a result of the gingivitis, recession of the gingival margin occurs and this may make the tooth crowns seem very long. Irregular incisor crown wear progressing to peg-like teeth is a common adjunct to gingivitis, especially late in the condition. Grass impaction around incisors is a further common finding which may also have a seasonal incidence, but its significance remains unclear. The course of broken mouth is variable; the time from the appearance of severe gingivitis to loss of tooth may span as little as a year but may extend over three years, the duration being farm dependent.

Pathologically, broken mouth is a chronic localized inflammatory response to local bacterial antigens in subgingival plaque.[4] A complex microbial flora develops in the gingival sulcus as soon as a tooth erupts into wear with an inflammatory response in the sulcal wall and associated lamina propria dominated by plasma cells and fewer polymorphs. The

character of response suggests that immune and autoimmune mechanisms may play a role in its development. On farms that remain free of broken mouth the amount of subgingival plaque remains small and the response localized, but where broken mouth occurs, the amount of plaque and its morphological complexity increases significantly and the local inflammatory response also intensifies. Up to 18 months prior to tooth loss, though usually about a year to 9 months before, the gingivitis generalizes to a periodontitis which may involve all the supporting periodontal ligament and even alveolar bone.

Incisors are lost following local tissue damage. While the mechanisms causing this damage have yet to be clarified, the results are clear. Under normal circumstances the morphology of the periodontal ligament and surrounding bone appear well suited to accepting the mechanical grazing forces applied to the incisors.[5] The tissue damage in broken mouth is associated with deepening of the sulcus round the tooth to form a pocket and the loss of collagen from within the ligament. Both destroy the functional morphology of the periodontium, particularly on the lingual aspect of the incisors, and make the tooth liable to loss from normal grazing forces.

Recent work now suggests that broken mouth has similarities in its pathogenesis to periodontal disease in man. Immune responses to subgingival plaque antigen components appear to be of primary importance to man and sheep though the responses in the sheep have yet to be clarified. Similarly, no specific plaque antigen or bacterial species has been implicated in either man or sheep. Rather, the current concept is one in which a number of micro-organisms are associated with tissue destruction and the induction of a plethora of host responses.

If these suggestions are correct then some explanation can be made for the apparently inconsistent role of the environmental factors mentioned earlier. In man it is well recognized that systemic and environmental factors can affect the rate of development of periodontitis and that they can work in different combinations; it is probable that a similar situation operates in sheep. It is postulated that a balance between periodontal health and disease prevails in the sheep and that any, or a combination, of those factors once considered causes of broken mouth may tip the balance towards disease rather than initiate the condition themselves; they may best be described as predisposing causes of broken mouth. However, many more environmental factors influence the health of the peridontium, its repair, nutrition and defences than those already mentioned. For example, good protein nutrition is central to periodontal health and many trace elements, e.g. copper, have been shown to have significant roles in maintaining host immune defences. While no evidence is available to link these factors to early incisor loss, their significance should not be overlooked in the investigation of dental problems; it may be wise to consider broken mouth a multi-factorial problem that may be unique to each farm.

There is no treatment or control for broken mouth. Management changes, to date, have been singularly unsuccessful in controlling it but this may be associated with the unrecognized long course of the condition; management changes may take 3–4 years to affect the incidence of broken mouth on a farm! A number of dental surgical procedures have also been tried, the most recent being a metal splint to aid tooth support. No earlier technique has given any more than temporary relief and the splint has, at the time of writing, yet to prove its economic and mechanical worth.

EXCESSIVE INCISOR WEAR

In New Zealand and Australia excessive incisor wear in which the incisor is worn down to gum level before 3 years of age, is a common and a significant economic condition.[6] In Britain what little information is available suggests that it is only significant very locally where sheep are grazed on very marginal sandy pasture, such as sand dunes. The tooth wear associated with broken mouth is not severe, the peg-shaped tooth mentioned above is never worn down to gum level.

HOLES, PITTING AND DISCOLOURATION OF INCISOR TEETH

Small, shallow pits in the enamel of the incisor crown, no larger than 1–2 mm across, are incidental findings during dental examinations. They appear sporadically, are associated with hypo-amelogenesis and appear to have little clinical significance; the suggestion that they are more prevalent in parasitized hoggs and yearlings is unsubstantiated.

Acid etching of enamel, caries and fluorosis are all characterized by enamel pitting, the last two being of occasional economic significance. Caries is the development of deep holes in incisor enamel by bacterial etching. The neck of the incisor at gum level is the most frequent site of attack (Fig. 6.2). Caries is quite common though few outbreaks are reported since malnutrition, the only significant consequence of caries, occurs only under very specific conditions of management. Crowns, weakened by caries, can snap off at gum level leaving a ragged stump; this can be a problem in hoggs fed on root crops during the winter, the stumps of their deciduous incisors are unable to cut up hard roots and body condition deteriorates as a result. Histologically the condition in lambs resembles that of man and, like that condition, is probably associated with diets high in soluble carbohydrate, e.g. concentrate feeding and some strains

Fig. 6.2 Dental caries on the anterior aspect of the fourth deciduous incisor (between arrows). A pit within the enamel is surrounded by brown enamel staining. The third incisor crown (stemmed arrow) has snapped off leaving a ragged, rounded crown.

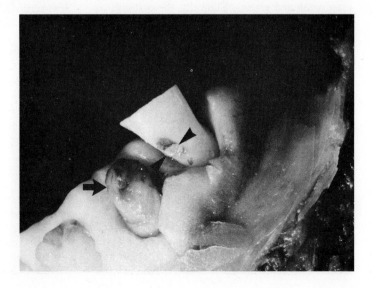

of root crops. Caries in adult animals has been reported in only one reference[7] but it is likely that it occurs sporadically and has little clinical significance.

Dental fluorosis is significant as the most usual way of recognizing fluoride intoxication, skeletal lesions appearing later than dental lesions. Pitting of enamel is extensive, the enamel is discoloured and chalky and the affected teeth wear much faster than normal. Fluorine ingestion may follow industrial contamination or the use of certain rock phosphate fertilizers, particularly those originating in North Africa. Where exposure to the contaminant is brief the resulting lesions may occur as an irregular groove around a pair of incisors which, with time, may be mistaken for carious pitting of enamel.

Discolouration of the visible incisor crown is a regular condition in sheep. Brown discolouration of the incisors, most pronounced at the gum margin and disappearing towards the incisive surface of the crown, is quite normal and is due to colouring of the porous cementum which covers the complete incisor crown at eruption. A thicker mineralized deposit over enamel, calculus, is occasionally seen but this is more common round the cheek teeth where it may be brown to black and occasionally have a metallic sheen. It is not clinically significant.

ANTERIOR SWELLING OF THE MANDIBLE

Localized unilateral swellings of the anterior portion of the horizontal ramus below the incisors in 2–4 year old ewes may occasionally become a flock problem[8] since the size is often sufficient to prevent normal mastication and grazing. Clinically, the swellings are bony and involve the incisor supports, one or more incisors always being displaced or missing. Histologically, the swellings are fluid-filled spaces lined with stratified epithelium, lying within a thin shell of alveolar bone (Fig. 6.3). One, or more, permanent incisor is found in close apposition to, or within, the cyst wall. The cause of these so-called dentigerous cysts is unknown. The sporadic occurrence within a flock does not suggest that they are the equivalent of the human congenital dentigerous cyst. It has been proposed that they arise as a result of abscessation of periodontal tissues during the development and eruption of permanent incisors followed by partial recovery and cyst formation from displaced epithelial fragments. In New Zealand, workers believe that nutritional factors such as a subclinical copper deficiency may play a role in their pathogenesis. Both of these hypotheses have yet to be substantiated.

DISEASES OF THE CHEEK TEETH

While conditions of the incisors have had some attention, on the whole, molar and premolar disease have been disregarded due to the difficulty of clinical examination. However, abattoir surveys[9] have shown that molar disease is more common than is realized and may be a severe flock problem for similar reasons to broken mouth. In fact, in one survey there was a better correlation between molar disease and body condition at slaughter than there was between the latter and incisor loss.

Molar disease is a problem of adult sheep but they may not show any clinical evidence of it until only a few normal cheek teeth remain. The first signs may

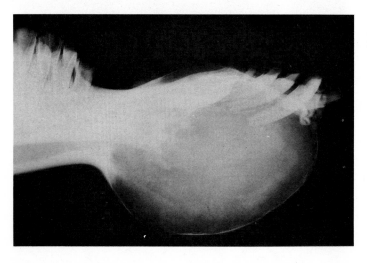

Fig. 6.3 Radiograph of anterior mandible showing so-called dentigerous cyst involving all the incisors.

be pregnancy toxaemia; ewes with molar problems and under metabolic stress are not able to achieve an adequate food intake. Otherwise, a picture is seen of slow deterioration in condition of a few ewes, occasional halitosis and some dribbling of rumen liquor during cudding. Digestive upsets with fluid intestinal contents may be a complication of inadequate food mastication. Manipulation of the mandible from the ventral aspect of the jaw may show hard bony swellings in the region of the cheek teeth of some ewes and the smoothing out of cheek tooth outlines in others due to impaction of grass between the teeth. Actual loss of mandibular or maxillary teeth may also be palpated from the outside.

These symptoms have been seen to accompany a number of apparently distinct syndromes (gingivitis, periodontitis, periostitis, tooth loss, mal-eruption, overgrowth and 'wavy mouth') but whether they are different manifestations of a single condition or a number of epidemiologically distinct diseases is unclear.

There is no information available to explain the pathogenesis of molar disease. However, both gingivitis and periodontitis are seen at necropsy, many cases being more acute and purulent than those seen in broken mouth; the lesions resemble those seen in the New Zealand syndromes of 'acute ulcerative ginvigitis'.[10] An alveolitis and periostitis probably develop as a consequence of the periodontitis and it is the new bone formation occurring as a result of the inflammation that is palpable externally. The sulcus between the third premolar and first molar is most often affected due to the interdental gap at this site.

On farms where molar disease is associated with broken mouth, gingivitis and periodontitis is more chronic than that just described. There is no purulent discharge and periostitis is rare. The relationship between the acute and chronic conditions is unknown. It is probable that the more chronic gingivitis discussed under broken mouth is aggravated to the more acute form by unrecognized, farm specific environmental factors. Similarly, many of the other syndromes appear to be secondary complications of the basic acute periodontitis. Thus if the alveolar bone is sufficiently rarified by alveolitis, then abscessation and fistulation to the ventral aspect of the jaw can ensue. In other cases all the ligamentous support to the molar tooth is lost but, because of its length and bony support, the tooth is not shed. Such teeth are likely to wear unevenly giving an irregular grinding table and shorter tooth. If a tooth is lost, the opposing tooth will overgrow into the empty socket and may cause ulceration and lacerations of the

tongue, gums and cheek. Packing of food material into pockets and sockets gives a severe halitosis and will exacerbate any infection still present. 'Shear mouth', the exaggerated peaking of the molar crowns, does not appear to be associated with these periodontal syndromes although it does produce similar lacerations of surrounding tissues.

There is no known cause for these conditions. Metabolic stress on the skeletal supports has been suggested but since molars are often affected whilst incisor teeth are clear of apparent disease this would appear unlikely. A pathogenesis similar to broken mouth may be a possibility in some cases, while in others the secondary contamination by bacterial pathogens of a site already weakened by chronic periodontitis may be postulated. A bacterial aetiology is supported by the observation that large doses of oxytetracycline given as a fluorochrome bone marker has improved the condition in one flock. However, all these suggestions require substantiation.

Treatment has never been attempted apart from the one situation mentioned. Once molar disease becomes severe enough to induce clinical symptoms it is probable that the condition is irreversible and the animal should be removed from the flock. Similarly, there is no recognized preventative measures for molar disease.

OCCLUSION

Correct occlusion, the meeting of the incisors within 1–3 mm of the front of the upper dental pad, is considered important in the selection and breeding of rams and ewes; good occlusion is said to minimize dental problems. While severely overshot (sow mouthed) or undershot (parrot mouthed) mandibles may be inefficient and predispose to dental problems, there is no proof that occlusion within these extremes has a marked effect on broken mouth[4] or other dental disease. Recent work has shown that selection for perfect occlusion is over-emphasized, the trait has a low heritability factor.

An undershot jaw associated with inability to close the mouth occasionally develops in hoggs. This is a rachitic syndrome and the section on osteodystrophic diseases of sheep should be referred to for this condition.

STOMATITIS

Apart from the 'exotic' conditions of foot-and-mouth disease and bluetongue, stomatitis is encountered in a number of local and generalized diseases. A catarrhal stomatitis, which may be associated with a gingivitis and mainly involves the posterior fauces of the soft palate, frequently develops in the course of many acute debilitating diseases and in local diseases of the mouth and pharynx, e.g. some chemicals from foodstuffs. There is oedema and local swelling, hyperplasia of local lymphoid tissues and later desquamation of epithelium. Recovery from the systemic condition also cures the stomatitis.

Severe contagious pustular dermatitis or 'orf' may present typical proliferative lesions on the tongue, palate and dental pad as well as the lips and skin. These lesions are raised, reddened foci with a severe peripheral hyperaemia; they do not keratinize.

TUMOURS

Fibrosarcomas of the jaw do occur. These are discussed in Chapter 45.

FUSOBACTERIUM NECROPHORUM AND OTHER BACTERIAL INFECTIONS

'Orf' lesions in lambs can become contaminated with secondary bacterial invaders, especially *Fusobacterium necrophorum*, giving a necrotic stomatitis which can be fatal. In fact necrotic stomatitis can be a sequel to any lesion which disrupts the integrity of the oral mucosa, e.g. traumatic lacerations following wavy mouth or displaced teeth, actinobacillosis and even eruption of permanent teeth. However, *Fusobacterium necrophorum* becomes a problem contaminant only where the animals environment is damp, dirty and overcrowded. Diagnosis can usually be made on the gross appearance of the lesion, a coagulative necrotic focus with surrounding hyperaemia and later fibrotic encapsulation, or following isolation of the causal organism on smear or culture. Sulphonamides or penicillin have been used in the treatment of necrotic stomatitis but it is only successful in the early stages of the disease. The condition can be prevented by attention to hygiene.

Flock outbreaks of actinobacillosis due to *Actinobacillus lignieresi* occur sporadically. Lesions consist of multiple fibrotic nodules in the subcutaneous tissues of the cheek, lips, nose and throat. These nodules progress to fibrous sinuses that fistulate to the oral cavity or outside, often releasing thick, smell-free, adherent, green-yellow pus. Similar granulomatous lesions can also occur with *Corynebacterium pyogenes*, although the pus here is foul-smelling, more fluid and yellow, thus specific diagnosis depends upon isolation or demonstration of the causal organism.

Epidemiologically, both granulomata are due to contamination of oral lacerations or wounds by grass awns or misuse of a drenching gun. A particularly common site for these lacerations is in front of the incisor teeth where one often gets a single shallow ulcer. *Actinobacillus lignieresi* is a facultative anaerobe inhabiting the normal gastrointestinal tract. The conditions can be treated with streptomycin intramuscularly but is best prevented by limiting the occurrence of oral lacerations and wounds.

REFERENCES

1 Herrtage M.E., Saunders R.W. & Terlecki S. (1974) Physical examination of cull ewes at point of slaughter. *Veterinary Record,* **95**, 257–60.

2 Ministry of Agriculture, Fisheries and Foods. (1977) Teeth Loss in Sheep. Agricultural Development and Advisory Service Ways and Means Panel Report.

3 Porter W.L. (1972) Premature Tooth Loss in Sheep. FRCVS Thesis, London.

4 Spence J.A., Aitchison G.U., Sykes A.R. & Atkinson P.J. (1980) Broken mouth (premature incisor loss) in sheep: the pathogenesis of periodontal disease. *Journal of Comparative Pathology,* **90**, 275–92.

5 Spence J.A. (1978) Functional morphology of the periodontal ligament in the incisor region of sheep. *Research in Veterinary Science,* **25**, 144–51.

6 Barnicoat C.R. (1957) Wear in sheep's teeth. *New Zealand Journal of Science and Technology,* **38A**, 583–632.

7 Colyer F. (1936) *Variations and Disease of the Teeth of Animals,* John Bale, Sons and Danielsson, Ltd., London.

8 Dyson D.A. & Spence J.A. (1980) A cystic jaw lesion in sheep. *Veterinary Record,* **105**, 467–648.

9 Richardson C., Richards M., Terlecki S. & Miller W.M. (1979) Jaws of adult culled ewes. *Journal of Agricultural Science,* **93**, 521–9.

10 Mackinnon M.M. (1959) A pathological study of an enzootic paradontal disease of mature sheep. *New Zealand Veterinary Journal,* **7**, 18–26.

7   CLOSTRIDIAL DISEASES

D. BUXTON

Clostridial diseases of sheep have long been recognized although their aetiology was in doubt. James Hogg, the Ettrick shepherd, writing in 1807 named four 'species' of braxy; 'bowel sickness, sickness in the flesh and blood, dry braxy and wet braxy'. All four 'species' resemble to a remarkable degree current descriptions of braxy, blackleg, pulpy kidney and struck respectively.

In the last quarter of the nineteenth century the bacterial nature of several of these conditions was realized and this allowed more accurate definitions of disease to be developed. Conditions such as blackleg, braxy, black disease and the enterotoxaemias, including lamb dysentery, struck and pulpy kidney, were found to be due to a group of anaerobic, spore-forming organisms which became known as the clostridia and which were shown to produce rapidly fatal disease by the secretion of potent toxins.[1]

ENTEROTOXAEMIA

Enterotoxaemia is the term used to describe disease caused by the toxins produced by *Clostridium perfringens* within the intestines.

Cause

There are five distinct types of *Cl. perfringens* (A, B, C, D and E), each of which produces various amounts and permutations of twelve toxins. Disease in sheep is caused principally by *Cl. perfringens* types B (lamb dysentery and haemorrhagic enteritis), C (struck and haemorrhagic enteritis) and D (pulpy kidney) although *Cl. perfringens* type A has been held responsible for a fatal haemolytic disease of lambs and for the occasional case of gas gangrene in sheep.

Clostridium perfringens is rod-shaped, 2–6 μm by 0.8–1.5 μm, with a capsule and is non-motile. Spores are usually central or subterminal, oval and wider than the bacterium. Young cultures stain Gram-positive but in older cultures Gram-negative organisms are present. The different types of *Cl. welchii* produce different amounts and types of toxin. Three major lethal toxins are alpha (lethal necrotizing lecithinase), beta (lethal necrotizing) and epsilon (lethal necrotizing), which largely dictate the type of disease which develops.[2]

LAMB DYSENTERY

Clinical signs

This disease which usually affects good strong lambs under two weeks old is caused by the beta and epsilon toxins of *Cl. perfringens* type B. The clinical course is seldom longer than a few hours,[3] the lamb has abdominal pain, stops sucking, collapses and dies, its faeces being semi-fluid and bloodstained. Occasional animals survive a few days and can display central nervous signs due to the development of a focal symmetrical encephalomalacia (FSE).

Pathology

At post-mortem examination serous, sometimes blood-stained, fluid is present in the peritoneal cavity and the intestines appear dark-red with discrete to confluent ulceration of the mucosa. In longer-standing cases with deeper more penetrating mucosal ulceration there may be an associated peritonitis with blood tinged fibrin and even adhesions. The ulcers which are apparent through the serosa are well-defined, yellow, necrotic, and surrounded by a rim of hyperaemic tissue. Large numbers of organisms with the morphology of *Cl. perfringens* can be seen in the ulcers by histological examination but inflammatory cells are generally sparse. The gut contents are blood-stained and the mesenteric lymph nodes oedematous. The liver is pale and friable, the kidneys enlarged, pale and sometimes soft and the lungs congested. The pericardial sac often contains straw-coloured fluid and the brain can show lesions of FSE due to the epsilon toxin but this is a more common manifestation of pulpy kidney disease.

STRUCK

Clinical signs

Originally recognized in adult sheep in the Romney Marsh district of Kent, struck is caused by the rapid multiplication of *Cl. perfringens* type C in the intestinal contents and with elaboration of alpha and beta toxin. Death is sudden and follows a short illness in which the animal might display abdominal pain.

Pathology

At post-mortem examination the peritoneal cavity contains a large amount of straw-coloured, sometimes pinkish fluid and the abdominal vessels are congested. The small intestines are hyperaemic, the luminal surface is often ulcerated, most frequently in the jejunum, while the large intestines appear normal. In addition there is excess pleural and pericardial fluid as well as haemorrhages on the surface of the heart. When autopsy is delayed the carcass can resemble a case of blackleg.

PULPY KIDNEY

Clinical signs

Pulpy kidney disease can occur in sheep of any age but is most frequent in young lambs from about 4–10 weeks of age and fattening stock 6 months to a year old. It is caused by the rapid multiplication of *Cl. perfringens* type D in the small intestine and the subsequent absorption of epsilon toxin, which is produced by the organism in the form of a non-toxic prototoxin and is converted to a lethal toxin by the action of trypsin.[4]

Pulpy kidney disease is peracute and of short duration with most cases being found dead. Those that are observed before death show hyperaesthesia, ataxia progressing to recumbency, opisthotonos, intermittent convulsions with or without nystagmus and stertorous respiration between convulsions. As well as neurological signs, diarrhoea is seen in cases which live longer. Affected animals do not recover.

Pathology

At post-mortem examination the animal is usually in good condition, there is excess straw-coloured pericardial fluid and haemorrhages are present in the heart. The liver is swollen and friable and some congestion of intestinal mucosa might be present, the contents being more fluid than normal. The appearance of the kidneys can be quite characteristic, in a fresh carcass they appear swollen and pale but they autolyse more rapidly than normal and after a few hours, if held under a gentle stream of water, the parenchyma is washed away leaving the frond-like cortical stroma. This feature gives the disease its name. Glucose in the urine and high levels in blood are often found in pulpy kidney disease and can assist in a diagnosis.[5]

Macroscopic evidence of swelling of the brain is sometimes apparent with flattening of cerebro-cortical gyri, tentorial herniation of the occipital poles of the cerebrum and cerebellar coning. In animals displaying clinical signs for longer than normal, evidence of FSE might be present and seen as symmetrical reddened foci in the corpus striatum, thalamus, midbrain and cerebellar peduncles.

Histological examination of the brains from cases of pulpy kidney disease can reveal up to three basic changes [6, 7, 8], perivascular oedema, vascular haemorrhages and focal malacia. Perivascular oedema, which arises from the direct action of epsilon toxin on vascular endothelium, is most common in the meninges but is also often seen in cerebral cortex, corpus striatum, thalamus, midbrain, cerebellar white-matter cores and cerebellar peduncles. Haemorrhages are common, widespread and most readily seen in cerebral cortex and corpus striatum. Malacia is focal and sharply demarcated from apparently normal surrounding tissue. The foci appear 'rarified', the neuropil is vacuolated, glial nuclei pyknotic and neurones degenerate with shrunken dark nuclei and eosinophilic cytoplasm containing little discernible Nissl substance. In more severe and long-standing cases haemorrhages can be present with substantial softening and infiltration by macrophages. Examination of the kidney from fresh cases reveals only cloudy swelling with some necrosis of proximal convoluted tubules and some oedema and congestion of renal cortex.

HAEMORRHAGIC ENTERITIS

Clinical signs

This condition affects lambs in their first few days of life and is caused either by *Cl. perfringens* type B or type C. Death follows a short clinical illness in which the lamb can shiver, show abdominal pain and pass blood-stained, fluid faeces.

Pathology

Post-mortem examination reveals fluid gut contents, usually haemorrhagic, with mucosal necrosis most severe in jejunum and ileum. Free blood is sometimes found in the lumen. In some cases changes are less severe with acute hyperaemia in the intestines, oedema of the gut wall and a few erosions of the mucosa. Further changes include hyperaemia of the mesentery, swollen, wet, mesenteric lymph nodes, blood-tinged peritoneal fluid, ecchymotic haemorrhages in serous membranes, excess pericardial fluid and a pale friable liver.

DIAGNOSIS OF ENTEROTOXAEMIAS

An initial diagnosis of the enterotoxaemias can often be made on the history, clinical signs and post-mortem findings, but a more precise diagnosis can be achieved by demonstrating the presence of specific toxins by means of a neutralization test.[9] The neutralization test can be used to identify toxins in isolated cultured bacteria, intestinal contents or tissue fluid. Culture fluids should be filtered. In the case of intestinal contents, which should be collected as soon after death as possible and a small amount of chloroform added as a preservative, they should be diluted with broth, centrifuged and their supernate removed for testing. The test fluid should be divided into two and one part incubated at 37°C for one hour with trypsin (0.05 per cent w/v) in order to activate any epsilon toxin that might be present. The two samples can then be tested by mouse inoculation, the mice usually dying within 12 hours if lethal toxin is present. The identity of the lethal toxin can be sought by attempting to neutralize the intestinal contents with specific antitoxins (see Table 7.1).

The neutralization test can also be performed intradermally in depilated albino guinea pigs when sites of inoculation are observed for 48 hours for signs of necrosis.

Intestinal contents are occasionally toxic for reasons other than the presence of clostridial toxins and in addition tests on cultures are sometimes inconclusive if the organism fails to produce sufficient quantities of diagnostic toxin or is insufficiently virulent. When tests are inconclusive conventional cultural and biochemical methods must be used.

To isolate and culture *Cl. perfringens* a loopful of fresh intestinal contents should be inoculated on to blood agar and incubated anaerobically for 24–48 hours. Alternatively the sample can be inoculated

Table 7.1 The neutralization reactions which occur between *Cl. perfringens* toxins and antitoxins

Cl. perfringens		Antitoxins			
Type	Major Lethal Toxins	A	B	C	D
A	alpha	+	+	+	+
B	alpha, beta, epsilon	−	+	−	−
C	alpha, beta	−	+	+	−
D	alpha, epsilon	−	+	−	+

+ neutralization of toxin
− no neutralization of toxin
N.B. trypsinization activates epsilon toxin but also destroys beta toxin

into cooked meat broth, which has had its free oxygen reduced by heating, and incubated at 37°C until bubbles of gas develop at the base of the medium. A loopful of culture should then be inoculated on to blood agar and incubated anaerobically. This method can yield an almost pure culture of *Cl. perfringens*.

Clostridium perfringens grows on ordinary laboratory media at 37–47°C under anaerobic conditions although glucose and blood will enhance growth.[9] It should be noted that in the diagnosis of *Cl. perfringens* enterotoxaemia, the demonstration of lethal toxin in intestinal contents and/or the isolation of *Cl. perfringens* are not sufficient on their own for a diagnosis as both can be found in normal animals. However, the demonstration of lethal toxin in tissue fluids is very signficant.

EPIDEMIOLOGY AND TRANSMISSION OF ENTEROTOXAEMIAS

The spores of *Cl. perfringens* types B, C and D are found in soil and faeces of normal animals in areas where disease is prevalent as well as in intestinal contents of infected sheep.[4] Their presence in the intestinal tract of normal animals is important because they are able to form the focus for a fatal infection as and when conditions alter to allow their rapid multiplication.

As *Cl. perfringens* is saccharolytic it can multiply extremely rapidly in the presence of high levels of carbohydrate when oxygen tension is low. Thus, in the case of lamb dysentery, disease is most prevalent in lambs which ingest large quantities of milk. Offspring of heavy-milking hill breeds are usually more prone to disease than lambs from lighter-milking breeds with a high lambing percentage where the intake of milk is more restricted. A similar situation exists with pulpy kidney disease. When ruminating sheep ingest carbohydrate it is entirely broken down in the forestomachs and none reaches the small intestines. However, when sheep are changed suddenly from a low to a high plane of nutrition the microbial flora of the rumen does not adjust sufficiently rapidly and partially digested food containing carbohydrate passes to the small intestines. This allows *Cl. perfringens* type D to multiply rapidly and produce large amounts of epsilon protoxin which is converted to lethal epsilon toxin by the action of trypsin. After conversion epsilon toxin must remain in the lumen at extremely high concentrations for several hours before it can be absorbed and produce signs of intoxication.

CONTROL PREVENTION AND TREATMENT OF ENTEROTOXAEMIAS

The most effective way of controlling enterotoxaemia is by vaccination. Sheep receiving an initial course should be injected with two doses of the toxoids of *Cl. perfringens* in adjuvant at least 6 weeks apart. Boosting injections should be given annually 2–4 weeks before lambing. Lambs should receive the first dose when 8–12 weeks old and a second dose 4–6 weeks later. Vaccines containing toxoids of *Cl. perfringens* made up in an oily adjuvant are also available. These should provide protection for two years after intraperitoneal injection and can be boosted with an aqueous version of the vaccine 12–2 weeks before lambing.

Antisera active against the toxins of *Cl. perfringens* offer temporary protection for about 4 weeks and can be given to young lambs at risk. They are of little therapeutic value however and no effective treatment exists for the enterotoxaemias.

Vaccination should be combined with careful management of stock. This should include the gradual introduction of sheep to high levels of nutrition in order to allow the rumen flora to adjust and similarly sheep should be removed from a high plane of nutrition when cases of enterotoxaemia occur.

BLACKLEG

Synonyms Blackquarter, blackquarter metritis, quarter-ill, quarter-evil, emphysematous gangrene, symptomatic anthrax.

Cause

Blackleg is caused by *Clostridium chauvoei (Cl. feseri, Cl. chauvoei* type B), and was the first naturally-occurring anaerobic infection to be identified. *Clostridium chauvoei* is a rod-shaped organism 3–8 μm long, 0.5 μm wide with rounded ends although it can be pleomorphic. It has peritrichous flagella, is motile, contains central or sub-terminal oval spores wider than the bacillus but has no capsule. It stains Gram-positive in fresh cultures and in tissues although pleomorphic forms usually stain unevenly and mature spores do not stain positive.

Clinical signs

Infection in sheep develops rapidly and commonly follows shearing, docking, parturition and castration wounds.[10] Within 48 hours there is a high fever and if limb muscles are involved the animal becomes stiff and unwilling to move. The skin over the affected area may be discoloured but subcutaneous oedema and gas production can be difficult to detect. In cases of blackquarter metritis, associated with parturition, severe perineal oedema is sometimes seen. Infections of the head can produce marked oedema and even bleeding from the nose (big head). In almost all cases a short period of anorexia, profound depression and prostration is followed by death.[3]

Pathology

Clostridium chauvoei produces alpha (oxygen stable haemolysin), beta (deoxyribonuclease), gamma (hyaluronidase) and delta (oxygen labile haemolysin) toxins and infection produces a severe necrotizing myositis, toxaemia and death.

After death the carcass often swells and bloats rapidly. Straw-coloured and sometimes blood-stained gelatinous fluid is present in the subcutaneous tissues and fascia around the lesion while affected muscle is dark red, contains gas bubbles, is moist with oedema fluid and smells of rancid butter.

At the onset of disease a cellulitis develops, initially with oedema and haemorrhage, followed by extension of infection to adjacent muscles which rapidly become necrotic. Inflammatory cell infiltrations into muscle are slight and most cells are degenerate due to toxin produced by *Cl. chauvoei*. In blackquarter metritis subcutaneous perineal oedema develops and extends to the thigh muscles which become dark and swollen. Oedema of vaginal wall and uterus occurs and in pregnant animals extends to the fetus which becomes rotten and bloated.

Lesions away from the primary site of infection can include congestion of the lungs, fibrino-haemorrhagic pleuritis and pericarditis and pale, round foci in liver and renal cortex.

Diagnosis

While clinical and pathological findings suggest a diagnosis of blackleg, a more certain diagnosis depends on the detection of the causal organism. Just after death Gram-positive rods should be visible in films taken from the lesion and oedematous fluid but in smears made several hours later greater numbers of spores and pleomorphic forms will be present.

Isolation of the organism can be attempted. Pure cultures can be obtained from the centre of the lesion and, as animals develop a septicaemia just before death, from heart blood.[9] *Clostridium chauvoei* must be grown under anaerobic conditions on ordinary laboratory media at 37°C with added liver extract or glucose.

All strains have one common somatic antigen but have one or other of two flagellar antigens. Their spore antigen is the same as that of *Cl. septicum*.

A rapid, accurate method of diagnosis is to apply specific *Cl. chauvoei* antiserum, coupled to fluorescein isothiocyanate, to smears or frozen sections of affected tissue and examine the slides for specific fluorescence with a microscope fitted with an ultraviolet light source. Guinea pigs are susceptible to inoculation and can be protected with specific antiserum.

A diagnosis should seek to exclude the possibility of anthrax, hypomagnesaemia, hypocalcaemia and fracture of a limb.

Epidemiology and transmission

The spores of *Cl. chauvoei* survive well in the soil and as a result, areas of high risk exist. In sheep the majority of cases arise from contamination of a wound.

Control, prevention and treatment

An attempt should be made to reduce all wounds to a minimum and to treat adequately those that occur. In endemic areas penicillin can be used prophylactically both to treat wounds and at assisted lambings. However, a more satisfactory means of control than the use of antibiotics is vaccination. Vaccines, which consist of formalin-killed whole cultures of *Cl. chauvoei* precipitated with alum, take 14 days to develop immunity after being injected. Pregnant ewes should be given one dose of vaccine 3 weeks before lambing and lambs born to these ewes and which receive colostrum will be protected against disease, including umbilical infection, for at least 3 weeks. Sheep less than a year old which are vaccinated before being moved to infected pasture develop only short-lived protection with one dose and require revaccination 6 weeks later. Annual revaccination of all animals before a period of risk is recommended. In practice vaccines often provide protection against up to eight clostridial diseases.

Treatment of clinical cases has met with little success but intravenous hyperimmune serum together with the injection of crystalline penicillin both intravenously and into the wound has been tried.

BRAXY

Synonyms Bradsot, Breckshuach, Cling, Bradapest

Cause

Clostridium septicum (Cl. chauvoei type A, *Vibrion septique, Bacillus oedematis maligni)* a rod-shaped organism 2–10 μm long and 0.4–1.1 μm wide, with rounded ends which can appear filamentous when chains of bacilli develop. It has peritrichous flagella and in young cultures is Gram-positive although after 3–4 days incubation Gram-negative forms occur. *Cl. septicum* has no capsule and pleomorphic forms stain irregularly with Gram's method.

Clinical signs

Braxy appears in late autumn and winter and usually affects lambs born in the previous spring. Clinical disease is of sudden onset and short duration with the animal having a high fever, appearing depressed and not eating. Abdominal pain with accumulations of abdominal gas can occur. Recumbency and coma are followed by death within a few hours of the onset of clinical signs.

Pathology

Cl. septicum produces alpha (lethal, necrotizing and haemolytic), beta (deoxyribonuclease), gamma (hyaluronidase) and delta (necrotizing and haemolytic) toxins and infection causes severe abomasitis and toxaemia which results in a high mortality.

Autolysis is rapid but in a fresh carcass, examination reveals acute inflammation of the abomasum with oedema and haemorrhages in the mucosa and submucosa and in some cases ulceration. In addition, excess straw-coloured serous fluid can be present in the pericardium with sero-sanguinous fluid in the abdominal cavity. In some cases the subcutaneous tissue over the abdominal cavity is oedematous and contains gas.

Diagnosis

Sudden deaths in young stock in autumn and winter together with post-mortem findings can suggest the possibility of braxy. However, for a definitive diagnosis it is necessary to demonstrate in smears made from abomasal lesions the presence of large numbers of Gram-positive rod-shaped or filamentous organisms which fluoresce, after treatment with specific antiserum to *Cl. septicum* conjugated with a fluorescent dye, and viewed with a microscope equipped with an ultraviolet light source.

Isolation and culture of *Cl. septicum* from the sheep is another important means of diagnosis.[9, 11]

Epidemiology and transmission

There is a greater incidence of braxy in Iceland, the Faroe Islands, Norway, Wales, Scotland, northern England and Ireland although it also occurs in other locations such as Australia and New Zealand. *Cl. septicum* survives well in the soil and is frequently present in faeces of herbivores. Disease is apparently triggered by the ingestion of frosted food which seems to produce a favourable micro-environment in which *Cl. septicum* can invade the tissues and multiply. Disease therefore occurs in unvaccinated flocks in late autumn and winter and affects lambs born the previous spring as older animals in enzootic areas appear to develop some immunity to the disease. Disease is also more common on mountainous farms and in young sheep wintered on hill pastures or old lowland grazing.

Control, prevention and treatment

While the incidence of braxy can be reduced by wintering sheep on new lowland pastures the most effective means of prevention is by vaccination. Commercial vaccines contain purified formol toxoid of *Cl. septicum* with an adjuvant. To provide protection against braxy in their first winter sheep should be injected in early autumn with two doses of vaccine, 3–4 weeks apart. Vaccines often also contain purified formalin-killed cultures of *Cl. chauvoei* to protect against blackleg and some also immunize against diseases caused by *Cl. tetani, Cl. oedematiens* and *Cl. perfringens.*

Treatment of clinical cases of braxy is seldom practical but injections of penicillin with or without hyperimmune serum have been tried.

BLACK DISEASE

Synonym Infectious necrotic hepatitis.

Black disease, a fatal peracute infectious condition, was known in Australia before the end of the nineteenth century and was first reported in the UK by Jamieson, Thompson and Brotherston in 1948.[12]

Cause

Black disease is caused by *Cl. oedematiens* type B *(Cl. novyi* type B, *Cl. gigas)* and is usually associated with the presence of infection by liver flukes *(Fasciola hepatica). Cl. oedematiens* type B is a large rod-shaped organism 3–10 μm long by 1–1.5 μm wide although some strains can grow to 20 μm in length. It has peritrichous flagella, is motile under anaerobic conditions, has no capsule and contains central or subterminal oval spores which are wider than the bacillus. It is Gram-positive although, after prolonged culture, it can be Gram-negative.

Clinical signs

The disease is characterized by the rapid onset of dullness, followed by an unsteady gait, an inability to move, collapse and quiet death within a few hours of the onset of clinical signs.

Pathology

Cl. oedematiens type B produces alpha and beta toxins. After death putrefaction is rapid. Subcutaneous blood vessels are engorged and can cause the carcass to blacken rapidly, giving the disease its name, and oedema is is often present in the abdominal wall. The liver is engorged and dark with yellow areas of necrosis up to 3 cm in diameter and surrounded by a red zone of hyperaemia. In addition there is often evidence of recent migration of liver flukes. Blood-tinged serous fluid is usually present in the thoracic and abdominal cavities and the pericardial sac and there are subendocardial and subepicardial haemorrhages.

Histology of liver has shown that migrating immature flukes create tracts of damage, which contain necrotic debris, blood and an eosinophil rich inflammatory exudate, surrounded by coagulative necrosis infiltrated by neutrophils. Spores of *Cl. oedematiens* type B which have been shown to be present in normal ovine livers are able to germinate in the damaged tissue and to produce vegetative forms which multiply rapidly and produce toxins, further tissue damage, toxaemia and death. Bacilli can be seen in tissue sections, most readily at the margin of the lesion just inside the zone infiltrated by neutrophils. They are rarely seen in surrounding viable tissues although at death they become dispersed to other organs from which they can be isolated.

Diagnosis

The clinical picture and post-mortem findings should suggest a diagnosis of black disease and the presence of fascioliasis in the flock would further support a diagnosis. However, a definitive diagnosis depends upon the positive identification of *Cl. oedematiens* type B in the tissue.[9] A rapid, accurate means of identification is to treat smears or sections of suspect tissue with antiserum to *Cl. oedematiens* coupled with a fluorescent dye and to examine them with a microscope fitted with an ultraviolet light source. In a positive case fluorescent bacteria are seen in the smear or section of the lesion.

Isolation of *Cl. oedematiens* by culture can be difficult especially from highly contaminated field samples.

Epidemiology and transmission

Both *F. hepatica* and *Cl. oedematiens* type B were implicated in the aetiology of black disease by Turner in 1930.[13] Since then it has been shown that the spores of *Cl. oedematiens* type B can be found both in the soil and in livers of normal sheep on farms where the disease occurs.[14] The disease has also been reproduced experimentally in sheep by the administration of metacercariae of *F. hepatica* followed two weeks later by *Cl. oedematiens* type B spores.[15] While it is generally accepted that the liver damage caused by fluke infection is an important step in the genesis of black disease, other forms of liver damage can trigger the condition.

Control, prevention and treatment

Control of fluke infection, both on the pasture and by treatment of the sheep, is an important and effective measure and should be instituted along with vaccination against *Cl. oedematiens*. Vaccines, which frequently offer protection against other disease, are prepared from toxoids of the organism and sometimes include formalized whole cultures of *Cl. oedematiens* type B as well as an adjuvant. At least two doses of vaccine not less than four weeks apart are adequate for primary immunization and normally gives protection for one year. Booster vaccinations are therefore required annually. Treatment of the disease is not a practical proposition.

BACILLARY HAEMOGLOBINURIA

Synonym Red water disease.

Cause

Bacillary haemoglobinuria is not a common condition of sheep being primarily a disease of cattle. It is an acute rapidly fatal disease caused by *Cl. oedematiens* type D *(Cl. novyi* type D, *Cl. haemolyticum)* a Gram-positive rod 5–10 μm long, 1–1.5 μm wide with similar properties to *Cl. oedematiens* type B (see black disease) although in culture it does not usually ferment maltose.

Clinical signs

It is a disease of sudden onset with fever and rumenal stasis with or without detectable abdominal pain. The animal breathes rapidly, has dark-red urine, becomes jaundiced and dies within a short time of the onset of clinical signs.

Pathology

Cl. oedematiens type D produces beta toxin which causes intravascular haemolysis, anaemia and haemoglobinuria. Thus at post-mortem examination the bladder contains red urine and the carcass is jaundiced. In addition there is subcutaneous gelatinous oedema and myocardial haemorrhages while the pleural and abdominal cavities and pericardial sac contain excess fluid, sometimes tinged with blood. However, the most characteristic change is in the liver, which is similar to that seen in black disease and consists of pale necrotic foci surrounded by a zone of hyperaemia. Histopathological examination of this lesion reveals the presence of many organisms with the morphology of *Cl. oedematiens*.

Diagnosis

The clinical and post-mortem findings suggest a diagnosis but to be more certain, isolation and culture of the causal organism, as described for black disease, or its identification by means of a fluorescent antibody test is necessary.[9]

Epidemiology and transmission

The disease is most common in summer and autumn and on irrigated or naturally damp pasture. The spores of *Cl. oedematiens* type D exist in the soil and can be found in the livers of normal healthy animals grazing on contaminated pasture. Bacillary haemoglobinuria depends upon some kind of hepatic injury, possibly liver fluke as in black disease, to trigger germination of the spores and their subsequent multiplication.

Control, prevention and treatment

Lasting protection can be achieved by vaccination as for black disease together with the treatment of pasture and stock to remove liver fluke infection. Treatment with penicillin can be attempted.

MALIGNANT OEDEMA

Synonym Big head.

Cause

Malignant oedema is an acute rapidly fatal wound infection caused by bacilli of the genus *Clostridium* (*Cl. septicum, Cl. chauvoei, Cl. perfringens, Cl. oedematiens* type A, *Cl. sordellii*).

Clinical signs

Clinical signs appear rapidly after infection. At the site of primary infection a swelling develops, which will 'pit' on pressure, and gas may be detected as the skin becomes darkened and more tense. A high fever is present, toxaemia develops and death usually occurs within 1–2 days.

Pathology

Malignant oedema is typically a cellulitis, in which infection spreads up and down fascial planes, rather than a myositis. Thus the skin over the lesion can appear gangrenous and there is subcutaneous oedema while the underlying muscle is not usually severely affected. Oedema fluid is blood tinged and can contain gas bubbles except when infection is caused by *Cl. oedematiens* when oedema fluid is gelatinous, clear and contains no gas.

Diagnosis

The clinical and post-mortem findings suggest a diagnosis of malignant oedema but confirmation requires laboratory identification of the causal organism.[9]

The differential diagnosis should seek to exclude other causes of sudden death, including blackleg, in which more direct involvement of muscle occurs.

Epidemiology, transmission, control, prevention and treatment

As the condition occurs when a wound becomes contaminated with dirt containing clostridial spores, the minimization of injury to sheep, particularly at shearing and lambing, together with antiseptic treatment of wounds reduces the risk of infection.

Prophylactic administration of penicillin with assisted lambings has been recommended and the proper use of clostridial vaccines ensures a high level of flock protection.

Once infection is recognized administration of antibiotics and local treatment of the wound with hydrogen peroxide or other oxidizing disinfectants can be attempted.

REFERENCES

1 Sterne M. (1981) Clostridial infections. *British Veterinary Journal*, **137**, 443–54.
2 Smith L.D. (1979) Virulence factors of *Clostridium perfringens*. Reviews of Infectious Diseases, **1**, 254–60.
3 Watt J.A.A. (1960) Sudden death in sheep. *Veterinary Record*, **72**, 998–1001.
4 Bullen J.J. (1970) Role of toxins in host-parasite relationships. In *Microbial Toxins* (Eds.Ajl S.J. & Montie T.C.) Vol.I, pp.233–76. Academic Press, New York, London.
5 Gardner D.E. (1973) Pathology of *Clostridium welchii* type D enterotoxaemia I. Biochemical and haematological alterations in lambs. II. Structural and ultrastructural alterations in the tissues of lambs and mice. III. Basis of the hyperglycaemic response. *Journal of Comparative Pathology*, **83**, 499–529.
6 Griner L.A. (1961) Enterotoxaemia of sheep. I. Effects of *Clostridium perfringens* type D toxin on the brains of sheep and mice. *American Journal of Veterinary Research*, **22**, 429–42.
7 Buxton D. & Morgan K.T. (1976) Studies of lesions produced in the brains of colostrum deprived lambs by *Clostridium welchii (Cl. perfringens)* type D toxin. *Journal of Comparative Pathology*, **86**, 435–47.
8 Buxton D., Linklater K.A. & Dyson D.A. (1978) Pulpy kidney disease and its diagnosis by histological examination. *Veterinary Record*, **102**, 241.
9 Buxton A. & Fraser G. (1977) *Animal Microbiology*, Vol.1.Ch.20 pp.205–28. Blackwell Scientific Publications, Oxford.
10 Wilson G.S. & Miles A.A. (1975) *Topley and Wilson's Principles of Bacteriology, Virology and Immunity*, 6e, Vol.2, p.2268, Edward Arnold, London.
11 MacLennan J.D. (1962) The histotoxic clostridial infections of man. *Bacteriological Reviews*, **26**,177–274.
12 Jamieson S., Thompson J.J. & Brotherston J.G. (1948) Studies in black disease. 1. The occurrence of the disease in sheep in the North of Scotland. *Veterinary Record*, **60**, 11–14.
13 Turner A.W. (1930) Black disease (infectious necrotic hepatitis) of sheep in Australia. Bulletin 46, Council for Scientific and Industrial Research, Commonwealth of Australia.
14 Bagadi H.O. & Sewell M.M.H. (1973) An epidemiological survey of infectious necrotic hepatitis (black disease) of sheep in southern Scotland. *Research in Veterinary Science*, **15**, 49–53.
15 Bagadi H.O. & Sewell M.M.H. (1973) Experimental studies on infectious necrotic hepatitis (black disease) of sheep. *Research in Veterinary Science*, **15**, 53–61.

8   ENTERITIS IN YOUNG LAMBS

D. R. SNODGRASS AND K. W. ANGUS

It has been estimated that between 2 and 4 million lambs die each year in the UK. Most of these losses are not due to infectious diseases of the young lamb but to abortions and stillbirths on the one hand and starvation or chilling of the lamb on the other. However, of the infectious conditions affecting the lamb, diseases of the alimentary tract are the most important.[1] Enteric diseases most commonly manifest as diarrhoea and can result in significant mortality or economic loss from reduced condition, drugs, and labour. Many flocks, even those on extensive husbandry systems throughout the rest of the year are kept intensively at lambing time. This produces ideal conditions for a build-up of infectious neonatal disease: close confinement, gradual increase of contamination and a continual through-put of susceptible young animals to amplify infectious agents. The fact that lambing takes place in winter or early spring means that adverse environmental factors such as cold, wet, or windy weather, are prevalent and these conditions are often associated with diarrhoea outbreaks.

Causes

No comprehensive surveys have been published to enable assessment of the relative importance of different infectious agents. However, studies have been made of individual micro-organisms, and these are described separately.

LAMB DYSENTERY

Caused by *Cl. perfringens (welchii)* type B (see chapter on Clostridial Diseases)

ESCHERICHIA COLI

E. coli is a normal inhabitant of the bowel of sheep. By the second day of life, the healthy lamb excretes 10^9–10^{10} *E. coli*/g of faeces, and this count decreases gradually with age until a plateau of about 10^6–10^7 *E. coli*/g is reached in the adult sheep. Only a minority of strains of *E. coli* are capable of causing diarrhoea and knowledge of the virulence factors involved is well documented. Early experiments to assess the enteropathogenicity of *E. coli* were performed by either oral inoculation of neonatal lambs, or by inoculation of ligated intestinal loops. Isolates capable of causing diarrhoea in the experimental lambs also caused fluid accumulation in isolated intestinal loops. Techniques to identify virulence factors have advanced, and it is recognized that *E. coli* capable of causing diarrhoea possess an antigenic pilus (known as K99 and common to calf and lamb isolates) which enables them to adhere to intestinal epithelium and produce an enterotoxin, which in calf and lamb strains is usually a heat-stable, low molecular weight, non-antigenic toxin (ST). Isolates dilating ligated gut loops, or possessing K99 and ST, are considered enterotoxigenic *E. coli* (ETEC).[2]

ETEC have been isolated from 20–43 per cent of scouring lambs in surveys in the UK and the USA. Affected lambs are usually less than one week old,

and exhibit fluid diarrhoea, dehydration, and weakness. Mortality rates up to 75 per cent have been reported.

SALMONELLAE

Salmonellosis is a sporadic cause of enteritis and loss in young lambs. Individual outbreaks are often very severe with fever, abortions, diarrhoea and high mortality in sheep in affected flocks. The major serotypes implicated are *S. dublin* and *S. typhimurium*, though more exotic serotypes e.g. *S. orienburg*, have been isolated in some incidents. Evidence from recent surveys suggests that in the UK *S. dublin* is the predominant serotype in ovine enteric salmonellosis.

The primary source of infection where *S. typhimurium* is involved is often difficult to identify, since this serotype can cause disease in a wide range of host species. With *S. dublin*, however, direct or indirect contact with infected cattle must be considered as a possible initiating cause.

The first indication of disease is frequently sudden death with no premonitory clinical illness, though dead lambs may display yellow staining of the perineum.[3] Sick animals are dull, febrile and refuse to suck; a greenish-yellow diarrhoea which may be blood-stained is commonly present. Older lambs are often thirsty, and may be found dead beside a source of water. The course of clinical illness is brief and death supervenes within 24 hours.

Necropsy findings in young lambs are often inconsistent. Abomasal and small intestinal contents are usually watery, and the macroscopic appearance of the intestinal mucosa varies from apprently normal to obviously inflamed over considerable lengths. The abomasal mucosa is often severely inflamed, with focal hyperaemia or haemorrhage in the plicae. Signs of dehydration may be present.[3] Mesenteric lymph nodes may be enlarged and oedematous, and salmonellae can normally be cultured from these, the small intestine, spleen and liver.

CAMPYLOBACTERS

All strains of *Campylobacter* associated with acute enteritis belong to a group with thermophilic characteristics, currently classified as *C. fetus* subs. *jejuni/ coli*. Although numerous strains of the organism have been found in sheep faeces, there is little evidence that they present a hazard to young lambs although regional ileitis resulting in unthriftiness has been reported in older lambs.[4] Attempts to induce clinical disease in lambs by oral transmission of *Campylobacter* strains from scouring calves have been unsuccessful, although infection was established. It is possible that persistent infection acquired in the neonatal period is responsible for a small proportion of unthrifty lambs at around weaning age.

ROTAVIRUS

Viral enteritis attracted much interest in the 1970s, and viruses are now regarded as significant causes of diarrhoea. Rotavirus in particular has been shown to occur globally in man and his domesticated mammals and birds.[5] Rotaviruses from different species are serologically related, but are usually distinguishable by genetic and antigenic analysis. Transmission of rotavirus from one host species to another has been frequently accomplished experimentally, but the existence of natural zoonotic transmission has not been confirmed. In particular, reports of diarrhoea in farm children at lambing time have not so far been adequately investigated.

The existence of a rotavirus from lambs has been confirmed and in the only published survey rotavirus infection was detected in 25 per cent of scouring lambs.[6] Antibody surveys suggest that rotavirus is endemic in the sheep population, as has been found in man, cattle and pigs.

Lamb rotavirus has not so far been adapted to grow in cell culture. However, experimental studies using gnotobiotic or colostrum-deprived lambs have enabled pathogenesis to be studied. As in other species, lamb rotavirus infects and destroys the mature absorptive villus epithelial cell of the small intestine, leading to villus atrophy and a malabsorptive diarrhoea.

OTHER VIRUSES

Conventional virological techniques based on tissue culture result in the isolation of enteroviruses, adenoviruses, and reoviruses from lamb faeces. Little significance is normally attached to these isolates, but reovirus type 1, ovine adenovirus type 1, and bovine adenovirus type 2 have all been associated with respiratory and enteric disease in lambs in large intensive farms in Hungary.[7, 8]

Studies based on electron microscopic examination of lamb faeces have been minor compared with the extensive investigations into diarrhoea in calves and piglets, with the result that comparatively little information is available. Astrovirus has been described from an outbreak of lamb diarrhoea, and has been shown experimentally to induce a mild diarrhoea in gnotobiotic lambs. However, there is no evidence that astrovirus is an important pathogen. Coronavirus-like viruses have also been described from lambs, but their morphology is distinct from true coronaviruses infecting calves and piglets and their significance is unknown.

CRYPTOSPORIDIUM

Endogenous stages of *Cryptosporidium*, a small protozoon of the enteric coccidia group, adhere to the microvillous borders of intestinal epithelial cells. The parasite differs from other coccidia in that it has a short (2–4 days) life-cycle, and apparently lacks host specificity.[9] Transmission is via infected faeces. Under experimental conditions, isolates from scouring calves, lambs and red deer infected gnotobiotic lambs, causing clinical diarrhoea with widespread cryptosporidial infection and severe damage in the intestine, particularly the ileum. An outbreak of cryptosporidiosis in artificially-reared lambs has been described, with severe diarrhoea and deaths[10] and more recent evidence suggests that serious outbreaks can occur in naturally-reared lambs at about 7–10 days old. However, the prevalence is unknown. Affected lambs quickly become dull and anorectic. They develop a 'tucked-up' appearance, and become stiff and reluctant to keep up with their dams. There is no febrile reaction. The length of clinical illness varies but is usually at least 7 days, before gradual recovery commences. Some lambs die after 2–3 days illness, others appear to recover, then relapse. Surviving lambs remain unthrifty for several weeks.

Necropsy findings are often vague: the carcass is usually thin and may be dehydrated. The intestines are flaccid, but the mucosal surface may appear normal, though in some instances there may be congestion in the distal small bowel. The caecum is often distended with khaki-coloured liquid contents, while the spiral colon is often empty. Histological examination shows widespread infection of the villous epithelium, particularly in the distal jejunum and ileum, with widespread atrophy and fusion of villi. In prolonged infections, the caecum and colon and sometimes the rectum, may be infected with the parasite, resulting in a severe typhlitis and colitis.

Diagnosis

To establish the aetiology of an outbreak of diarrhoea in lambs requires laboratory investigation, and
the possibility of lamb dysentery should not be
ignored. Salmonellosis typically presents as a wider
syndrome than neonatal diarrhoea alone, and if
salmonellosis is suspected, conventional bacteriological techniques to isolate *Salmonellae* should be
employed.

ETEC may be isolated from faeces on blood agar
and McConkey agar. No correlation exists between
haemolysin production and enteropathogenicity of
calf and lamb strains of *E.coli*. K99 expression can be
prevented by excess polysaccharide production, so
passage of *E.coli* on minimal medium such as Minca
+ 1% Isovitalex prior to testing for K99 by slide
agglutination is preferred. Heat stable toxin is normally assayed by inoculating cell-free culture fluids
to infant mice. Ideally, to confirm a diagnosis of
ETEC infection, the bacteria should be shown to be
present at abnormally high titres in the small
intestine i.e. $> 10^8/g$. To demonstrate this, an acutely-
ill untreated lamb should be sacrificed, *E. coli* counts
made, and the presence of K99-producing *E. coli*
adhering to intestinal epithelium demonstrated by
immunofluorescence of cryostat sections of small
intestine.

Rotavirus infections are more readily diagnosed,
and the technique most widely used is ELISA
performed directly on faeces samples. Sections of
intestine obtained as above can also be stained for
rotavirus immunofluorescence. In the absence of
specific immunological reagents, examination by
electron microscopy is satisfactory and offers the
added advantage of a catch-all technique for several
other enteric viruses. Briefly, an approximate 20 per
cent suspension of faeces in saline is made and the
coarse debris allowed to settle. A drop of supernatant
fluid is transferred to an electron microscope grid,
stained with a negative stain to delineate virus
particles, and examined in an electron microscope at
a magnification of 40 000–50 000. Alternatively, the
faecal suspension may be clarified by lowspeed
centrifugation and the supernatant fluid centrifuged
at high speed to pellet viruses. The pellet is then
resuspended in a small volume and examined.
However, no consistent increase in sensitivity has
been shown with this method. Rotaviruses have a
characteristic morphology, and can often readily be
recognized in large numbers from specimens prepared as described (Fig. 8.1).

To detect cryptosporidia in faeces smears on glass

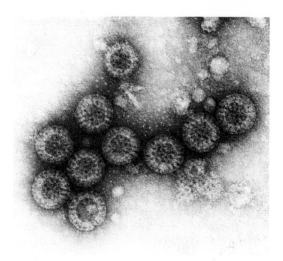

Fig. 8.1 Rotavirus particles in faeces. The characteristic
arrangements of the outer capsomeres giving the
impression of spokes radiating from a wheel hub led to
the adoption of the name rotavirus. (x 102 000)

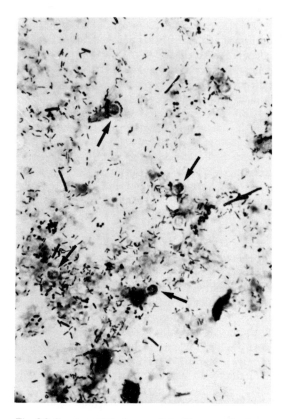

Fig. 8.2 Cryptosporidia (arrowed) in Giemsa-stained
faecal smear (x 4250)

slides are air-dried, fixed in methanol, and stained by Giemsa's method. Examination under x100 objective lens is necessary to detect the cryptosporidia, which are circular structures stained blue with no obvious capsule 3–4 μm in diameter, with often several eccentrically-placed pink-staining granules, and characteristically an unstained halo around the organism (Fig. 8.2). Faeces containing cryptosporidia are also infectious for suckling mice. Cryptosporidia can be readily demonstrated in histological sections of both small and large intestine from samples fixed shortly after death (Fig. 8.3). Up to 24 hours after death, cryptosporidia may be observed in mucosal scrapings.

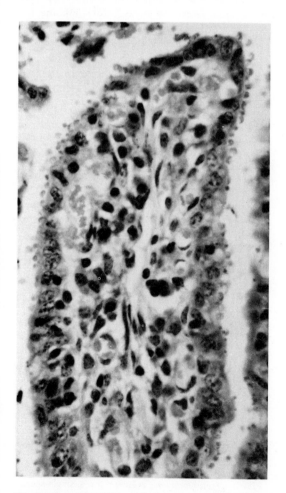

Fig. 8.3 Cryptosporidia attached to enterocytes in ileum of a naturally-infected calf. (x 1000)

Treatment and control

It is important to confirm if lamb dysentery is involved, so that appropriate prophylactic measures can be carried out, (see Chapter 7). Immunization against salmonellosis is unsatisfactory, although killed vaccines are useful on premises where a particular serotype has already been diagnosed. In the absence of lamb dysentery and salmonellosis, disease produced by the other micro-organisms involved can probably not be distinguished clinically. However, as there is no specific vaccine or therapy for any of these micro-organisms, control depends on non-specific measures. The lambing area should immediately be moved to a clean site to attempt to break the build-up of infection. If lambing takes place indoors it is frequently difficult to change lambing areas, but in that case young lambs should be moved outside as quickly as weather permits. Extra attention should be paid to the nutrition of the ewe, to ensure adequate colostral production, and to the care of young lambs to ensure adequate colostral intake.

No controlled trials of the use of antibiotics in lambs have been made, but by analogy with other species their use is probably of little value. Effective fluid replacement therapy and warming of sick lambs are the most appropriate remedies. Although sulpho-namides have been occasionally reported effective in bovine cryptosporidiosis, this has not been confirmed.

It is possible that vaccines being developed for use in calf diarrhoea could eventually be applied in sheep. These vaccines are intended for use in pregnant cows, to stimulate production of antibody in colostrum and milk to rotavirus and ETEC. Experiments in ewes and lambs have shown the validity of a similar approach, and an effective K99-based ETEC vaccine for ewes is feasible.[11] Whether a calf rotavirus vaccine given to ewes passively protects lambs against lamb rotavirus is not known, although antibodies to calf rotavirus in cows' milk have been shown to protect piglets against pig rotavirus infections. A separate lamb rotavirus vaccine is not at present feasible as lamb rotavirus has not been adapted to grow in cell culture.

'WATERY MOUTH'

Watery mouth is a colloquial name for a clinical entity responsible for sporadic but sometimes quite heavy losses in young lambs in the immediate post-natal period. Recent reports based on close observation of flocks where watery mouth is a recurrent annual problem[12, 13, 14] give a perspective on the clinical features and circumstances surrounding occurrences.

Lambs 12 h to 7 days old may be affected, but the highest incidence is in the 12–48 h age group. The condition is commonest in multiple births, but can occur in singles. Frequency of occurrence is greater in lambs born towards the end of the lambing period than in early lambs. The clinical expression of the condition varies from farm to farm and from year to year on the same farm, but may be so severe as almost to reach epidemic proportions.

Affected lambs do not suck and quickly become dull, lethargic and comatose. The muzzle and lower jaw are often wet with saliva or regurgitated stomach contents, and the mouth feels cold. The abdomen is often distended: lambs have a bloated appearance. Diarrhoea is absent; rather, lambs appear constipated, with no passage of the meconium. On handling the abdomen, a marked gurgling sound due to accumulation of gas in the intestines may be heard, known locally as 'rattle belly'. Death usually takes place within 12–24 h.

At necropsy features consistently found include retained meconium and the presence of excessive amounts of thin clear mucin (fetal mucin) in the abomasum. If colostrum is present, this is unclotted. In slightly older lambs, the abomasum may be distended with milk. *E. coli* is usually isolated from the gut and other organs (septicaemic colibacillosis).

Treatment is aimed at overcoming intestinal stasis by removing the meconium using enemas and laxatives, and controlling infection with oral and parenteral antibiotics. In mild cases, laxatives alone may be sufficient to effect a cure, but in severe outbreaks oral antibiotics may have to be given to lambs at birth, and possibly repeated later, to prevent further losses. *E. coli* antiserum may be of some value in controlling losses.

Analysis of possible predisposing causes indicates that these are multifactorial. Undoubtedly, inadequate colostral intake is an important contributory factor. As this is related to diminished sucking drive, such contributing factors as hypothermia, hypoglycaemia, or prenatal infections, e.g. border disease, toxoplasmosis, or enzootic abortion have been suggested. Irrespective of cause, adequate colostral intake must be ensured, if these lambs are to survive.

Management factors are probably of equal importance on some farms. Penning of the lambs with their mothers for up to 12 h is an important predisposing factor, particularly when it seems likely that build-up of pathogenic enterobacteria is occurring as lambing proceeds. Prolonged penning should only be considered if the weather is particularly inclement. In the most adverse circumstances, the lambing-pens may have to be abandoned for the season, and disinfected during the summer. There is also evidence that early castration or docking with rubber rings predisposes to watery mouth.

Preventive measures thus depend on individual circumstances. Attention to the management factors discussed above almost certainly helps to reduce the incidence in most circumstances. Chemoprophylaxis with oral antibiotics, though doubtless satisfactory in the short term, can only lead to further development of transmissible drug resistance in strains of enterobacteria, and is fundamentally undesirable. Vaccination of the ewes with a combined *Pasteurella haemolytica/E. coli* vaccine 4 weeks before lambing has been used to control the condition.

REFERENCES

1 Hughes L.E. (1964) An analysis of the results of postmortem examinations of sheep. Animal Disease Surveys Report No.3, MAFF, HMSO.

2 Sojka W.J., Morris J.A. & Wray C. (1978) Enteric colibacillosis in lambs with special reference to passive protection of lambs against experimental infection by colostral transfer of antibodies from ewes vaccinated with K99. In *Abstracts of the XII International Congress of Microbiology*, pp. 52–63. Munich.

3 Hunter A.G., Corrigall W., Mathieson A.O. & Scott J.A. (1976) An outbreak of *S. typhimurium* in sheep and its consequences. *Veterinary Record*, **98**,126–30.

4 Vandenberghe J. & Hoorens J. (1980) Campylobacter species and regional enteritis in lambs. *Research in Veterinary Science*, **29**, 390–1.

5 McNulty M.S. (1978) Rotaviruses. *Journal of General Virology*, **40**,1–18.

6 Snodgrass D.R., Herring J.A., Linklater K.A. & Dyson D.A. (1977) A survey of rotaviruses in sheep in Scotland. *Veterinary Record*, **100**, 341.

7 Belak S. & Palfi V. (1974) Isolation of reovirus type 1 from lambs showing respiratory and intestinal symptoms. *Archiv fur die gesamte Virusforschung*, **44**, 177–83

8 Belak S., Palfi V. & Palya V. (1976) Adenovirus infection in lambs. 1. Epizootiology of the disease. *Zentralblatt fur Veterinarmedizin*, **23B**, 320–30.

9 Tzipori S., Angus K.W., Campbell I. & Gray E.W. (1980) *Cryptosporidium:* evidence for a single-species genus. *Infection and Immunity*, **30**, 884–6.

10 Tzipori S., Angus K.W., Campbell I. & Clerihew L.W.

(1981) Diarrhoea due to *Cryptosporidium* infection in artificially reared lambs. *Journal of Clinical Microbiology,***14,** 100–05.

11 Sojka W.J., Wray C. & Morris J.A. (1978b) Passive protection of lambs against experimental enteric colibacillosis by colostral transfer of antibodies from K99-vaccinated ewes. *Journal of Medical Microbiology,* **11,** 493–9.

12 Shaw W.B. (1981) Lamb survival. *Veterinary Record,* **108,** 265.

13 Haig G. (1981) Lamb survival. *Veterinary Record,* **108,**195.

14 Collins R.O. (1981) Lamb survival. *Veterinary Record,* **109,** 43–44

9   COCCIDIOSIS IN LAMBS

L.P. JOYNER

With the increasing application of intensive husbandry methods the various causes of ill-thrift in young lambs have attracted increasing attention. Among these, enteritis in lambs aged 4–6 weeks is a common source of loss. Coccidiosis has been recognized as one of the causes of this syndrome but the relationship between the parasites and clinical disease is often difficult to establish. This is because most sheep are infected with several species of coccidia and most lambs at some time produce large numbers of faecal oocysts without showing any detrimental effects. Furthermore, there may be more than one factor contributing to the enteropathy at any one time.

The causal parasites

Coccidiosis occurs in many species of animals, generally due to species of the genus *Eimeria*. They are highly host-specific so that a given species is only found in one kind of animal and does not infect any others. There are at least eleven species found in sheep. It was thought that several were common to both sheep and goats but this has been found to be incorrect.[1,2] All the parasites of this genus in sheep parasitize the epithelium lining the alimentary tract. They are acquired by oral ingestion of the resistant oocyst and follow a life-cycle characterized by several cycles of multiple division so that ultimately they have the potential to parasitize many cells of the intestinal epithelium. Each species is generally located in a characteristic region of the gut. Multiplication however, is limited and the final number of parasites ultimately depends upon the size of the infective dose and the species involved. The pathogenic potential of a given species depends among other things on the activity and size of the multiplicative stages (schizonts), their location in the alimentary tract, the degree of tissue invasion and the rapidity with which the host becomes immune.

The immune status of the flock and the distribution of oocysts within the environment are keys to the epidemiology of the disease. Very little is known about the immune mechanism in coccidiosis. It is probably cell-mediated and as yet there are no tests which can be used to assess the immunity of individual animals. The young, previously unexposed lamb is highly susceptible, but from birth in a conventional flock it quickly acquires infection and by hygiene and careful management to avoid early crowding, the animal can be given an opportunity to develop immunity before the overall infection level builds up.

Development in the host terminates with the formation of oocysts which are voided with the faeces. At first they are not infective but after a few days under ordinary conditions of moisture and temperature they become sporulated and able to infect further animals when ingested with the food. They are resistant to most disinfectants but they are destroyed by desiccation and exposure to ammonia.

Recent research has been directed to the identification of the different species of *Eimeria* in sheep. The sporulated oocysts can be differentiated morphologically but the unsporulated ones frequently do not present sufficient characters for this to be done (Fig 9.1). To prepare sporulated oocysts for examination in the laboratory, the faeces are first emulsified in water, strained through muslin and the oocysts separated by centrifugation. The oocysts are resuspended in 2 per cent potassium bichromate solution and shallow layers of the suspension are kept in glass dishes at room temperature or 28°C. Generally

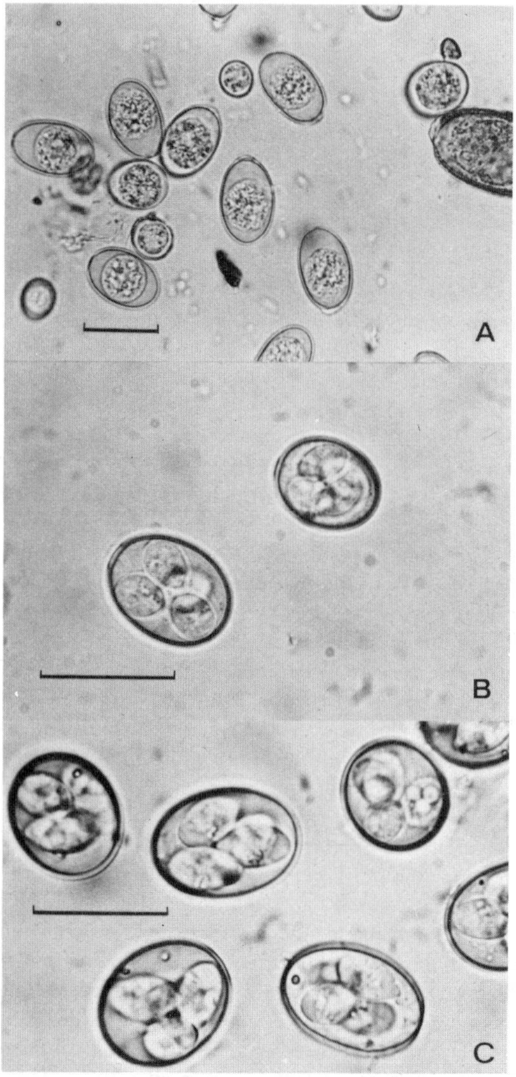

Fig. 9.1 A Unsporulated oocysts of ovine coccidia B Sporulated oocysts of *Eimeria crandallis* C Sporulated oocysts of *Eimeria ovinoidalis* The scale bar represents 20 microns in each case. (Crown Copyright)

sporulation is complete after 4 or 5 days.[3] Morphological features which can be used for the identification of individual species can be seen under the high-power microscope and have been described by several authors.[1, 4, 5, 6] A set of diagrams has also been prepared for diagnostic laboratories.[3]

The species of Eimeria recognized in sheep are listed in Table 9.1 and the oocysts of the two species recognized as pathogenic are shown in Fig. 9.1. *E. ovinoidalis* (formerly known as *E. ninakohlyakimovae*) (Fig. 9.1c) is especially pathogenic and relatively light infections kill lambs with severe haemorrhagic diarrhoea. Lesions containing developmental stages of the parasites which penetrate into the submucosal tissues are found in the ileum, caecum or colon. *E. crandallis* (Fig. 9.1b) is not so acutely pathogenic but can cause extensive diarrhoea and weight loss.

Surveys have shown that faecal oocyst counts may be high in lambs aged 6–10 weeks and any of the species may occur. There often are no symptoms and the counts progressively fall to a few hundreds per gram which is characteristic of the mature animal.

Most species respond to treatment with sulphonamides administered either orally or by injection. Oocyst numbers may fall and diarrhoea be relieved but often there is a partial relapse after the treatment.

Clinical signs and pathology

Coccidia, initially, are parasites of intestinal epithelium and the first response to infection is usually a loss of epithelial cells. This is generally accompanied by villus atrophy, an extension of the glandular part of the epithelium and the ratio of villus height to total mucosal thickness becomes reduced. The resultant smooth gut surface is a common response to infections but it results in a reduced efficiency of food absorption which gives rise to the ill-thrift which is usually the first sign of the disease.

Eimeria crandallis and *Eimeria ovinoidalis* characteristically are found in the ileum. The latter species also affects the caecum in its later stages and lesions may extend into the colon. Most ovine coccidia produce at least one generation of large schizonts which develop in the lamina propria, often forming whitish colonies which can be seen by the naked eye through the mucosal surface. When smears of this tissue are examined under the microscope they can be seen to contain masses of merozoites. In large numbers the schizonts can disrupt the mucosa causing haemorrhage which is the basis of the haemorrhagic diarrhoea of acute cases. In extreme

Table 9.1   Characters of oocysts of *Eimeria* spp. from sheep

Species	Oocyst Dimensions (Average in microns)	Morphological Characters
E. ahsata	25 × 39	Prominent polar cap. Yellowish brown
E. ovina	20 × 31	Polar cap. Ellipsoidal
E. crandallis	17 × 24	Shallow polar cap. Sometimes absent. Broad sporocysts
E. faurei	21 × 29	No polar cap—micropyle at narrow end. Pale yellowish-brown colour
E. granulosa	21 × 29	Urn-shaped, large polar cap at broad end. Yellowish-brown
E. intricata	32 × 47	Large, thick-walled oocysts, brownish. Polar cap
E. ovinoidalis	18 × 23	No polar cap. Thin-walled colourless. Ellipsoidal
E. pallida	10 × 14	No polar cap. Thin-walled. Ellipsoidal—pale yellow
E. parva	14 × 16	No polar cap. Nearly spherical. Colourless
E. marsica	13 × 19	Inconspicuous polar cap. Ellipsoidal, v. pale yellow
E. weybridgensis	17 × 24	Shallow polar cap. Sometimes absent. Elongate sporocysts. (c.f. *E. crandallis*)

cases this can be fatal before any oocysts appear in the faeces. Severe damage can still be caused by the latest stages in the life-cycle, the sexual stages, because they are the most numerous and develop in the epithelial cells to be shed as oocysts into the gut lumen to pass out with the faeces. The epithelial lining of the gut may thus become stripped away causing loss of fluids, blood or permit entry of other infections or toxins.

The regeneration process is complicated, depending for instance upon the degree of damage to the intestinal crypts, or the nutritional status of the animal. Full recovery to a normal growth pattern may take quite a long time.

Diagnosis

Diagnosis of coccidiosis in sheep is often difficult because there is rarely a clear relationship between unthriftiness and coccidial infections. Lambs may show diarrhoea, reduced growth, high faecal oocyst counts and low helminth burdens often with a good response to sulphonamide therapy. Such criteria do not necessarily give an unequivocal diagnosis because as previously stated, normal lambs may pro-

duce large number of oocysts and in acute infections lambs may die before any oocysts are shed and those which are produced may be atypical and deformed. Following acute infections oocyst production may fall rapidly leaving a sick animal with non-specific lesions in the gut. The response to sulphonamide therapy may also be misleading because these compounds have a wide spectrum of activity which includes a range of infections beside coccidia. The flock history and records of laboratory investigations of a number of cases are often useful guides to the diagnostician.

From a study of cases from British flocks submitted for laboratory examination during 1978–1979 when the incidence of diagnosed coccidiosis was particularly high Gregory and others[7] concluded that the diagnosis of individual cases of coccidiosis in lambs can be complicated. Generally, severe diarrhoea with high oocyst counts with either *E. ovinoidalis* or *E. crandallis* predominating or numbers of deformed oocysts of these species, together with lesions associated with developmental stages of the parasites in the ileum and/or caecum are good evidence for a specific diagnosis.

On a flock basis it was concluded that the disease typically presents the following features: many lambs aged about 4–6 weeks with diarrhoea, some of which give high faecal oocyst counts in which *E. crandallis* or *E. ovinoidalis* predominate. Animals which die show post-mortem lesions in the ileum or caecum.

Control

Epidemiological studies indicate that disease may be related to the numbers of parasites acquired in early life and the rate of transmission between very young animals. Hygiene in the lambing sheds is therefore important, as is the provision of adequate clean litter and bedding in subsequent housing. The quicker the lambs can be transferred and maintained on clean grazing the better. Treatment of established infections is generally by means of sulphonamides. Sulphamezathine is commonly given orally or by injection; the long-acting formulations given parenterally, or amprolium in the drinking water may also be effective. Some flockmasters anticipate the problem and give treatment routinely when lambs are at the age when from previous experience trouble is expected.

REFERENCES

1 Levine N.D. & Ivens V. (1970) The coccidian parasites (*Protozoa, Sporozoa*) of ruminants. *Illinois Biological Monographs*, Vol.44. University of Illinois Press.
2 McDougald L.R. (1979) Attempted cross-transmission of coccidia between sheep and goats and description of *Eimeria ovinoidalis* sp. n. *Journal of Protozoology*, **26**, 109–13.
3 Ministry of Agriculture, Fisheries and Food. (1977) Technical Bulletin No. 18, 2e. *Manual of Veterinary Parasitological Laboratory Techniques*. HM Stationery Office, London.
4 Joyner L.P., Norton C.C., Davies S.F.M. & Watkins C.V. (1966) The species of coccidia occurring in cattle and sheep in the South-West of England. *Parasitology*, **56**, 531–41.
5 Norton C.C., Joyner L.P. & Catchpole J. (1974) *Eimeria weybridgensis*. sp. nov. and *Eimeria ovina* from the domestic sheep. *Parasitology*, **69**, 87–95.
6 Catchpole J., Norton C.C. & Joyner L.P. (1975) The occurrence of *Eimeria weybridgensis* and other species of coccidia in lambs in England and Wales. *British Veterinary Journal*, **131**, 392–401.
7 Gregory M.W., Joyner L.P., Catchpole J. & Norton C.C. (1980) Ovine coccidiosis in England and Wales 1978–1979. *Veterinary Record*, **106**, 461–2.

10   JOHNE'S DISEASE

N.J.L. GILMOUR AND K.W. ANGUS

Synonym Paratuberculosis

Johne's disease is a chronic enteritis of ruminants caused by the bacterium *Mycobacterium johnei* (syn. *M. paratuberculosis*). The disease was first described in cattle in Germany in 1895 but the causal organism was not grown in the laboratory until 1912. Johne's disease is of world-wide distribution and is economically important, not only as a cause of death but also losses which result from reduced productive capacity during the lengthy preclinical stage of the disease. Johne's disease has remained a problem for so long because of the absence of a simple, accurate diagnostic test to detect subclinical infection.

The distribution of Johne's disease in cattle is world-wide. There have been fewer reports of the disease in sheep but it has been reported from Britain, Iceland, Germany, Spain, Italy, Yugoslavia, Egypt, Israel, Iraq, South Africa and New Zealand. Its introduction into Iceland by 5 rams from Germany in 1933 and its subsequent spread to sheep and cattle has been well documented.[1]

Cause

The cause of Johne's disease is infection with *Mycobacterium johnei*. This bacterium is aerobic, non-motile, Gram-positive and acid-fast, 1–2 μm in length and 0.5 μm broad. It will grow on artificial laboratory media only in the presence of a growth factor (mycobactin) derived from mycobacteria. After up to 6 weeks incubation at 37–39°C *M. johnei* forms small, raised, dull-white, rough colonies.

A pigmented strain of *M. johnei* has been isolated from Johne's disease in sheep.[2] The pigmentation is due to a bright yellow, non-diffusable pigment. Pigmented strains grow very slowly and sparsely on artificial media. Colonies may only be visible after 9 months' incubation. Non-pigmented sheep strains are culturally and antigenically indistinguishable from strains from cattle. *M. johnei* will survive in faeces, soil and water for more than a year.

Clinical signs

The predominant clinical sign of Johne's disease in sheep is loss of bodily condition due to muscle wasting and there are no specific signs. Although the disease is always chronic there is considerable variation in the course in sheep. In some instances the disease progresses relatively rapidly with the interval between the appearance of wasting and death measured in months. In other cases, after the initial loss of condition, there may be no clinical deterioration for long periods. In sheep, diarrhoea is not a feature of the disease. This is probably due to the sheep's ability to reabsorb water in the large intestine. In advanced cases the faeces become soft and unformed.

Pathology

It must be emphasized that gross changes in sheep with Johne's disease are often difficult to detect, and do not resemble those of the disease in cattle.[3, 4] However, the bright yellow colour of the intestinal mucosa in sheep infected with pigmented strains of *M. johnei* is pathognomonic.

In advanced cases there is wasting with gelatinous atrophy of fat depots and serous effusion into body cavities.[4] In the intestine, macroscopic changes are confined to the posterior jejunum and ileum, and the mesenteric lymph nodes. The ileum may be thickened and may feel doughy when handled, but more usually the only visible change in the lining is a slight fleshy or velvety thickening,[4] or a faint granularity of the surface, perhaps with slight congestion.[3] These changes can only be recognized with confidence if the observer has considerable experience of the appear-ance of the normal sheep bowel. Occasionally, there may be a tendency for the mucosa to undergo fissuring when bent over the fingers. Where infection with pigmented strains has occurred, the mucosal lining takes on a bright yellow colour, due to the presence of numerous pigmented *M. johnei* in the lamina propria.

The afferent lymphatic vessels in the mesentery may be thickened and convoluted, and may even contain numerous small (1–4 mm) whitish nodules, which may be caseous or even calcified. Similar nodules or white flecks may be seen on the peritoneal surface of the ileum, or the cut surface of the intestinal wall or the mesenteric lymph nodes. The latter are almost invariably enlarged and may be very prominent at necropsy.

The microscopic lesions of Johne's disease in sheep vary according to the strain of the organism and the resistance of the host. Lesions associated with pigmented strains of the organism consist either of focal, or more frequently, of diffuse accumulations of epithelioid macrophages, which replace the normal structures in the mucosa. The crypts of Lieberkuhn become atrophic and disappear, leaving a lamina propria packed with pale cells. Similar infiltrates of epithelioid cells may be found in the submucosa and the lymphoid follicles of the Peyer's patches, but they seldom extend to the mesenteric lymph nodes.[3] The epithelioid cells comprising the lesions often contain numerous *M. johnei* bacteria, readily demonstrated by Ziehl-Neelsen staining of histological sections.

The reactions to infections with non-pigmented strains vary considerably. Infiltration with epithelioid macrophages may be diffuse, but more likely focal. Discrete masses of epithelioid cells are seen in the villi just beneath the surface epithelium, in the submucosa, the lymphoid follicles of Peyer's patches, the afferent lymphatics to the mesenteric lymph nodes, and the nodes themselves. In all these sites, the epithelioid cells are usually associated with pleomorphic multinucleated cells, mainly of the Langhans type, and the lesions are heavily invested by other mononuclear cells.[3] The focal lesions may take on a distinctive tuberculoid character in some sheep, or they may exhibit caseation, calcification or fibrosis, the resultant nodular lesions being visible macroscopically. This tendency for lesions to undergo caseation or calcification is an important point of difference from the lesions of Johne's disease in cattle.[4] Acid-fast organisms can be demonstrated only in very small numbers in tuberculoid lesions, and are usually absent from caseous or calcified foci.

The tendency for lesions to undergo caseation or

calcification is believed to be a manifestation of local resistance or hypersensitivity. This supposition has been at least partly confirmed experimentally by examination of lesions in sheep infected with *M. johnei* which exhibited strong or weak skin hypersensitivity to intradermal inoculation of PPD avian tuberculin.[5] Electron microscopy of epithelioid macrophages in lesions showed that *M. johnei* were undergoing degeneration in these cells in sheep with strong skin hypersensitivity, whereas the bacteria seemed to be able to multiply in epithelioid cells in sheep with weak skin hypersensitivity. Thus, skin hypersensitivity appears to be linked to the capability of the host to reduce the weight of intestinal infection.

Diagnosis

Natural Johne's disease occurs in sheep of any age over about one year although the disease may occur in younger sheep after experimental infection. However, most cases occur in breeding ewes. The problems of diagnosis centre on the confirmation of the presence of Johne's disease in the wasting sheep. There is no evidence that allergic and serological tests are of value in detecting infection in the preclinical stage of the disease or even in the clinical stage. The history of flock losses due to sheep dying after a wasting illness may be helpful in diagnosis but deaths may often be sporadic and vary in numbers from year to year. Diagnosis on a flock basis is best achieved after post-mortem examination of a wasting sheep. Confirmation of the presence of the disease is obtained by the finding of acid-fast bacilli with the morphology of *M. johnei* in smears from the ileal mucosa or mesenteric lymph nodes, stained with Ziehl-Neelsen. However, acid-fast organisms may be absent even although enteric lesions are present.

Histology of the ileal mucosa and the caecum and mesenteric lymph nodes is of most significance especially since thickening of the intestines or enlargement of the mesenteric lymph nodes are not constantly present. Culture of these tissues is also of value if taken into consideration with the histology but it is possible to recover *M. johnei* from these tissues in the absence of lesions sufficient to cause the actual disease.

In the live animal the finding of clumps of acid-fast organisms in the faeces provides evidence of the presence of the disease but here again a negative result does not rule out the possibility of the disease being present. Faecal culture is of some diagnostic value but again there are problems. The long time (3–4 months) required for a result to be obtained and the overgrowth of the cultures with contaminants limit the usefulness of this method. Biopsy of the terminal ileum and ileo-caecal lymph nodes is also a useful diagnostic aid. Tissues taken at biopsy can be examined histologically and cultured for the presence of *M. johnei*

The diagnosis of the pigmented form of Johne's disease presents few problems at necropsy. The bright yellow pigmentation of the ileum and the presence of enormous numbers of acid-fast bacilli in smears from the intestinal lesions and usually enlarged mesenteric lymph nodes provide confirmation of the diagnosis.

Tissues taken at necropsy or intestinal biopsy specimens must be treated with oxalic acid and inoculated onto a medium which contains antibiotic to control the bacteria, yeasts and fungi normally found there. The medium described by Brotherston et al[6] is satisfactory for this purpose and has the added advantage that it is transparent. This allows the presence of colonies of *M. johnei* to be detected as early as 3–4 weeks after inoculation. In differential diagnosis, all wasting diseases of sheep must be considered, especially those resulting from intestinal helminths and liver flukes.

Epidemiology and transmission

Infection is spread by way of the faeces. *M. johnei* is excreted in the faeces of infected sheep before clinical signs of the disease are apparent and usually in larger numbers once clinical signs have developed. Young sheep are more susceptible than adults, i.e. if infected as lambs they are more likely to develop clinical signs, if infected as adults overt disease is less common.

The cycle of infection in a flock starts with the lambs becoming infected by faeces containing *M. johnei*. There are ample opportunities for this to occur, e.g. faecal-contaminated udders. Infection can also be acquired *in utero* if the disease in the ewe is advanced. It has been shown experimentally that as few as 1000 *M. johnei* will infect lambs.[6] Infection is established when *M. johnei* invades the lymphatic tissue in the mucosa of the small intestine, where it multiplies over the next 2–3 months and spreads to the draining mesenteric lymph nodes. After this initial phase of infection a number of different sequelae occur. The outcome of infection seems to depend on two things; first the ability of the host to mount a cell-mediated immune response with delayed-type hypersensitivity and secondly the dose of the initial infection, as a heavy initial infection is likely to be overcome than a light one. Depending on the host's response one of three possible events develops: the initial infection is overcome, the infection persists for many months or years (this category comprises the carrier which does not develop the clinically-apparent disease) or the intestinal lesions slowly progress till the cellular changes in the intestine interfere with its functions. The principal biochemical lesions are a loss of plasma proteins into the lumen and concurrent malabsorption of amino acids [7] and enhancement of protein metabolism in the liver, the substrates for which are probably mobilized from the muscles which are wasted as a result. It has been shown experimentally that reinfection does not influence the course of the disease and it is the initial infection which determines the final outcome.

Control

Johne's disease as a flock problem tends to be cyclic in that the disease often appears then disappears after a few years in the absence of control measures. There is no treatment for Johne's disease. Sheep in which the disease has been confirmed are likely to be excretors of *M. johnei* in the faeces and as such are a danger to other sheep. It is therefore important that they and any wasting sheep should be culled. Lambs born from ewes which develop the disease should not be retained in the flock as they are most likely to be infected either from their dams' faeces or in utero. Vaccination with killed *M. johnei* in oil adjuvant prevents the development of clinical disease but does not eliminate the infection.[8] Vaccines prepared for

use in cattle are suitable for sheep. Vaccination should be confined to lambs up to about 3 months of age and revaccination is not advised. As vaccination causes allergy to avian and mammalian tuberculin and antibodies detectable by the complement fixation test for Johne's disease, it cannot be used in flocks which are liable to be tested with these for export certification purposes. It is not known if it is possible to vaccinate against the pigmented form of Johne's disease.

REFERENCES

1 Sigurdsson B. (1956) Paratuberculosis of sheep and methods for controlling it. *Organisation for European Economic Co-operation.* Paris, 169–88.

2 Taylor A.W. (1951) Varieties of *Mycobacterium johnei* isolated from sheep. *Journal of Pathology and Bacteriology,* **63**, 333–6.

3 Stamp J.T. & Watt J.A. (1954) Johne's disease in sheep. *Journal of Comparative Pathology,* **64**, 26–40.

4 Jubb K.V.F. & Kennedy P.C. (1963) *Pathology of Domestic Animals.* 1e Vol. 2. pp.119–22. Academic Press, London, New York.

5 Gilmour N.J.L., Angus K.W. & Mitchell B. (1978) Intestinal infection and host response to oral administration of *Mycobacterium johnei* in sheep. *Veterinary Microbiology,* **2**, 223–35.

6 Brotherston J.G., Gilmour N.J.L., & Samuel J.McA. (1961) Quantitative studies of *Mycobacterium johnei* in the tissues of sheep. 1. Routes of infection and assay of viable *M. johnei. Journal of Comparative Pathology,* **71**, 286–99.

7 Paterson D.S.P., Allen W.M. & Lloyd M.K. (1967) Clinical Johne's disease as a protein losing enteropathy. *Veterinary Record,* **81**, 717– 18.

8 Gilmour N.J.L. (1976) The pathogenesis, diagnosis and control of Johne's disease. *Veterinary Record,* **99**, 433–4.

PARASITIC GASTROENTERITIS 11

R.L. COOP AND M.G. CHRISTIE

Parasitic gastroenteritis (PGE) is mainly a disease of young sheep and is caused by infection with mixed trichostrongylid nematodes particularly those of the genera *Ostertagia* and *Trichostrongylus* (Table 11.1).

Table 11.1 Common parasites of sheep in the United Kingdom

Abomasum	*Ostertagia circumcincta*
	Ostertagia trifurcata
	*Trichostronglus axei
	*Haemonchus contortus
Small intestine	*Trichostrongylus vitrinus
	*Trichostrongylus colubriformis
	*Nematodirus battus
	Nematodirus filicollis
	Cooperia curticei
	Strongyloides papillosus
	Bunostomum trigonocephalum
	Monezia expansa
Large intestine	Chabertia ovina
	Oesophagostomum venulosum
	Trichuris ovis

*The main contributors to outbreaks of parasitic gastroenteritis

The majority of grazing sheep become infected with gastrointestinal parasites but whether or not their performance is adversely affected depends on the rate and level of intake of infective larvae, the age of the animal, the plane of nutrition and previous experience of helminth infection.

ABOMASAL PARASITISM

Ostertagia infection

In Britain *O. circumcincta* is the most prevalent species, with small numbers of *O. trifurcata* often present.

Ostertagia is the major contributor to PGE in young lambs and can cause acute disease (Type I ostertagiasis) leading to a rapid decline in growth rate. It most frequently occurs in intensively-grazed lambs during the summer months. Clinical signs are watery diarrhoea with soiling of the fleece, dehydration and failure to gain weight. Severe gastritis is often seen at post-mortem examination with elevated gastric pH and many (10 000–15 000) adult *Ostertagia* spp. may be present.

When the intake of infective larvae is relatively low, outbreaks of clinical disease may not occur. However, subclinical *Ostertagia* infections can lower food intake and impair growth rate.[1] A subacute or chronic form (Type II ostertagiasis) is occasionally seen in both housed and out-wintered ewes and yearlings in the late winter/early spring which is similar to Type II ostertagiasis in cattle. It is caused by the emergence of hypobiotic larvae which were picked up the previous autumn and over-wintered in the gastric mucosa. Affected sheep often show intermittent diarrhoea and progressive loss of body-weight and condition.

In moderate to heavy infections the abomasal wall is thickened and inflamed with marked cellular infiltration and numerous raised nodules are present on the surface of the abomasal folds. In severe cases the nodules may coalesce to form the characteristic 'Morocco leather-like' appearance which occurs with severe hyperplastic gastritis. As the larvae develop within the gastric glands they distend the lumen and stretch the cellular lining causing dedifferentiation and a reduction in the numbers of parietal and peptic cells.[2] Adjacent non-parasitized glands are also affected, parietal cells being replaced by non-functional cells and as a consequence the pH of the abomasal contents becomes more alkaline. A proportion of larvae develop into adults in the mucosa causing the septa between parasitized glands to rupture, forming multilocular spaces.

Haemonchus infection

Haemonchus contortus is usually associated with warm temperate and tropical regions but outbreaks of haemonchosis occasionally occur in Britain particularly in the southern counties. Clinical features of haemonchosis are pallor of the mucous membranes, hyperpnoea and tachycardia. Pathogenesis is associated with blood loss caused by the feeding activities of larvae and adult worms and diarrhoea is not normally seen. Acute disease results from large intakes of infective larvae and animals rapidly become unthrifty and weak. Anaemia and oedema occur and morbidity is frequently high. Chronic haemonchosis is due to a more gradual intake of larvae and results in a general wasting condition often resembling a state of malnutrition. Growth rate gradually declines and the fleece may be open and lack-lustre. Chronic haemonchosis with severe anaemia can occasionally occur in ewes in the spring and results from maturation of hypobiotic larvae ingested the previous season.

At necropsy, the carcass is frequently pallid and watery and the liver pale and friable. Abomasal contents may be dark brown and the mucous lining pale, mucoid and covered with dark red spots. A few raised pale discoid nodules, 2–5 mm in size, may be present in the plicae. Microscopically, the main features relate to the local effects of individual adult worms or larvae, with the more diffuse effects resulting in hypertrophy of the mucosa. Adult worms usually lie between the surface epithelium and a thick layer of surface mucus, and local lacerations or erosions with haemorrhage from superficial capillaries are associated with their presence. Developing larvae may be seen coiled in the gastric glands where they cause local distention, cytolysis and loss of parietal cells. Overall hypertrophy of the mucosa is partly due to an increase in the upper mucus-secreting capacity of the glands and partly to cellular infiltration of the lamina propria. Abomasal lymph nodes are often enlarged and hyperplastic.

Trichostrongylus infection

Trichostrongylus axei can cause depression of appetite and a check in growth rate. *T. axei* is often accompanied by *Ostertagia* and other species of *Trichostrongylus* and diarrhoea may or may not be present. Larval penetration causes inflammation and hypertrophy of the abomasal mucosa and erosion of the mucous membrane can occur with the formation of small 'ring-worm'-like lesions.

SMALL INTESTINAL PARASITISM

Trichostrongylus infection

Trichostrongylus spp. are important contributors to ovine PGE and in Britain *T. vitrinus* is the more common species. Trichostrongylosis is usually seen as a chronic wasting condition, often in hoggs and ewes in the early winter, but acute disease can sometimes occur in lambs. Clinical features include sudden loss of appetite and a marked decline in growth rate. Dark coloured diarrhoea is frequently seen in the more severely affected animals and the fleece is often open and brittle. At necropsy large numbers (20 000–30 000) of *Trichostrongylus spp.* may be present. Subclinical infections can also impair the utilization of nutrients.

Gross lesions are those of enteritis with increase in mucus production, inflammation of the anterior small intestine and hypertrophy of the duodenal and

Fig. 11.1 Section of duodenum from a lamb infected with *T. vitrinus* showing total villous atrophy and elongation of the crypts.

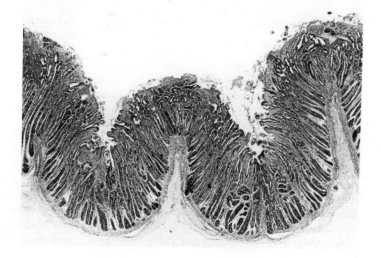

jejunal mucosa. The villi are frequently short and distorted and in severe cases may be completely absent leaving the mucosa covered with small rounded protruberances (Fig. 11.1). The intestinal crypts become dilated and elongated and the lamina propria is usually heavily infiltrated with inflammatory cells. In *T. vitrinus* infections of long duration the affected areas tend to be more localized, forming the

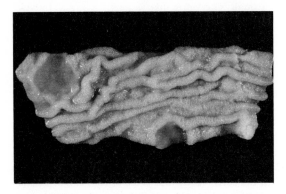

Fig. 11.2 Portion of small intestine from a lamb infected with *T. vitrinus* showing the flattened focal areas ('finger-print' lesions).
The photographs are reproduced by permission of the British Veterinary Association.

characteristic 'finger-print' lesions (Fig. 11.2). *Trichostrongylus spp.* can easily be overlooked at necropsy as the thread-like worms (5–8 mm in length) are not readily visible in gut washings, and only become apparent when the majority of the digesta has been removed on a fine sieve.

Nematodirus infection

Nematodirus battus can cause significant losses in young lambs around May and June in Britain. Nematodiriasis is caused by a sudden mass hatch of infective larvae on the pasture and the first signs are scouring in a proportion of the flock. Disease is rare in lambs over 3 months of age. Lambs show acute enteritis with profuse watery diarrhoea, often associated with lethargy and loss of appetite. The fleece frequently becomes rough and the lambs may show the typical 'tucked-up belly' appearance. Weight loss can be rapid, with severe dehydration and if the infection is unchecked, a proportion of the lambs may die within a few days. Lambs which survive usually suffer a check in growth rate and it will often

be a couple of months before they return to reasonable condition.

Nematodirus filicollis is less pathogenic as the infective larvae hatch over an extended period.

At necropsy there may be masses of worms coiled together in the intestine or very few parasites may be present, the majority having been expelled during the diarrhoeic phase. Developing stages and adult worms are involved in the pathogenesis. Gross lesions are normally those associated with catarrhal enteritis and acute inflammation of the alimentary tract. The mesenteric lymph glands may be enlarged. Microscopically, the mucosa shows superficial lesions with local distortion and compression of the villi in contact with parasites, leading to necrosis of the surface epithelium with the formation of local erosions. The microvilli may be sparse and stunted and the lamina propria infiltrated by inflammatory cells.

Other parasite infections

Cooperia curticei, Strongyloides papillosus and *Bunostomum trigonocephalum* are occasionally found in the small intestine at necropsy but are generally not present in sufficient numbers to be pathogenic. *Monezia expansa* is the commonest tapeworm in young lambs but although they can be large (1.5–2 m) and fill the lumen of the alimentary tract they are considered to be of low pathogenicity and are usually eliminated from the host after a few months.

The main parasites in the large intestine are *Oesophagostomum venulosum, Chabertia ovina* and *Trichuris ovis*. They are normally present in small numbers and cause very little pathogenic effect. *O. venulosum* feeds on small plugs of tissue and may leave small ulcers on the intestinal mucosa. *C. ovina* can cause enteritis if present in sufficient numbers, producing small haemorrhages in the wall of the colon.

Epidemiology of gastrointestinal parasitism

Veterinary epidemiology has two tasks: the appreciation of the natural history of diseases and quantifying disease in populations to give information on incidence and prevalence.[3] The following discussion of the epidemiology of gastrointestinal parasitism refers to nematode (round worm) parasites. In the area of gastrointestinal parasitism much further progress has been made with the first task than with the second.

The starting point for an understanding of the natural history of gastrointestinal parasitism and of the methods which have been developed to prevent damage to lambs is a consideration of the life cycle. In a susceptible host, infective larvae which are eaten with the grass normally reach maturity in their chosen part of the alimentary tract in 2–3 weeks. When they reach maturity the female worms are fertilized by male worms and start laying eggs into the surrounding digesta. These eggs are carried down the gut and pass to the exterior in the faeces thus starting the free-living stage of the life cycle. The essence of this free-living phase is that the developing worm passes through the following stages, egg, first stage larva (L1), second stage larva (L2), third stage or infective larva (L3). The L1 and L2 are stages of active growth but the L3 is a stage at which further development is suspended until the larva is eaten by a sheep. L3 larvae can survive for several months on pasture.

The pattern of development described in the previous paragraph is common to all species of round worm parasites of sheep but there is an important difference between *Nematodirus* and other genera in respect of the point on the cycle at which larvae hatch and leave the protection of the egg shell. In *Ostertagia, Trichostrongylus spp.* and all other genera, except *Nematodirus*, the very small L1 (about 0.2 mm long) hatches from the egg and feeds on bacteria in the dung, as does the L2 which in its turn grows and moults to become the L3, which retains the moulted skin of the L2 as a protective sheath. In *Nematodirus* the egg is much larger than in other species and the embryo is equipped with sufficient food reserves to carry it through development within the protection of the egg shell and it is the infective L3 which hatches from the egg. This difference between *Nematodirus* and the other species of sheep round worms in the manner of their free-living development is reflected in the time taken for its completion. In *Nematodirus battus* this is normally 10–12 months, the lambs of one year infecting those of the next year. With the other genera, given adequate moisture and the conditions of summer, free-living development can be completed in less than 2 weeks but at cooler times of the year the time can be as long as 3 months.

Suspended development

The L3, although motile, is a stage at which further development is suspended. Suspension of development can also take place in the parasitic phase of the life cycle. *Ostertagia* L3's ingested by ewes in autumn or winter may become arrested early in their parasitic development. Such arrested forms may become activated around the time of lambing or during winter housing.

Application of knowledge of the life cycle in the control of worm damage

The currently accepted account of the epidemiology of *Nematodirus* emphasizes the timing of the hatching of the L3's from the eggs as the deciding factor in making any year one of high incidence of this disease. The disease is readily recognizable and has a sharply defined seasonal pattern (late April to June). These features, taken together with the unusual form of the free-living development, have made it possible to develop a reasonably complete account of its epidemiology and the way in which this knowledge can be applied in practice is described in Chapter 50.

The situation is much less clear with other genera of gastrointestinal nematodes, the most important of which is *Ostertagia*. Where pasture is used repeatedly for sheep the numbers of *Ostertagia* L3 on the pasture reaches a peak in July–August and this contamination is derived in part from the ewes which excrete worm eggs during lactation and in part from the lambs themselves. Even if all sheep are withdrawn from a pasture in late summer or autumn, sufficient L3 survive the winter to infect ewes grazing the pasture early in their lactation the next year, and the lambs, as they start to graze take in enough of these overwintered larvae to develop an egg output in their faeces which adds substantially to the contamination of the pasture. Safe grazing practices are designed to prevent exposure of lambs to the mid-summer peak of contamination.

Measuring parasite damage

With regard to the second task of epidemiology, quantifying disease, an interesting attempt to correlate outbreaks of parasitic gastroenteritis with climatic events has led to the suggestion that years of high incidence of disease are correlated with weather conditions in the previous year.[4] While further examination of this suggestion would involve a detailed discussion of possible mechanisms, it is notable that this approach has to start with an attempt to collate and classify information on the incidence of parasitic gastroenteritis into the four subjectively assessed categories: low, below average, above average, high. The need to use such a simplified classification indicates the extent of the task still facing epidemiology in quantifying even the incidence of frank disease. The problem of quantifying the damage to production caused by subclinical parasitism is even greater. The description given in this chapter of the effects which can be demonstrated under controlled experimental conditions make it seem likely that some degree of impairment of performance due to *Ostertagia* is widespread where lambs graze pasture contaminated by their dams and later by themselves, but currently there is no means of distinguishing in the field between parasitic and nutritional causes of suboptimal performance. The recorded success of clean grazing systems suggests that a degree of parasite damage is almost universal in conventional grazing systems, but it must be remembered that implementation of a clean grazing system normally involves a total reappraisal of the existing system and the resulting improvements in production may result from other changes unconnected with the control of parasites.

EFFECTS OF GASTROINTESTINAL HELMINTHS ON PRODUCTION

A common feature of gastrointestinal parasitism in sheep is a poor growth rate and in acute cases, actual loss of weight and debilitation. Meat and wool production are inevitably lowered and where prophylactic treatment is minimal a greater proportion of stock may be retained as stores. The economic impact of gastrointestinal parasitism to the livestock industry has prompted many investigations into the underlying mechanisms responsible for the loss of production.[5, 6]

Abomasal parasitism

Depressed productivity in lambs infected with *Ostertagia* is largely due to reductions in food intake.[1] Similarly, lowered food intake is a feature of *T. axei* infections. However, as animals develop resistance to infection, appetite and growth rate gradually return to normal. Experimental studies have shown that the deposition of muscle protein in the carcass can be reduced in *Ostertagia*-infected lambs, the greater part of the reduction being attributable to the lowered food intake, although the apparent digestibility of

food and utilization of nutrients is also depressed. Endogenous protein loss, which results from increased sloughing of epithelial cells, increased mucus production and increased leakage of plasma proteins from the damaged abomasum, contributes to the poor retention and utilization of nitrogen in the parasitized animal.

Ostertagia infection can lower mineral deposition in the body and impair the growth of the skeleton. The amount of bone matrix is also reduced and these effects on the skeleton are thought to arise from changes in energy and protein availability rather than the malabsorption of minerals. The effect on skeletal growth may be important in fat lamb production as the smaller framework limits to some extent the amount of muscle deposited.

Decreased digestive efficiency, poor retention of nitrogen and increased leakage of macromolecules into the abomasum also occurs in *T. axei* and *Haemonchus* infections. With *Haemonchus* a further drain on the animals' reserves results from the haematophoragic activity of the worms, and this aspect has been studied in considerable detail.[7]

Intestinal parasitism

Although *Trichostrongylus spp.* can cause extensive lesions in the intestine, malabsorption per se is not a major factor contributing to impaired productivity, any local effects being compensated by digestion and absorption from the remainder of the alimentary tract. Indeed, it has been shown recently that the true digestibility and absorption of protein is unaffected in sheep infected with intestinal *Trichostrongylus* spp. Poor growth rate of infected animals mainly results from a reduction in food intake. Experimental studies, using pair-feeding, have shown that the efficiency of feed utilization is also lowered in intestinal parasitism.[8] In addition, retention of nitrogen is reduced, leading to poor carcass quality even with subclinical levels of infection. Part of the reduced efficiency of food conversion of parasitized animals is associated with increased rates of blood and intestinal protein turnover which must limit the availability of aminoacids for synthesis of muscle and wool protein.

Gastrointestinal parasitism can reduce wool growth and decrease the length and diameter of the fibres. The wool tends to be more brittle which lowers its processing qualities. It has been suggested that these changes may be associated with an increase in the requirement for sulphur-containing aminoacids which are essential for the formation of wool protein.

Intestinal helminth infections interfere with bone mineral metabolism leading to poor retention of calcium and phosphorus. Skeletal growth is reduced and the quality of the bones is poor due to reductions in the density of the bone matrix and its degree of mineralization. The basic cause of the skeletal lesions is thought to be induced mineral deficiency coupled with impaired nitrogen and energy metabolism. *Trichostrongylus* infection may be a contributing factor in the aetiology of osteoporosis seen in growing lambs.

Plane of nutrition

The nutritional status of the host influences the course and pathogenicity of gastrointestinal infections by affecting the initial establishment of the parasite and/or development of the infection. The level of nutrition may influence immunological competence and also affect the ability of the animal to repair the gastrointestinal mucosa and maintain the increased turnover of blood proteins which frequently accompanies infection. Generally, the effects of helminth infestation are more severe when sheep are maintained on a low plane of nutrition. These interactions are obviously complex and are influenced by several factors such as the age of the animal at the time of infection, the extent of previous exposure to parasites and the type of diet.

REFERENCES

1 Sykes A.R. & Coop R.L. (1977) Intake and utilization of food by growing sheep with abomasal damage caused by daily dosing with *Ostertagia circumcincta* larvae. *Journal of Agricultural Science,* **88**, 671– 7.

2 Armour J., Jarrett W.F.H. & Jennings F.W. (1966) Experimental *Ostertagia circumcincta* infections in sheep: development and pathogenesis of a single infection. *American Journal of Veterinary Research,* **27**, 1267–78.

3 Thrusfield M.V. (1980) The scope and content of epidemiology courses in veterinary curricula. Proceedings of the Second International Symposium on Veterinary Epidemiology, pp.303–14. Australian Government Publishing Service, Canberra.

4 Ollerenshaw C.B., Graham E.G. & Smith L.P. (1978) Forecasting the incidence of parasitic gastroenteritis in lambs in England and Wales. *Veterinary Record,* **103**, 461–5.

5 Symons L.E.A. & Steel J.W. (1978) Pathogenesis of the loss of production in gastro-intestinal parasitism. In *The Epidemiology and Control of Gastro-intestinal Parasites of Sheep in Australia,* (Ed.Donald A.D. Southcott W.H. & Dineen J.K.) Division of Animal Health CSIRO, Australia.

6 Dargie J.D. (1980) Pathophysiology and helminth parasites. In *Digestive Physiology and Metabolism of Ruminants.* (Eds.Ruckebusch Y. & Thivend P.) MTP Press, Lancaster.

7 Dargie J.D. (1975) Applications of radioisotopic techniques to the study of red cell and plasma protein metabolism in helminth diseases of sheep. Symposium, British Society Parasitology, **13**, 1–26.

8 Sykes A.R. & Coop R.L. (1976) Intake and utilization of food by growing lambs with parasitic damage to the small intestine caused by daily dosing with *Trichostrongylus colubriformis* larvae. *Journal of Agricultural Science,* **86**, 507–15.

LIVER FLUKE

12

A. WHITELAW

Synonyms Ovine fascioliasis, Liver fluke diseases, Liver rot

Fasciola hepatica is the most important trematode of domestic ruminants and exists in all countries where environments suitable for the intermediate snail host prevail. According to Dunn[1] 'Fascioliasis is one of the last difficult problems in helminthology besetting the veterinarian'. Economically the estimates of loss to the agricultural industry attributable to deaths, ill-health, condemnation of livers, the costs of treatment and prophylaxis are substantial.

Cause

Fasciola hepatica is a member of the class Trematoda, sub-order Digenea, family Fasciolidae, genus *Fascioloides*. It is of prime importance in ruminants, but can infect most other mammals including man. Its site as an adult within the host is the liver. Morphologically *F. hepatica* is a leaf-shaped worm 2–5 cm in length. The eggs are operculate, oval 150 μm long by 90 μm wide, and yellowish-brown in colour.

Life cycle of Fasciola hepatica. Eggs from the adult fluke in the bile ducts of the host pass into the intestine to be voided in the faeces. One fluke can pass between 5000–20 000 eggs per day. At temperatures above 10°C a miracidium develops in the egg. The time taken for development is temperature dependent, being about 6 weeks at 15°C and 10 days at 22°C. Light is essential for hatching and eggs must be free from faeces, and also must have a complete film of moisture surrounding the egg.

When conditions are suitable, the miracidia usually hatch as a mass over a few hours, and propelled by cilia, they actively seek a snail host of the genus *Lymnaea*. Their maximum infective period is three hours. In the digestive tract of the snail they become young sporocysts, then follow one of two courses. The first is for the sporocysts to develop into rediae which upon entry to the liver of the snail produce daughter rediae; the second is to produce second generation sporocysts which cannot produce rediae. Therefore from rediae, daughter rediae, or second

generation sporocysts, the final stage in the snail, the cercaria, is produced. Six hundred or more cercariae can emerge from one miracidium. At optimal temperatures the cercariae are shed from about five weeks onwards after the entry of the miracidium into the snail. Snails will retain cercariae in the absence of sufficient moisture, but once conditions are favourable mass emergence of cercariae occurs and they swim to attach themselves to herbage where they lose their tails and secrete a tough cyst wall to become metacercariae. Under optimal conditions they can survive for a year, but many can be lost by desiccation, or during a prolonged period of freezing. Once ingested by the host the young flukes excyst in the intestine, eat through the mucosa, and traverse the peritoneal cavity to reach the liver capsule which they penetrate. They are voracious feeders and migrate through the liver parenchyma to reach a bile duct, where they mature to become adult flukes. Egg laying takes place some 10–12 weeks after the initial infection.

Life cycle of the intermediate snail host. The most important species of *Lymnea* involved in the transmission of fascioliasis vary in their distribution in the world, and of these *Lymnaea truncatula*, occurring in the UK, is the most widespread. *Lymnaea truncatula*, brown in colour, requires moisture, preferably slow moving water and mud with a slightly acid pH. If dry conditions prevail they can descend into mud, undergo aestivation (shutting down their metabolic activity) to re-emerge when more suitable conditions of moisture occur. This is important in the epidemiology of the fluke. The optimum temperature range for reproduction of the snail is between 15 and 26°C when rapid production of snail egg masses occurs. At temperatures below 10°C the snails can survive over the winter months to produce a new generation of snails in the following spring. One snail can produce several thousand descendants in favourable conditions.

Clinical signs

Fascioliasis in sheep is recognized as occurring in three distinct forms which are related to the weight of infection, the duration of the period of infection and thus to the time of year and the availability of metacercariae.

Acute fascioliasis Metacercariae shed by the summer infection of snails peak in large numbers in the herbage in September and October. At this time wastage by detachment or desiccation has not occurred and sheep can ingest large numbers over a short period of time resulting in a massive invasion of the liver parenchyma by more than 1000 immature flukes almost simultaneously. The ensuing pathological damage to the liver results in sudden deaths from September to December. Occasionally if stock are not placed on dangerous pastures until the winter months the sudden deaths can occur from December to February. Sheep may be in good body condition and the first sign of trouble may be several sudden deaths.

Examination of the remainder of the flock shows lethargic animals, with pale mucosae, exhibiting dyspnoea when made to move. Palpation of the abdomen causes a 'guarding' action, and evidence of liver enlargement and possibly ascites is present. Diagnosis is confirmed by post-mortem, parasitological, and blood examinations. Rough handling can cause liver rupture and death.

Subacute fascioliasis The pathogenesis of this condition presents clinical signs, compatible with acquisition of infection over a longer period of time, which are related to parenchymal damage and the presence of adult flukes in the bile ducts. Deaths occur later in the year than in acute fascioliasis, usually between December and February. A distinguishing feature in infected animals is loss of condition, with the flock as a whole showing ill-thrift, a degree of lethargy, poor fleeces and low body condition scores. Examination of individuals shows pale mucosae, the presence of palpable liver enlargement without the 'guarding' of the abdomen seen in the acute form. The submandibular oedema of the chronic form is absent, and ascites is not usually present. This form may be seen in the wake of an outbreak of the acute form, when sheep have only been treated once and have remained on the pasture.

Chronic fascioliasis This is the commonest form and can be acquired from either the summer infection of snails when the peak of metacercarial availability has passed or from the winter infection of snails producing fewer metacercariae in the spring. It therefore occurs either in February and March or June and July and results in a progressive loss of condition due in the main to the presence of adult flukes in the bile ducts. Chronic fascioliasis is exacerbated by poor nutrition or where the requirements of the ewe are at a peak, as in late pregnancy or during lactation and shows as a progressive loss of condition developing to emaciation with typical sub-mandibular oedema. Anaemia is frequently severe and mucosae are extremely pale. Although deaths occur, many sheep show no more than ill-thrift when fluke burdens are low or when adequate nutrition is available. Sheep dying from other causes may reveal the presence of chronic fascioliasis. Immune reactions as seen in cattle are not featured in sheep and many flukes have a life-span equivalent to their host.

Pathology

The overall pathogenicity of fascioliasis centres around the damage to the liver and the degree is dictated by the parenchymal damage due to migrating flukes and the effects of adult flukes in the bile ducts. During the migration of young flukes through the intestinal wall and within the peritoneal cavity there is little evidence of pathological damage. Once penetration of the liver capsule occurs evidence of petechiation and blood oozing from pinpoint punctures arise. In the liver parenchyma hepatic cells are destroyed, haemorrhage from hepatic vessels and tissue reaction in the track of the burrowing fluke occurs. Hepatic cells show necrosis and infiltration by leucocytes into the damaged areas is visible as dark red or yellowish lines on or in the hepatic substance. The severe liver damage and haemorrhage resulting from large numbers of migrating flukes produce a severe anaemia resulting in sudden death. In the subacute and chronic forms the lesser numbers of migrating immature flukes do not produce the same degree of liver damage, and clinical signs may be absent because adequate undamaged liver parenchyma is present. In the subacute and chronic forms after eight weeks the immature flukes enter the bile ducts and develop to the adult stage. Once the bile ducts have been entered the pathological processes associated with the active ingestion of blood by the flukes produce a macrocytic hypochromic anaemia the severity of which is dependent on the numbers of flukes present. A degree of fibrosis in the bile duct walls occurs in sheep but the reaction is very much less than the fibrosis and calcification featured in cattle infections. Distension of the bile ducts and a roughening of the liver surface are readily visible.

In the subacute form ascites is not always present. Fluke burdens are of the order of 1000 ± 300. In the chronic form lack of body fat is linked with the impairment of metabolism. Fluke burdens are from 50–200 or more adult flukes. Old migratory tracts may be seen as yellow or red foci on the liver surface. Oedema, ascites, and hepatic lymphadenitis are present. There are several detailed reports in the literature concerning the immunological mechanisms and the histological changes associated with hepatic fibrosis in fascioliasis.[2, 3] The influence of the plane of nutrition upon the severity of anaemia produced and on the albumin metabolism was recently reported.[4, 5] Black disease may be a complication of fascioliasis. For a description of this see Chapter 7.

Diagnosis

Acute fascioliasis Sudden deaths in sheep grazing wet marshy areas usually between late September to December, occasionally later, in an area of endemic fluke disease, possibly with prior knowledge of a high incidence forecast, should alert the practitioner to suspect acute fascioliasis as the problem.

Post-mortem examination reveals the characteristic lesions, and squeezing a section of the liver into water followed by sieving reveals the small immature flukes. The clinical signs in other members of the flock are fairly pathognomonic. Animals may be in good condition. Faecal examination for fluke eggs is negative and the use of blood parameters and enzyme tests such as serum glutamic oxaloacetic transminase (SGOT) and plasma dehydrogenase (GDH) may help to confirm the diagnosis.

Differentiation from other causes of sudden death include clostridial disease, acute pasteurellosis, acidosis, whilst the coexistence of diseases such as parasitic gastroenteritis and cobalt deficiency should be borne in mind.

Subacute fascioliasis Rapid loss of condition is seen with deaths occurring less rapidly than in the acute form. It can follow an outbreak of the acute form earlier in the season. Again the history, environment, and time of year are indicative, with a forecast of higher risk of fascioliasis than normal. Post-mortem examination showing some evidence of migratory tracts, with large numbers of adult flukes in the bile ducts is often all that is required for a diagnosis of subacute fascioliasis. In live sheep a tentative diagnosis on clinical signs of anaemia, liver enlargement and poor body condition can be confirmed by positive faecal fluke egg counts and by the examination of blood parameters which show hypochromic macrocytic anaemia, low packed cell volume and hypoalbuminaemia. Enzymes such as SGOT and gamma-glutamyl transferase (GGT) will be elevated and can be helpful. A heavy worm burden, poor nutrition and cobalt deficiency are possible differential diagnoses which can also coexist with fascioliasis.

Chronic fascioliasis. Diagnosis is made on a post-mortem examination, clinical signs of severe progressive loss of body condition, the presence of anaemia and submandibular oedema. Fluke eggs are present in the faeces of a representative number of animals in the flock. History and area are useful indicators, allied to the presence of unthrifty sheep from February to March resulting from the summer infection of snails, or from June onwards from the winter infection of snails. Post-mortem signs are diagnostic with low to moderate numbers of adult flukes, and signs of liver fibrosis and ascites. Confirmation can be aided by blood examination and enzyme estimations for SGOT and possibly GGT. Differential diagnosis should include roundworm infection, malnutrition and cobalt deficiency; with less likely candidates being scrapie, Johne's disease and jaagsiekte, although these are much more likely to occur as individual cases.

Epidemiology

The studies carried out by many research workers, have played a major role in the understanding of the epidemiology of fascioliasis.[6, 7, 8] Ollerenshaw[6] in a paper on the ecology of the liver fluke brought forward some extremely important facets which are highly relevant in considering the epidemiology of the disease. He pointed out that the usual diagram of the life cycle, showing the successive stages, was an idealistic one and did not sufficiently stress the potential hazards to the fluke at each of six phases of the cycle listed by him, whereby each stage had not only to develop in a particular environment but having done so, had to pass from that environment to the next where the succeeding stage had to develop. Thus varying degrees of mortality at each development and transport stage meant that the biotic potential of the parasite was counterbalanced by this mortality.

The critical determinants for the completion and the time taken for the completion of the successive stages, from the passage of a fluke egg to the presence of an egg-laying adult fluke, are moisture and temperature. Adequate moisture is essential to break up the infected faeces on the pasture, to surround the egg in a film and to prevent desiccation of the developing egg. Moisture is essential to allow the miracidium to swim and enter the snail and after development there, to encourage the emergence of cercariae and allow these to swim and encyst on herbage as metacercariae. If metacercariae are to be protected against desiccation, moisture must still be adequate. The development and survival of snail populations also depends on moisture.

The importance of temperature is paramount. Below 10°C fluke eggs do not develop and hatch, and reproduction and development of the snail is halted. In practice this means that in the UK the period between April to November is the only time when adequate moisture and appropriate temperatures synchronize to allow completion of the liver fluke cycle. The 'summer' infection of snails in late spring and early summer results when miracidia from fluke eggs, deposited in the previous autumn and winter or in the spring of the current year, arise to enter the early snail population. Cercariae from this infection encyst as metacercariae in late summer and autumn. The 'winter' infection of snails is produced from miracidia hatching from fluke eggs in late summer or early autumn. Fluke eggs deposited later do not hatch till the following spring because of the constraints of lower temperatures. The 'winter' infection of snails does not produce cercariae until the following spring. Thus the parasite achieves a carry over from one year to the next.

Treatment, control and prevention

Two significant factors in the last two decades have been the development of accurate forecasting systems and the advent of chemotherapeutic agents with high and safe activity against both immature and adult flukes.

Forecasting systems The work of Ollerenshaw and others in the ecology of the liver fluke and intermediate snail host led to an extension of the epidemiological knowledge linking climatic conditions and the liver fluke cycle into forecasting systems. The system of Ollerenshaw used in England and Wales, and that of Ross in Northern Ireland and Scotland both relate meteorological data to the probable incidence of fascioliasis in particular years, allowing appropriate measures to be implemented.

For detailed information regarding forecasting systems the reader should refer to Ollerenshaw[9] and Ross.[10]

Flukicidal drugs The new range of these is characterized by high efficiency, good safety margins and the ability of some to kill immature flukes, as well as adults. In general the more efficient the drug in killing very young immature flukes the less its efficiency in killing adults, but within the range of products the choice most appropriate to the form of fascioliasis encountered can be made with confidence.

Treatment

Where outbreaks occur the clinical signs and diagnosis of the form of fascioliasis present indicates the choice of drug.

Acute fascioliasis. The superiority of the drug diamphenethide (Coriban; Wellcome Ltd.) in killing extremely young flukes makes it the drug of choice and a single dose combined with a move to clean pasture is the best regime. If a move is not possible continuous dosing at three-week intervals is advisable. If other drugs such as rafoxanide (Flukanide; Merck Sharp & Dohme.) or nitroxynil (Trodax; May & Baker Ltd.) are used, two doses and a move to clean pasture are recommended because their activity against immature flukes below 4–6 weeks of age is not high. Where a move is not possible, repeated doses at intervals of 3 weeks is advocated.

In treating acute fascioliasis the farmer should be warned that further deaths can occur despite therapy and handling of clinical cases should be gentle.

Subacute or chronic fascioliasis In these forms the greater activity of rafoxanide and nitroxynil against adults compared with diamphenethide favours their use.

Where loss of condition has occurred it is valuable to improve the nutrition of the sheep.

Control and prevention

In endemic fluke areas routine use of flukicidal drugs is employed, the frequency of dosing being based on the current forecasts. However, certain regions and farms, in high rainfall areas with established snail habitats, require above average dosing schedules, relaxation of which could prove disastrous. A suitable regime in normal or low incidence years is to dose all stock in May, October and January. In years of above average forecasts or in areas with a high rainfall additional doses in June, November and February are indicated. While these methods give good control, the progression of some flukes to maturity ensure that eggs continue to be deposited on pastures for variable periods and the necessity for annual routine dosing is not removed.

An alternative method, termed 'strategic dosing' based on work by Armour, Corba and Bruce[11], has been investigated by the Hill Farming Research Organisation on their farm in the west of Scotland since 1973.[12] Based on the rationale that by killing flukes in the immature stage, egg deposition on the pasture would be minimized or eliminated and that the presence of snails would be irrelevant, near eradication of fascioliasis at that farm has been achieved.

Apart from the use of routine or strategic flukicidal drugs, prevention can be achieved by other managemental methods and some of these, if applicable, should be the methods of choice in arriving at freedom from the disease. These managemental methods may themselves be combined or used in combination with therapy.

Flocks   The aim here is to ensure that sheep do not graze an area where metacercariae are available. This can be achieved by fencing off well-defined snail habitats so as to exclude stock. This is only practicable when these are small and few in number otherwise it is not economic to do so. An alternative is to exclude stock from fluke areas from the end of August till February. This is not always possible and still has an element of risk.

Pastures   The most desirable and efficient method is drainage. Where this is possible long-term prevention is assured by eliminating snail habitats. The cost, however, can be high and unfortunately many snail habitats do exist in undrainable high rainfall areas. Drainage by open ditches is contra-indicated, as unless well maintained, these can fill in and create pools of slow moving water and mud ideal for the proliferation of snails.

The use of molluscicides on snail habitats received much attention at one time but would not appear to have been adopted to any great extent in the UK. Designed to remove the snail host, potential disadvantages are prohibitive costs, if large areas require treating, the remarkable ability of snails to repopu-

late an area, the potential risks to stock, operators and fish. However, molluscicides are worthy of consideration in certain circumstances and if proposed, specialist advice should be sought.

A suggested check list summary should include:
1. Be aware of endemic fascioliasis in your area.
2. Be suspicious of sudden deaths or loss of condition in sheep on wet reedy pastures. Determine the form of fascioliasis by clinical and post-mortem examination. Check parasitological and blood parameters from representative numbers of sheep. Obtain feedback from abattoirs.
3. Look for snail habitats, rushes, wet flushes and the presence of snails.
4. Type of flock (open or closed) and previous history of fascioliasis. Are bought in cattle and sheep dosed?
5. Fluke forecast for current year.
6. Treatment. Choose drug appropriate to type of fascioliasis. Move to clean pasture.
7. Control and prevention. Drainage; fencing; molluscicides; routine or strategic dosing schedules.
8. Health programmes. Dosing schedules; parasitological examinations; abattoir checks; map snail habitats.

Other trematodes of sheep

Whilst *Fasciola hepatica* is the most important trematode, in some areas of the world two other species can assume importance and it is pertinent to include a brief account of these in this chapter.

Fasciola gigantica   This is a member of the sub-class Digenea, family Fasciolidae, genus *Fascioloides*. This fluke has a similar morphology to *F. hepatica* but is larger; 5–7 cm long by approximately 1.2 cm wide. Characteristically it has less well-defined shoulders and long straight sides in comparison with *F. hepatica*. Its distribution includes Africa, Asia, Southern USA, Spain, Southern Russia and the Middle East. The egg is larger than that of *F. hepatica*. *L. auricularia* is the most important intermediate host. The biology of this snail and the others which can act as intermediate with hosts of *F. gigantica* is similar to that of snails associated with *F. hepatica*.

F. gigantica has a similar life cycle to *F. hepatica*, although the different phases are longer, and it has a closer association with water. The prepatent period is from 13–16 weeks.

The pathogenesis of this fluke is similar to that of *F. hepatica*. In sheep both the acute and chronic forms occur. Control measures based on those applicable to *F. hepatica* infection are recommended.

Dicrocoelium dendriticum   This trematode is a member of the sub-class Digenea, family Dicrocoeliidae, genus *Dicrocoelium*. It has a world-wide distribution and in the UK it is important only in the islands off the west of Scotland. It is a small lanceolate fluke approximately 1.2 cm long by 0.25 cm wide. The egg is small, (45 by 30 μm) dark brown, with an operculum and characteristically the egg has one flat side. When passed in the faeces it already contains miracidia.

This fluke does not require water. The egg is ingested by various species of land snail, then the miracidium emerges to reach the hepatopancreas of the snail. Two generations of sporocysts are produced from which cercariae, in masses held in a form known as slime-balls, are expelled by the snail. This phase takes approximately three months. Thereafter, further development is achieved when the second intermediate host, an ant of the genus *Formica* ingests the slime-ball. Whole ant colonies can become infected in endemic areas. In the ant the cercariae migrate to the abdominal cavity to become metacercariae. Some cercariae migrate to the CNS of the ant and this aberrant form makes the ant climb up herbage, increasing the availability to grazing animals. Ingestion of the ant releases the metacercariae which then migrate to the liver via the bile duct where, spreading through the biliary system, they pass throughout the liver.

Very large numbers of these flukes can be acquired by individual sheep. Diagnosis requires finding the typical eggs. Control cannot be achieved in a similar manner to *F. hepatica* because snail sites cannot be adequately defined, and anthelmintics used for *F. hepatica* are not usually effective unless at dosage rates considerably larger than those employed with *F. hepatica*. It has been reported from Greece that albendazole (Valbazen; Smith Kline & French), at dosage rates of 20 mg and 15 mg/kg, was safe and efficacious against *D. dendriticum* in sheep, whilst in Norway two doses of 10–12 mg/kg given a week apart reduced faecal egg counts by about 90 per cent.

There are also reports from Russia showing that chickens folded on to infected pastures at a stocking density of 500 per hectare provide a degree of control.

REFERENCES

1 Dunn A.M. (1978) *Veterinary Helminthology,* 2e, p.196. Heinemann Medical Books Ltd., London.
2 Dargie J.D., Armour J. & Murray M. (1974) Immunological mechanisms in fascioliasis. Proceedings of the Third International Congress of Parasitology. p. 495. Munich.
3 Rushton B., Murray M., Armour J. & Dargie J.D. (1974) The pathology of primary and reinfection lesions of fascioliasis in the ovine liver. *Ibid.* pp.498–9.
4 Berry C.I. & Dargie J.D. (1978) Pathophysiology of ovine fascioliasis: the influence of dietary protein and iron on the erythrokinetics of sheep experimentally infected with *Fasciola hepatica. Veterinary Parasitology,* **4**, 327–39.
5 Dargie J.D. & Berry C.I. (1979) The hypoalbuminaemia of ovine fascioliasis: the influence of protein intake on the albumin metabolism of infected and of pair-fed control sheep. *International Journal for Parasitology,* **9**, 17–25.
6 Ollerenshaw C.B. (1959) The ecology of the liver fluke *(Fasciola hepatica). Veterinary Record,* **71**, 957–65.
7 Armour J., Urquhart G.M., Jennings F.W. & Reid J.F.S. (1970) Studies on ovine fascioliasis. II. The relationship between the availability of metacercariae of *Fasciola hepatica* on pastures and the development of the clinical disease. *Veterinary Record,* **86**, 274–7.
8 Ross J.G. (1967) An epidemiological study of fascioliasis in sheep. *Veterinary Record,* **80**, 214–17.
9 Ollerenshaw C.B. & Rowlands W.T. (1959) A method of forecasting the incidence of fascioliasis in Anglesey. *Veterinary Record,* **71**, 591–8.
10 Ross J.G. (1970) The Stormont 'wet day' forecasting system for fascioliasis. *British Veterinary Journal,* **126**, 401–8.
11 Armour J., Corba J. & Bruce R.G. (1973) The prophylaxis of ovine fascioliasis by the strategic use of rafoxanide. *Veterinary Record,* **92**, 83–89.
12 Whitelaw A. & Fawcett A.R. (1981) Further studies in the control of ovine fascioliasis by strategic dosing. *Veterinary Record,* **109**, 118–19.

SECTION 3

NERVOUS SYSTEM AND LOCOMOTION

13

SCRAPIE

B. MITCHELL AND J.T. STAMP

Synonyms La tremblante (Fr: trembling); traber-krankheit (Ger: trotting disease); Rida (in Iceland).

Scrapie is a non-febrile, progressive, chronic, degenerative disorder of the central nervous system of sheep and goats which is inevitably fatal and occurs as a natural infection throughout the world. It has been present in Europe for 250 years, perhaps longer, and has been recorded in India, Africa, Asia and in both South and North America. It has on occasion been imported into Australia, New Zealand and South Africa but following eradication by slaughter these countries presently appear to be free of the disease.

Sheep scrapie is the best known member of a larger group of diseases that includes transmissible mink encephalopathy (TME) and two diseases of man, Kuru and Creutzfeldt-Jakob disease, all of which have a similar brain pathology.

Cause

The aetiology of scrapie is, as yet, not defined but much information has been established about its transmissible scrapie agent which has a number of biologically defined strains. It is equally widely accepted that there is a genetically controlled interaction between the host species and the infective agent determining the susceptibility of individual animals to the clinical manifestation of disease.

Experimental disease has been established following inoculation by a number of routes of homogenates of tissues and of tissue fluids from affected animals to sheep, goats, mice, rats, golden hamsters, voles, gerbels and New World monkeys. The agent can withstand heat, formalin, ultraviolet light and ionizing irradiation to an extent that one would not expect of a conventional animal virus. The exact molecular form and size are not known but infectivity has been demonstrated in 50 nm pore size membrane filtrates. The molecular weight has been assessed at between 150 000 and 1 500 000 daltons.

Clinical signs

The lesions of the disease are essentially vacuolation in the CNS and the clinical signs of scrapie precipi-tated by these lesions can be considered under three headings:

Itch, rubbing and nibbling of the skin Intense pruritus leads to sheep rubbing against fixed objects, gates, fences, etc, often resulting in extensive wool loss and even damage to the skin particularly round the base of the tail (Fig. 13.1) and along the chest

Fig. 13.1 Sheep affected with scrapie showing loss of wool over the hindquarter as a result of rubbing because of pruritis.

wall. Hair-covered skin is also rubbed leading to denuded patches on the poll and face and biting of the lower legs. During the rubbing movements the sheep display exaggerated signs of ecstasy and even on occasion become demented making nibbling movements of the lips and chewing adjacent objects. When affected sheep are being assessed clinically this 'nibbling reflex' can be elicited by deep massage of the muscles along the back.

Hyperexcitability, trembling and stupor Change of temperament can be very variable, some sheep responding by hyperexcitability and trembling on being approached and caught. Some collapse and remain unconscious for up to a minute as ultimate manifestation of this excitation. A contrasting manifestation in some sheep is stupor and lassitude.

Incoordination of gait   This is often the first indication of scrapie abnormality and is observed by the diligent shepherd, who knows his animals well, when the flock comes forward to the feeding trough, etc. At first there is a high-stepping of the front legs which are somewhat incoordinated with the movements of the back legs. As the disease progresses the back legs become affected and what initially is seen as weakness when the animal turns, progresses to incoordination, missed steps and a trotting action over the next 2–6 weeks. In more advanced cases animals may be unable to stand in the morning but after being helped up will walk in a relatively coordinated way during the rest of the day.

Emaciation is a common feature of the terminal case and progression to death usually takes about 6 weeks. However, this is not a predictable entity; some animals becoming very fat, others dying relatively suddenly after a few days of observed signs. No authenticated cases have been known to recover although some may exhibit signs for many months.

Diagnosis

The diagnosis of scrapie at present is determined by (i) the clinical signs in the live animal and by (ii) the demonstration of vacuoles in the brain post-mortem.

There is in scrapie disease a complete absence of any demonstrable specific immune process and the agent has neither been shown to be antigenic nor to evoke any inflammatory response. (Paradoxically, mice treated by the anti-inflammatory steroid prednisone acetate at the time of challenge by scrapie agent had very prolonged incubation periods and about 20 per cent survived a challenge which was 100 per cent lethal to control mice.)[1] Standard tissue culture techniques in the laboratory have failed to demonstrate the agent. In consequence there are no laboratory tests available other than neuropathological examination.

The clinical diagnosis of scrapie is not easy. Animals show numerous combinations of the described clinical signs and breed (or strain of agent) differences are considerable. It is noticeable that in the experimental disease the principle manifestation differs depending on the strain of agent with which the animal has been inoculated. This is particularly noticeable in goats where the response to one scrapie agent is exclusively pruritus whilst to another it is ataxia. In the natural field disease the signs in a given flock tend to be constant and are readily identifiable by the shepherd associated with that flock.

It is important to exclude confusion with other diseases. Rubbing may be caused by ectoparasites and hind-limb incoordination is associated with other diseases of the CNS, for example, maedi-visna and polioencephalomalacia. Because of these difficulties it is essential to confirm a clinical diagnosis histologically.

Pathology

The only consistent lesions of scrapie occur in the CNS. Microscopical lesions invariably present in clinical cases of scrapie are progressive vacuolation and degeneration of neurones mainly in the medulla, pons and mid-brain. Interstitial spongy degeneration is often found in the same areas as neuronal vacuolation and occasionally there may be neuronal loss. The character and distribution of the lesions have been clearly described in a number of publications.[2, 3, 4] There are no inflammatory changes in scrapie.

Epidemiology and transmission

Scrapie has been studied in depth for more than 50 years but precise knowledge of its epidemiology and transmission is still incomplete. Information has been derived from two sources: field observation and experimental studies.

In the natural disease death occurs most frequently in sheep between 2 and 5 years old, the most common age being 36–40 months. It has been estimated that if culling for commercial reasons was not practised about 20 per cent of scrapie cases would occur in animals over 5 years of age. Natural disease is atypical if it occurs in sheep below 1 year of age.

There is considerable evidence that the majority of sheep developing scrapie have been infected since early life and the scrapie agent has replicated within the sheep during the whole of this period without causing any apparent signs. It is well established that maternal transmission (vertical spread) of scrapie agents accounts for much of the familial pattern seen in outbreaks of natural scrapie. The progeny born to young ewes which in later life die from scrapie, are frequently affected.

There is no information, as yet, whether maternal transmission depends upon infection of the egg, the embryo or the lamb at the time of birth but fetal fluids have been used successfully to transmit infection as has placenta. It is widely believed that lambs acquire prenatal infection. Milk does not seem to be involved. In terms of natural transmission to in-contact animals (lateral spread) dissemination of infection at parturition is thought to be the most serious risk for contamination of the environment.

There is good evidence that pasture contamination is an important source of infection but, other than with placenta, little work has been done on the excretion of the agent from infected animals. Oral infection is a proven route of transmission. Little is known about the persistance of infectivity on the pasture but the agent is very stable when subjected to physical insult. It is thought that lambs are more readily infected than adults. Goats are readily infected by mixing with scrapied sheep.

Experimental transmission studies have yielded information on a number of important issues. Scrapie can be induced in sheep and in other species by a transmissible agent. There are many strains of scrapie agent (possibly in excess of 20). Their properties have been defined biologically by serial passage through inbred strains of mice of defined genotypes and in sheep and goats because some strains have not, as yet, produced disease in mice. The strain type is characterized by (1) the incubation interval between inoculation and the onset of the signs of scrapie and by (2) brain 'lesion profiles' in which the numbers and distribution of vacuoles in brain sections taken from predetermined slices of the mouse brain are scored.[5] The pattern of stability on passage has resulted in the strains of agent each being grouped into three main classes. Class I (which includes the agent, ME7, most frequently recovered from field cases) is most stable, agents retaining their identity irrespective of the genotype or even the species of host through which they are passaged. Class II agents (which include the components of the inoculum most frequently used in experimental sheep studies, SSBP/1: sheep scrapie brain pool No 1) are completely stable when passed in the mouse genotype originally used for their isolation but their properties change if they are subsequently passaged in another genotype. Class III agents are unstable in that they display a single discontinuous change at some stage in passage within a single mouse genotype. Following the change the acquired properties are similar in some respects to those manifested by agents in Class I.

The genetic aspects of scrapie in mice are mainly in relation to the control of the incubation period. A gene was designated 'sinc' (scrapie incubation) with two alleles which, in the original studies with scrapie agent ME7, were termed s7 and p7 because of their influence in shortening or prolonging the incubation period. With some strains of agent the failure to induce scrapie in defined genotypes of mice is attributed to the prolonged incubation period being greater than the life span of the mice. This is a very important concept because it raises the spectre of certain genotypes, after infection, continuing to harbour the agent but unable to manifest the signs of scrapie before their natural death. These animals which could include sheep, might carry infection and possibly be a source of contamination to others rather than being 'resistant' to infection as is often so with conventional diseases.

Sheep transmission experiments have been of two main types—pathogenesis studies to elucidate sources of infective material and genetic studies. Breeding programmes in Britain using Cheviots and Herdwicks have been directed over about 20 years at selecting for increased or decreased susceptibility to scrapie induced by the subcutaneous injection of a defined test inoculum (SSBP/1). Animals were as-cribed a positive (+ve) or negative (−ve) status in response to the manifestation of clinical disease, confirmed histologically, within 24 months post-inoculation. These projects have indicated that the susceptibility (within 24 months) of sheep to this defined challenge (SSBP/1) is controlled genetically by a gene designated 'Sip' (scrapie incubation per-iod). Furthermore, by crossing studies between the +ve line and −ve line animals it has been demon-strated that 'susceptibility' is inherited as a dominant trait so that animals of homozygous (Sip/Sip) or heterozygous (Sip/sip) status are equally susceptible in response to this defined experimental challenge. Homozygous recessive animals (sip/sip) do not develop scrapie within the 24 month period but some, if injected intracerebrally, develop the disease after very long incubation periods. As with the mouse models it may be more correct with this SSBP/1 inoculum to look on the genetic control as being a prolongation of incubation interval rather than absolute 'resistance'. In practical field terms extreme prolongation, beyond slaughter age, is a commer-cially acceptable aim but in terms of international traffic and certification of freedom from infective agent this concept poses many problems.

A second important finding emanating from these breeding projects was that −ve line animals, if challenged by an agent (CH1641) different from the SSBP/1 inoculum, developed scrapie as frequently as did the +ve line sheep. Thus sheep bred for 'resistance' to one agent can be susceptible to experimental challenge by another strain of scrapie agent. This has important implications for field attempts to breed sheep having resistance to scrapie. If sheep are transferred to an environment in which a different strain of agent is present they could be susceptible.

Studies on the natural disease were extensive in the early part of this century when transmission was believed to be hereditary. Over the last 30 years most work has been directed at elucidating the relative contributions to transmission of a finite scrapie infective agent and of a specific genetic component. In this research it has been very difficult to design environmental situations which make interpretation of results unequivocal. The long incubation period of the disease has exacerbated these difficulties because it is virtually impossible in agricultural management to avoid changes in pasture locations, of flock groupings and of personnel over the extended number of years during which studies must be pursued. There has been much acrimony about the interpretation of results.

One firmly held opinion, evolved over 20 years of meticulous recording of a large breeding group of preponderantly Suffolk sheep, is that scrapie is naturally determined by an autosomal recessive gene and does not depend on a naturally communicable scrapie agent.[6, 7] There is overwhelming alternative evidence that when sheep, themselves healthy and from a flock background of freedom from scrapie extending over decades, are introduced to an environment including contact with scrapie-affected sheep a variable proportion contract the natural disease.[8, 9, 10] It must be concluded that this is evidence for a naturally-transmissible scrapie agent. However, susceptibility is not absolute and the development of the disease may depend on many factors. Some of these are the weight of the contaminating infection, the age of the sheep at the time of exposure and, on the basis of data interpreted from experimental work, the genotype of the animal which may very well prolong the incubation period beyond the commercial life span of the animal. Manifestation of clinical disease following the natural transmission of field strains of scrapie agent may well be confined preponderantly to animals having a homozygous recessive genotype. This is in contradistinction to the dominant genetic control of incubation period demonstrated in experimental transmission studies with one strain of agent (SSBP/1) but with another agent (CH1641) the genetic control was different and as yet its dominance has not been ascertained. The hereditary control of scrapie and whether this is manifest as a dominant or a recessive trait may well depend on the strain of scrapie agent which is present in the environment or in the sheep population.

Field evidence suggests two situations. One in which sheep of all genotype permutations remain healthy in the absence of infective agent but when transferred to a contaminated environment display the interaction of genotype and susceptibility to infection by developing scrapie. The second is where scrapie is endemic to a lesser or greater degree in a sheep population. The familial manifestation of scrapie is predominantly, although not exclusively, associated with vertical transmission of the disease from affected ewes to their offspring. These affected ewes contribute half of the genotype to their offspring and abundant opportunity for infection.

Throughout the history of scrapie research there has been much debate regarding the role of affected rams in the spread of scrapie. There is no evidence that a ram transmits infection by coitus, in semen or through the sperm. There have been, however, many claims that the progeny groups of some rams have a much higher prevalence of scrapie than that in the progeny groups of contemporaries. The prevalence of scrapie in affected flocks usually is little higher than 1–2 per cent per annum. However, in badly affected flocks this figure may be as high as 10–15 per cent with up to 50 per cent of the 2–3 year old age group being affected. It is in these situations that specific rams are incriminated as transmitters of infection but the interpretation of the epidemiology can be rationalized equally well on the basis of the flock having a ubiquitously high distribution of scrapie agent and the ram effect a consequence of the genetically-determined susceptibility rather than direct transmission.

Control

There is no treatment for scrapie. Control measures have to rely on harnessing the currently available knowledge of the epidemiology of the disease and should be approached with clearly defined objectives—total eradication of the disease or its containment within economically viable limits. The former is practicable only in international situations where the environment is uncontaminated and imported livestock are subject to stringent quarantine regulations beyond the incubation period of scrapie. This may be 8–10 years and involves several generations of sheep. It is impossible to issue veterinary certification of freedom from scrapie in the absence of any diagnostic test. Farmer certification of flock *freedom* from the disease is not particularly meaningful. The farmer is unlikely to be conversant with the disease unless it has been a problem and his flock would require to be entirely closed with no introduction of stock rams or other commercial sheep.

Pragmatically control measures should aim at culling those animals most likely to develop the disease. Transmission studies have indicated that all progeny of affected females should be culled and, to minimize contamination of the environment, affected females killed before they lamb. In-contact spread from affected rams does not constitute too high a risk conditional on their being isolated as early as possible after the onset of clinical signs. If doubt arises about the precision of a diagnosis the animals should be segregated till the evidence is unequivocal.

Genetic control is at present imprecise. The genetic control of susceptibility in any flock to the prevailing strain of agent may well be precise but available evidence suggests that the genotypes 'resistant' to one strain of agent may be fully susceptible to a different strain. Against this background any recommendation for the complete culling of the progeny of nominated affected sires could be devastating to the commercial selection process within the flock to an extent much more damaging than a low annual loss from scrapie. Progeny groups generally do not derive infection from their sire, merely the genetic susceptibility. This they could do from perhaps three-quarters of the contemporary sires in a breed but in the absence of widespread environmental contamination the status of these contemporaries is not demonstrated.

Attempting to eliminate scrapie by identifying sires whose genotype should leave exclusively 'resistant' progeny is inevitably expensive and laborious. Progeny testing is necessary and the interval until the status of progeny can be classified is 3–5 years. A reservoir of ewes with a predictably high risk of taking scrapie is required to allow test matings of 'negative' rams. An environment continuously being contaminated in the course of this progeny testing may well be a greater hazard to eliminating the disease than the benefits likely to accrue from testing the rams. If, however, in the course of normal breeding programmes rams are identified as leaving progeny apparently 'resistant' to the scrapie agent prevalent on a farm then they should be exploited.

Flocks having a very high prevalence of scrapie are best sold for slaughter. Regrettably some are dispersed as breeding stock with tragic consequences for the flocks of the unsuspecting purchasers and the breed at large. The interval necessary before restocking can be safely achieved is not known. Evidence from Iceland indicates that scrapie can recur but it is not clear if this emanates from persistence of infection in the environment or by contamination from birds, rodents or other vectors.

In summary, the main policies in control should be the removal of affected ewes and all their progeny. The progeny of affected sires are not necessarily infected but it is prudent to apply more stringently than usual, the commercial or breed selection criteria when assessing flock replacements so that only the minimum numbers having a potentially scrapie-susceptible genotype are retained.

REFERENCES

1 Outram G.W., Dickinson A.G. & Fraser H. (1974) Reduced susceptibility to scrapie in mice after steroid administration. *Nature, Lond.,* **249**, 855–6.

2 Zlotnik I. (1958) The histopathology of the brain stem of sheep affected with natural scrapie. *Journal of Comparative Pathology,* **68**, 148–66.

3 McDaniel H.A. & Morehouse L.G. (1964) The diagnosis of scrapie. In: Proceedings of the 67th Annual Meeting of the United States Livestock Sanitary Association, Albuquerque, 1963. pp.550–64. Ann Arbor Michigan US Livestock. Sanitary Association.

4 Kimberlin R.H. (1981) Scrapie. *British Veterinary Journal,* **137**, 105–12.

5 Fraser H. & Dickinson A.G. (1968) The sequential development of the brain lesions of scrapie in three strains of mice. *Journal of Comparative Pathology,* **78**, 301–11.

6 Parry H.B. (1979) Elimination of natural scrapie in sheep by sire genotype selection. *Nature, Lond.,* **277**, 127–9.

7 Parry H.B. (1979) Aetiology of natural scrapie. *Nature, Lond.,* **280**, 12.

8 Brotherston J.G., Renwick C.C., Stamp J.T., Zlotnik I. & Pattison I.H. (1968) Spread of scrapie by contact to goats and sheep. *Journal of Comparative Pathology,* **78**, 9–17.

9 Dickinson A.G., Stamp J.T. & Renwick C.C. (1974) Maternal and lateral transmission of scrapie in sheep. *Journal of Comparative Pathology,* **84**, 19–25.

10 Hourrigan J.L. (1981) (Epidemiology of scrapie). Proceedings of the Canada Department of Agriculture, Scrapie Consultative Meeting, February 24 and 25 1981. pp. 21–32. Ottawa.

LOUPING-ILL

14

H.W. REID

Synonyms Trembling

Louping-ill is an acute virus disease of the central nervous system (CNS) affecting most species of domestic animals as well as man.

A disease of sheep associated with ataxia and incoordination has been recognized in the upland grazings of Scotland for many years. However, despite numerous investigations the identity of the causal agent remained elusive until 1930 when workers in Edinburgh isolated virus from the brains of affected sheep and demonstrated its transmission by the sheep tick *(Ixodes ricinus)* shortly thereafter. By 1934 a vaccine prepared from formalinized infected sheep brains had been developed and further research was minimal until the late 1960s.

Cause

Louping-ill virus belongs to the genus *Flavivirus* and is antigenically closely related to viruses of the tick-borne encephalitis complex which are found throughout the northern latitudes of the world where their principle disease association is with man. Like all *Flaviviruses* the virus is fragile and readily destroyed by heat, disinfectants, and acidic conditions. The capacity of the virus to agglutinate goose red blood cells is exploited in establishing the identity of virus isolates and as a haemagglutination-inhibition (HI) test for detecting antibody. In the laboratory, virus is propagated either in mice by intracerebral inoculation or in tissue culture cells including baby hamster kidney (BHK-21), pig kidney, chick embryo and sheep kidney cells.

Clinical signs

All ages of sheep appear to be equally susceptible although in endemic areas louping-ill is most frequently diagnosed in lambs and yearlings. Following infection between 5 and 60 per cent of animals develop clinical signs which vary from slight transient ataxia to sudden death. In affected animals incoordination progressing to paralysis, convulsions, coma and death within 24–48 hours is the general course. In a proportion of non-fatal cases residual torticollis or posterior paralysis may remain for weeks or months. Fever is not consistently present during the clinical phase of infection.

Following subcutaneous injection of virus initial replication occurs in the drainage lymph-node. Thereafter virus is disseminated by the circulation attaining high titres in blood and lymphatic tissues. During this viraemic phase there are few clinical signs apart from an elevated rectal temperature. After 3–5 days, virus can no longer be detected in the blood or lymphatic tissues but persists in the CNS and clinical signs may develop. Serum antibodies may be detected 5–6 days after infection which at first is largely of the IgM class and is progressively replaced by IgG antibody over the following 7–10 days.

At one time it was considered that clinical signs only developed in those animals in which virus gained access to the brain and great emphasis was placed on factors which might predispose to this. However, it is now recognized that following infection virus invariably replicates in the CNS. In subclinical infections only limited neuronal damage occurs before viral replication is interrupted by the immune response.[1, 2]

Pathology

Pathological changes directly attributable to louping-ill virus are restricted to the CNS, although terminally, secondary pneumonic lesions may develop. Swelling of the brain and hyperaemia of the meningeal blood vessels are the only gross post-mortem changes.

The principle histological lesion is widespread non-suppurative meningoencephalomyelitis. Neurone necrosis and neuronophagia are most prominent in the motor neurones, cord, medulla, pons, midbrain and the Purkinje cells of the cerebellum. The distribution of neurones containing cytoplasmic viral antigen also follows a similar pattern. Focal gliosis and lymphoid perivascular cuffs are most prominent in the hind brain, brain stem and meninges. These inflammatory cells appear to be predominantly B-lymphocytes specific for louping-ill virus antigen and it is concluded that the intrathecal synthesis of antibody to louping-ill virus by these cells is responsible for limiting viral replication in the CNS of the non-fatal cases.[1]

Diagnosis

A diagnosis of louping-ill should be considered in animals exhibiting signs of neurological dysfunction or that have died suddenly in areas where ticks are active. In areas where ticks are absent louping-ill may also affect sheep with a history of recent movement. When louping-ill is suspected laboratory confirmation should be sought as the clinical signs are similar to those of other diseases of the CNS.

As indicated in the above sections is it clear that by the time clinical signs are apparent virus can only be detected in the CNS. In dead or moribund animals the brain should be removed carefully and small pieces (approximately 1–2 cm^3) of brain stem placed in 50 per cent glycerol saline while the rest of the brain is placed in 10 per cent formal saline. Details of the method employed and the precautions that must be taken can be found in Chapter 53. Histological examination of the formal-fixed brain provides evidence of a non-suppurative encephalomyelitis but a definite diagnosis relies on isolation of virus. The pieces of brain in the glycerol saline are removed, washed and homogenized as a 10 per cent w/v suspension prior to intracerebral inoculation into mice. The presence of louping-ill virus can be confirmed in mice that die or become paralysed or comatose, by homogenizing the brains and inoculating on to pig kidney cell cultures alone and in the presence of immune serum to louping-ill virus.

Where it is not feasible to obtain brain material, serological confirmation may be sought. A variety of tests has been described for detecting antibody including complement fixation, gel precipitation, neutralization and HI. In practice however, the HI test has proved of greatest value. Serum which is collected from a clinically affected animal is tested for HI antibody as unheated serum and after heating to 64°C for 30 minutes. A four-fold or greater reduction in titre in the heated sample indicates that much of the antibody activity is due to IgM from which it may be inferred that the serum has been collected from a recently-infected animal. However, as IgG may largely replace IgM before clinical signs are manifest the absence of readily detectable IgM antibody in positive sera does not preclude the possibility of recent louping-ill virus infection.[2]

Epidemiology

Transmission of louping-ill is dependent on the sheep tick *Ixodes ricinus* and thus the epidemiology of the disease is intimately linked to the vector. The two fundamental requirements of the tick are a moist micro-climate for its survival when the tick is not feeding and the availability of large mammals such as sheep, cattle, hares, deer etc., on which adult ticks rely for their blood meals.[3, 4]

During the three year life cycle of the tick, feeding occurs for only approximately 17 days the remaining time being spent on the ground where it is essential that the relative humidity remains close to saturation. Such conditions are present in the vegetational mat of the upland grazings of the UK but are absent in the relatively well-drained, intensively-farmed lowland areas. The distribution of the tick is therefore largely restricted to the upland sheep grazings, sheep being the principle vertebrate host. Likewise louping-ill may occur in the majority of hill sheep flocks although the extent of challenge varies widely.

The geographical distribution of louping-ill is thus seen to be determined by that of the tick and similarly the annual periodicity of tick activity determines the incidence of infection.

Ticks are generally inactive during the winter months becoming active in the spring when the mean maximum weekly temperature exceeds 7°C. The period during which a tick quests before it becomes finally exhausted and desicated is governed by the temperature and relative humidity but seldom exceeds 4–8 weeks. Thus the time when ticks are active varies throughout the country depending on the prevailing local climatic conditions. In all locations exhausted ticks that have failed to feed will die, which explains the relative absence of ticks during the summer months. Spring-fed ticks will not have moulted and be ready to feed until the following spring. However, in certain regions there is a second period of tick activity in the autumn. After feeding these ticks do not moult and feed until the following autumn, remaining a distinct and separate population from ticks which feed in the spring.[4]

It is self-evident that the incidence of louping-ill virus infection follows closely the periodicity of the tick, losses occurring in two peaks: one in the spring followed in some areas by a second peak in the autumn.

Two patterns of mortality are seen. The disease may affect all ages of animals when either infection is present at a low incidence or where infection has only recently been introduced to a farm. All ages of bought-in sheep may also be susceptible. On farms where louping-ill is endemic losses occur primarily in lambs and replacement breeding stock, older animals being immune. Colostrum-derived antibody does, however, provide highly efficient protection in lambs of immune ewes which are therefore unlikely to become infected during their first spring. Thus on endemic farms only lambs acquiring insufficient colostral antibody are likely to die from louping-ill. Lambs protected by colostrum in their first year of life are fully susceptible to infection the following spring and it is therefore in the ewe lambs retained for breeding that the heaviest mortality is frequently seen.

The sheep tick is not a fastidious feeder and while adult ticks are restricted to larger mammals the larval and nymphal stages feed on any category of vertebrate. It is therefore, not surprising that infection with louping-ill virus has been reported in a wide variety of domestic and wild species of vertebrate. Most illuminating, however, have been the studies of infection of red grouse. Experimental infection of this species generally proved fatal and field studies indicated that in some areas where louping-ill is endemic a high mortality due to virus infection occurred in the grouse. Studies of infection in other vertebrates indicated that apart from sheep the intensity of viraemia that developed was insufficient to transmit virus to the tick. Thus contrary to the previously held view that louping-ill virus is primarily maintained in a wild vertebrate-tick cycle the converse appears to be true and present evidence indicates that the maintenance of louping-ill in nature is essentially a sheep-tick cycle. Infection of grouse only contributes occasionally as grouse die out in louping-ill endemic areas and infection of other vertebrates is of no biological importance.[5, 6]

Control

Control of louping-ill may be achieved either by immunizing sheep by vaccination against louping-ill virus infection or by preventing sheep from becoming infected by control of the tick.

Since the original isolation of the virus a number of prophylactic methods have been advocated including the use of both live virus and formalin inactivated vaccines derived from infected sheep brains, inacti-

vated vaccines prepared from infected mouse brain, or chick, sheep and BHK 21 cell cultures. The value of administrating immune serum has also been assessed in an extensive field trial. However, of these methods, vaccination using inactivated tissue culture propagated virus incorporated in oil adjuvant has proved to be of greatest value. Initially such vaccines were prepared from virus antigen concentrated by precipitation using methanol. Difficulty was experienced in producing batches of vaccine of consistantly high potency and it was found necessary to administer two injections to ensure protection. Subsequent developments have resulted in the production of vaccine, prepared from virus antigen concentrated by membrane filtration, which has better qualities of stability and immunogenicity and is effective following the administration of a single injection.

To control the disease on farms where it is endemic it is generally the practice to vaccinate all the ewe lambs, which are to be retained for breeding, either in the autumn or in the following spring before ticks become active. In addition, all purchased sheep should be vaccinated at least 28 days before exposure to tick-infested pasture. Finally, where the disease occurs for the first time it may be considered prudent to vaccinate the whole breeding flock.

Another control strategy is suggested from a consideration of the epidemiology of louping-ill. Provided the conclusion is valid that the maintenance of louping-ill depends on the availability of susceptible sheep, it follows that elimination of such animals will result in the eradication of louping-ill. The tick represents only a temporary reservoir of infection as transtadial but not transovarial transmission occurs and a tick population becomes clean of louping-ill in a period of two years in the absence of reinfection. Two methods of eradication are thus suggested, namely the physical removal of sheep from the environment or systematic mass vaccination over a period of two years. There is evidence from islands off the west coast of Scotland to suggest that both methods are effective. It is, however, generally not feasible to remove sheep entirely from a grazing for a period of two years and mass vaccination also is probably not justified unless the possibility of the reintroduction of infection by movement of tick infested wild or domestic animals can be ensured.

A final method of control is directed at reducing the prevalence or eradication of the tick. The attraction of such an approach is that not only would louping-ill be controlled but other tick-associated conditions would also be eliminated. In the absence of dips that have long residual properties, dipping alone is unlikely to achieve a marked effect unless the interval between dipping is short. However, a programme in which sheep are grazed on improved pasture, where ticks do not survive, during the periods of greatest tick activity and are then dipped and placed on the unimproved grazings following the peak of tick activity results in a progressive reduction in tick numbers and the disappearance of diseases associated with tick parasitism.

Such approaches to tick control succeed only if there are few alternative hosts available on which the adult stage of the tick can feed. In some areas wild species such as hares and deer host large numbers of ticks, thus unless they too can be removed from the pasture these methods are inappropriate. It is also essential that vigilence in tick control procedures is maintained for regeneration of a large tick population results in the reintroduction of all the tick-associated diseases into an entirely susceptible sheep population which may have devastating results.

It is clear, therefore, that care must be exercised in selecting the appropriate control strategy or combination for each individual situation.

REFERENCES

1 Doherty P.C. & Reid H.W. (1971) Louping-ill encephalomyelitis in the sheep. II. Distribution of virus and lesions in nervous tissue. *Journal of Comparative Pathology*, **81**, 531–6.
2 Reid H.W. & Doherty P.C. (1971) Louping-ill encephalomyelitis in the sheep. I. The relationship of viraemia and the antibody response to susceptibility. *Journal of Comparative Pathology*, **81**, 521–9.
3 Arthur D.R. (1973) Host and tick relationships: a review. *Journal of Wildlife Diseases*, **9**, 74–84.
4 Campbell J.A. (1952) Recent work on the ecology of the pasture-tick *Ixodes ricinus* in Britain. Report of the 14th International Veterinary Congress. (1949) 2, pp.113–19. London.
5 Reid H.W., Duncan J.S., Phillips J.D.P., Moss R. & Watson A. (1978) Studies on louping-ill virus (flavivirus group) in wild red grouse *(Lagopus lagopus scoticus)*. *Journal of Hygiene, Cambridge*, **81**, 321–9.
6 Reid H.W., Moss R., Pow I. & Buxton D. (1980) The response of three grouse species *(Tetrao urogallus, Lagopus mutus, Lagopus lagopus)* to louping-ill virus. *Journal of Comparative Pathology*, **90**, 257–63.

15

LISTERIOSIS

R.M. BARLOW

Synonym Circling disease

First described in New Zealand in 1931[1], listeriosis occurs world-wide and is prevalent in temperate climates. It is a bacterial infection with several unusual characteristics, the mechanisms of which are incompletely understood. Listeriosis may present in three forms of disease which only rarely overlap. These are, abortion in pregnant ewes, septicaemia in lambs, and meningoencephalitis in adult sheep. Only the last-mentioned is considered in detail here, the other two being considered in passing.

Cause

The cause, *Listeria monocytogenes* is a pleiomorphic organism which usually appears as a slightly curved Gram-positive rod. It is an aerobic or micro-aerophilic organism 1–2 μm in length and according to some authorities should be classified in the genus Erysipelothrix. *L. monocytogenes* however, can be readily distinguished from *E. insidiosa* by phage typing and by the increased zone of haemolysis it produces on incubation with *Corynebacterium equi* on sheep blood agar.

Man and most animal species can be symptomless carriers of infection excreting the organism in faeces and milk. *L. monocytogenes* survives in soil, faeces and fodder for many months under moist conditions. Primary isolation of the bacterium in culture may require a period of 'enrichment', i.e. maintenance of the source material, kept moist with buffered saline, at 4°C for up to a year. Storage of tissue in 50 per cent glycerol may also facilitate the isolation of *L. monocytogenes* from contaminated samples.

Clinical signs

In the abortifacient and septicaemic forms of the disease the premonitory signs include inappetance, lassitude and fever, but these may pass unnoticed, the first signs being respectively, abortion or the death of the host. Abortions most commonly occur during the last third of pregnancy and may be associated with metritis and retention of the placenta. Abortions occurring close to term are more likely to be accompanied by dystocia, maternal systemic disease and death, than are abortions at an earlier stage.

In listerial meningoencephalitis usually the first abnormality observed is that the head is turned and tilted to one side. The neck is held stiffly and the sheep tends to walk in circles towards the tilted side. Movements become incoordinated and affected animals may be found with the affected side leaning against a fence or wall. There is drooping of the ear and eyelids on the affected side and there is often conjunctivitis which may be bilateral. The mouth gapes and the tongue protrudes and there is difficulty in swallowing with drooling of saliva. There may be a mucopurulent nasal discharge and snoring respiration. The incoordination progresses to recumbancy with intermittent paddling movements which may be accompanied by opithotonus and nystagmus. Recovery is rare, the clinical course occupying up to 5 days.

Pathology and pathogenesis

In the abortifacient form the placenta shows pinpoint, yellowish, necrotic, foci involving the tips of the cotyledonary villi and histologically, a purulent exudate may also be evident. Miliary necrotic foci of similar appearance are present in the fetal viscera especially liver and spleen. The causal organism can be isolated from the lesions and also from the fetal stomach contents.

The septicaemic form, which occurs principally in lambs but also in some late aborting ewes, is also characterized by miliary foci of necrosis and microabscesses in liver and other viscera. Organisms are usually plentiful and readily isolated from fresh specimens.

The lesions in fetuses and lambs are consistent with there having been a bacteraemia. Presumably fetal infections arise from the placentitis, the lesions of which appear to be of greater age. This implies a preceding maternal bacteraemia but evidence for this is usually absent.

The encephalitic form of listeriosis is one of the very few examples of a true bacterial encephalitis. The lesions are of two types: parenchymal microabscesses and vaso-meningeal cellular infiltrations.[2] The former are localized to the brain stem and are usually unilaterally disposed. They may involve the region of the oculomotor nucleus in the midbrain, but are usually most intense in the regions of the medial and lateral vestibular and the trigeminal nuclei in the pons and medulla. They appear to arise from mobilization of microglia in small foci of parenchymal necrosis in which the bacteria can usually be demonstrated. The lesions become sparsely infiltrated with neutrophils and increase in size. Mononuclear cells are recruited to the periphery of the lesion and adjacent micro-abscesses may coalesce to form lesions which may be discernible macroscopically as areas of discoloration and softening a few millimetres in diameter.

The vaso-meningeal infiltrates are characteristically composed of mononuclear cells of moderately large size. There may be a few histiocytes and neutrophils admixed. Bacteria are rarely demonstrated in this type of lesion. The vascular meningeal infiltrates extend from regions of microabscess formation out along the Virchow-Robin spaces to the meninges, within which they spread considerable distances rostrally and caudally. The cellular components of the infiltrate become more lymphocytic as distance from micro-abscesses increases.

Though the question of a haematogenous route of infection must be considered, the restricted localization of bacteria-containing micro-abscesses suggests other possibilities. There is evidence that the cranial nerves are infected[3, 4] and peripheral neuritis with centripetal movement of infection may well be the source of the bacteria found in brain stem neurones.[5] It has not been proven that this is always the route of infection in naturally occurring listerial meningoencephalitis but since the disease almost invariably affects adult sheep with no evidence of systemic infection, it is reasonable to speculate that loss of integrity of buccal mucosa during shedding of deciduous teeth or in periodontitis would provide a satisfactory portal of entry adjacent to branches of the trigeminal nerve along which the infection could travel centripetally to produce localized brain stem encephalitis.

Diagnosis

In the abortifacient and septicaemic forms, diagnosis of listeriosis depends upon recognition of the miliary abscesses in visceral organs and should be supported by isolation of the causal organism.

Listerial meningoencephalitis may be suspected clinically on the basis of neurological signs but it must be differentiated from rabies, coenurosis and otitis media. Confirmation of diagnosis is readily obtained in most instances by histological examination of the brain and this may be quicker than bacteriological confirmation as 'enrichment' is often required before the organism can be demonstrated in culture.

Epidemiology and transmission

Individuals of many species can be symptomless carriers of *L. monocytogenes* and their faeces may contaminate soil and forage and provide opportunities for sporadic clinical infections throughout the year under suitable conditions of temperature and moisture.

Serious outbreaks of listeriosis mainly happen towards the end of winter and are usually associated with silage feeding. Grass ensiled from fields infested with moles *(Talpa europaea)* contains quantities of soil from the mole heaves which may be contaminated with *L. monocytogenes*. The organism can persist in silage and at pH > 5 multiplication may occur especially in the more aerobic conditions at the surface of the silo. Further aerobic opportunities for bacterial multiplication occur if the silage is spread before feeding and not immediately consumed. In one study[6] of an inwintered flock of 116 sheep which had not had a case of listeriosis in 3 years, 2 animals were excreting *L. monocytogenes* at the commence-

ment of housing. Silage was fed as part of the ration and at lambing time *L. monocytogenes* was isolated from faeces of 64 per cent and milk of 41 per cent of ewes. Paradoxically specific antibody (indirect haem-agglutination test) titres fell during the housed period, but no cases of listeriosis occurred. These observations suggest a complex aetiology and it may be that most mature sheep have a level of immunity related to the degree of environmental contamination but that some may develop disease under conditions of immunosuppression or if the infection reaches a site such as the pregnant uterus or a nerve trunk which is inaccessible to immune mechanisms.

Control, prevention and treatment

Insufficient is known of the pathogenesis and epidemiology for effective control measures to be advocated. Care should be taken in the preparation and feeding of silage if the chance of a major outbreak is to be minimized.

In sheep, treatment has little practical application. Tetracyclines may be used to control infection in valuable animals though they do not repair the neurological deficit resulting from listerial meningoencephalitis.

REFERENCES

1 Gill D.A. (1931) 'Circling' disease of sheep in New Zealand. *Veterinary Journal*, **87**, 60–74.
2 Cordy D.R. & Osebold J.W. (1959) The neuropathogenesis of listeria encephalomyelitis in sheep and mice. *Journal of Infectious Diseases*, **104**, 164–73.
3 Asahi O., Hosoda T. & Akiyama Y. (1957) Studies on the mechanism of infection of the brain with *Listeria monocytogenes. American Journal of Veterinary Research*, **18**, 147–57.
4 Charlton K.M. & Garcia M.M. (1977) Spontaneous listeric encephalitis and neuritis in sheep. Light microscopic studies. *Veterinary Pathology*, **14**, 297–313.
5 Charlton K.M. (1977) Spontaneous listeric encephalitis in sheep. *Veterinary Pathology*, **14**, 429–34.
6 Grønstøl H. (1979) Listeriosis in sheep. *Listeria monocytogenes* excretion and immunological state in healthy sheep. *Acta Veterinaria Scandinavica*, **20**, 168–79.

SWAYBACK

16

R.M. BARLOW

Synonyms Swingback, swingleback, jinkback, bent-back, evil, belland, bride, gefa-gwan (Wales), reng-uerra (Peru), pateleta (Argentina and Patagonia), enzootic ataxia (Australia).

Swayback is an ataxic disorder of newborn and young lambs which has been known for very many years and, as the plethora of synonyms indicates, the disease has a wide geographical distribution. It is possible that in earlier years these terms embraced more than one disease entity as there are no scientific descriptions prior to the 1920s and it was not until 1937 that the diagnosis gained a degree of aetiological specificity with the demonstration of the prophylactic value of copper supplements.[1]

Cause

Swayback is associated with a low copper status of the affected lamb and its mother. Unless prophylactic measures are taken the disease tends to be enzootic within reasonably well-defined areas. In these areas the soil may have a low copper content or the uptake of copper by herbage and its utilization by the animal may be inhibited by other factors. The trace element interactions are complex and probably varied. The best documented interaction is the copper-depressing effect of molybdenum in the presence of sulphate, but interactions with other metallic ions such as lead and silver have also been suggested from time to time.[2]

Clinical signs

In the congenital form of the disease, severely affected lambs may be still-born or live-born but small, dull, weakly and limp, with nystagmus and depressed corneal and pupillary reflexes. Some show fine head tremor or intermittent bruxism (grinding of the teeth).

Less severely affected lambs are often bright and alert with good pedal reflexes, though in the hind limbs the response to stimulus may be delayed. Such lambs are able to rise unaided but lack power and coordination. They feed well but have difficulty following the mother as exercise exacerbates the weakness of the hind limbs which sag to either side to produce a swaying, stumbling, ataxic gait.[3] Lambs of this type are usually fine-boned and may show incomplete extension of the carpal joints. They are liable to navel infections, hepatic necrobacillosis, joint-ill and bacterial meningoencephalitis.

In the delayed form of swayback the lambs appear normal at birth but develop clinical ataxia at any time up to 3 months of age or even later. Open surgical castration and docking at 4–6 weeks of age may precipitate fresh cases. It is often considered that the fast growing, better thriving lambs of the flock are more susceptible. The first sign of delayed swayback is an instability of the hock joints which with exercise rapidly progresses to the characteristic swaying ataxic gait. Lambs with delayed swayback often have unusually fine long bones and the fleece in some breeds has a lacklustre, staring appearance. Once clinical signs are established such lambs cease to thrive.

Occasionally a variant of the delayed form of swayback—the Roberts type[4]—is encountered. In this form of swayback lambs up to 6 weeks of age which were thriving and strong 24 hours previously are found twitching, trembling and wandering aimlessly. The condition progresses rapidly to recumbency with bruxism, apparent blindness with pupillary dilatation, extensor spasms, coma and death.

Pathology and pathogenesis

Macroscopic lesions in swayback are inconstant. In a proportion of congenital cases the cerebral white matter may contain sharply demarcated fluid-filled cavities or gelatinous transformations with indistinct boundaries. These lesions occur most frequently at the tips of the white matter cores of parietal and occipital gyri.

In the Roberts variant there is severe swelling of one or both cerebral hemispheres with flattening of gyri and herniation of the cerebellar vermis through the foramen magnum. The affected areas of cerebral cortex appear thickened due to oedema of the intracellular compartment, cellular degeneration and dispersal which are most pronounced in the deep cortical laminae and which also affect the subjacent white matter. Yellowish areas of necrosis may be present in the depths of affected sulci.

The pathognomonic lesions of swayback are microscopic and are found in the brain stem and spinal cord. They involve both nerve cells and fibre tracts. Nerve cell changes are most evident in the large neurones of the red and vestibular nuclei, the reticular formation and the ventral horns of the spinal cord.[5] The changes consist of swelling, vacuolation and chromatolysis proceeding to a 'hyaline'-like necrosis which rarely stimulates neuronophagia.

Altered nerve fibres in the brain stem have a rather scattered distribution but in the spinal cord they occupy chiefly the peripheral zones of the dorsal and ventral parts of the lateral funiculi and the sulco-marginal funiculi. Less commonly the septomarginal funiculus may also contain altered fibres. In the affected areas there is pallor of myelin and a positive Marchi reaction. There is evidence of Wallerian-type degeneration[6], myelin 'degradation'[7] and reduced myelin synthesis.[8]

Copper forms the prosthetic group of the enzyme cytochrome oxidase which is important in tissue respiration. In swayback it has been proposed that there may be a threshold of enzyme activity below which the cell is unable to support the metabolic requirements of structure, growth and function. The nerve cell may be unable to support its most peripheral cytoplasm, the axon, which degenerates, removing a stimulus for myelination in as yet unmyelinated regions of the CNS and causing Wallerian-like degeneration in already myelinated areas.[9] In a similar manner the peripheral cytoplasm of the oligodendroglial cell, (which forms the myelin sheath) may experience difficulty in fulfilling its biological role resulting in reduced synthesis or degradation of myelin.

Recently a distinctly different pathogenetic mechanism has been proposed.[10] In the Robe district of South Australia, copper-unsupplemented ewes invariably produce swayback lambs; examination of fetuses has indicated that myelin developed normally, but also gave the characteristic Marchi positive reaction of swayback. Cytochrome oxidase activity however, was similar to that in controls. It was suggested that the function of copper is to protect the lipids of myelin from damage (peroxidation) by circulating molecular oxygen. In the copper sufficient fetus such molecular oxygen would be effectively removed by the action of the copper-containing proteins, superoxide dismutase and caeruloplasmin. The distribution of the spinal lesions was attributed to their proximity to the 3 major spinal arteries in which abundant molecular oxygen might be expected. However, on this assessment it is difficult to account for the brain lesions which tend to occupy the terminal territories of the cerebral vasculature. Thus further work is necessary before the pathogenesis of swayback is fully understood.

Diagnosis

Clinical diagnosis of swayback in a flock is based upon the characteristic ataxia in newborn and young lambs. In the differential diagnosis it is necessary to consider border disease, navel infections and their sequelae and aspects of infectious abortion particularly chlamydia and *Toxoplasma gondii*. A diagnosis of swayback may be confirmed by histological examination of the brain and spinal cord, supported by copper assay of the tissues, (blood and liver), of the affected lamb and its mother. Values for blood < 9.5 μmol/l and liver values for lambs < 80 mg/kgDM indicate a low copper status; not infrequently swayback lambs and their mothers may have values < 3.0 μmol/l in blood and 10 mg/kgDM in liver.[3]

Epidemiology

As stated, swayback is associated with a simple or a conditioned deficiency of copper and thus has a fairly defined geographical distribution. In marginally copper-deficient areas swayback may follow upon pasture improvement procedures such as liming, ploughing and reseeding which respectively raise soil pH and destroy perennial, copper-rich plants. The disease tends to occur in cycles, being particularly prevalent following mild open winters when pasture is available and supplementary feeding, which usually contains adequate copper, is kept to a minimum. Swayback is most prevalent amongst the progeny of older ewes whose copper stores have been depleted over the years. Conversely bought-in ewe lambs or gimmers bred for the first time usually produce normal progeny. Adult sheep on 'swayback' farms often appear clinically normal though their well-being and productivity is suboptimal. In Western Australia however pregnant ewes on affected pastures are reported to develop a 'steely' or 'stringy' fleece and anaemia.[1] There is also evidence that breeds may differ in their susceptibility to copper deficiency.[11]

Control, prevention and treatment

Swayback can be effectively controlled by the prophylactic administration of copper to the ewe from mid-gestation onward. This may be given as a drench (1.0 g copper sulphate in 30 ml of water: this dose being repeated 2 or 3 times at 3 week intervals) or by injection with one of the commercially available copper edetate or copper methionine preparations. Recently, promising results have been obtained from oral administration of a bolus of cupric oxide 'needles' which lodge between the abomasal plicae and slowly release cupric chloride which is readily absorbed. Other forms of prophylaxis include the spraying of pasture with solutions of copper sulphate or the provision of copper-rich mineral licks. Neither of these methods is applicable to all situations and local professional advice should be sought before these methods are applied.

Treatment of lambs with congenital swayback, i.e. those which may have serious brain damage is useless. However, there is some evidence to suggest that the progress of delayed cases or the development of new ones may be arrested by the administration of copper to lambs. However, it must be emphasized that sheep are very susceptible to copper poisoning and great care must be exercised especially in the dosing of lambs. Small doses repeated are safer than a single large dose and oral administration is to be preferred. If copper sulphate solutions are used the dose should be given in milk to protect the delicate gastric mucosa from the astringent effects of the drench. Initially only a few lambs should be treated so that the safety of the dose can be assessed. Deaths from acute toxicity occur within 1–3 days and are manifest by pulmonary congestion and oedema and pleural effusion.

REFERENCES

1 Bennetts H.W. & Chapman F.E. (1937) Copper deficiency in sheep in Western Australia: a preliminary account of the aetiology of enzootic ataxia of lambs and an anaemia of ewes. *Australian Veterinary Journal,* **13**, 138–49.

2 Innes J.R.M. & Shearer G.D. (1940) 'Swayback' a demyelinating disease of lambs with affinities to Schilders encephalitis in man. *Journal of Comparative Pathology and Therapeutics,* **53**, 1–41.

3 Barlow R.M., Purves D., Butler E.J. & Macintyre I.J. (1960) Swayback in south east Scotland. II. Clinical, pathological and biochemical aspects. *Journal of Comparative Pathology,* **70**, 411–28.

4 Roberts H.E., Williams B.M. & Harvard A. (1966). Cerebral oedema in lambs associated with hypocuprosis, and its relation to swayback. I. Field clinical, gross anatomical and biochemical observations. II. Histopathological findings. *Journal of Comparative Pathology,* **76**, 279–83, 285–90.

5 Barlow R.M. (1963) Further observations on swayback. I. Transitional pathology. *Journal of Comparative Pathology,* **73**, 51–60.

6 Cancilla P.A. & Barlow R.M. (1966) Structural changes of the central nervous sytem in swayback (enzootic ataxia) of lambs. II. Electron microscopy of the lower motor neuron. *Acta Neuropathologica,* **6**, 251–9.

7 Smith R.M., Fraser F.J., Russell G.R. & Robertson J.S. (1977) Enzootic ataxia in lambs: appearance of lesions in the spinal cord during foetal development. *Journal of Comparative Pathology,* **87**, 119–28.

8 Howell J.McC., Davison A.N. & Oxberry J. (1969) Observations on the lesions in the white matter of the spinal cord of swayback sheep. *Acta Neuropathologica,* **12**, 33–41.

9 Barlow R.M. (1970) Contributions to Comparative Neuropathology. DSc., Thesis. University of Edinburgh.

10 Smith R.M., King R.A., Osborne-White W.S. & Fraser F.J. (1981) Copper and the pathogenesis of enzootic ataxia. In *Trace Element Metabolism in Man and Animals (TEMA-4)* (Eds. Howell J.McC., Cawthorne J.M. & White C.L.) ACT 2601, pp. 294–7. Australian Academy of Science, Canberra City.

11 Wiener G. & Field A.C. (1969) Copper concentration in the liver and blood of sheep of different breeds in relation to swayback history. *Journal of Comparative Pathology,* **79**, 7–14.

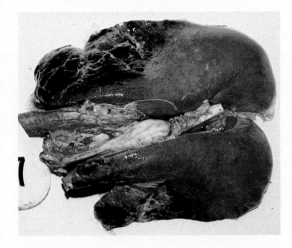

Plate 1 Peracute pneumonic pasteurellosis. Dark-red consolidated anterior lobes are covered by a gelatinous pleuritic exudate.

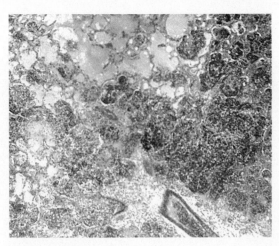

Plate 2 Necrotic lung in pneumonic pasteurellosis. Alveolar spaces are filled with a pink-stained protein-rich exudate, or are packed with intensely-stained spindle-shaped leucocytes which 'stream' from one alveolus to another. (H & E x 150)

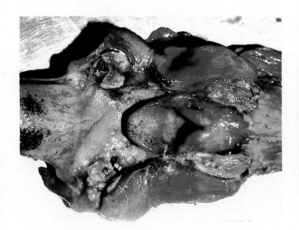

Plate 3 Necrotic lesions in the pharynx and tonsillar crypts of a sheep with systemic pasteurellosis. Photograph reproduced by kind permission of the Editor, *Journal of Medical Microbiology*.

Plate 4 Enzootic abortion: necrosis of the placental cotyledons and adjacent tissue.

17 POLIOENCEPHALOMALACIA

R.M. BARLOW

Synonyms Cerebrocortical necrosis (CCN)

This is an acute neurological disease of ruminants characterized clinically by head pressing, blindness, opisthotonos and convulsions, and pathologically by brain swelling and laminar necrosis of the cerebral cortex. It was first described in the USA in 1956[1] since which time there have been numerous reports world-wide. At present it is one of the most common neurological diagnoses and in feedlot sheep it has been estimated to account for 19 per cent of all deaths.

Cause

Polioencephalomalacia (PE) is a pathological change resulting from interference with cerebral metabolism at cell level. The naturally-occurring disease in ruminants is frequently associated with a deficiency of thiamine[2], a vitamin of the vitamin B complex.

Clinical signs

The incidence of PE is highest in the younger age groups and the disease has been recorded in sucking lambs. Cases usually occur sporadically but outbreaks involving several animals are not infrequent. Premonitary signs are rarely observed but include a brief period of scour when the affected animal may wander aimlessly away from the main flock. Within 24 hours affected animals appear blind and exhibit fine tremors. The head is held high and there may be salivation. The animal is reluctant to walk and soon becomes recumbent with opisthotonos, nystagmus or strabismus and paddling movements of the limbs. Convulsions may be precipitated by mild acoustic or tactile stimuli. Death rapidly supervenes.

Pathology

The consistent pathological changes are confined to the brain and are usually macroscopically recognizable. The cerebral hemispheres are swollen, pale and soft with yellow discoloration of some gyri especially in the frontal, dorsolateral and dorsomedial areas of cortex. There is swelling of the parahippocampal and cingulate gyri which may herneate beneath the tentorium cerebelli. The posterior vermis may be cone-shaped due to herneation through the foramen magnum.

The cut surface of the cerebrum reveals a laminar distribution of necrosis in the cortex of affected gyri with separation from the subjacent white matter. The tips of white matter cores may show liquefactive changes. When viewed in ultraviolet light (365 nm) affected areas of cortex may exhibit a bright white autofluorescence which has been attributed to accumulation of ceroid lipofuscin in macrophages.[3] This procedure readily demonstrates the bilateral, nearly symmetrical distribution of the lesions. It is a useful aid to diagnosis, but not all cases of PE fluoresce.

Histologically the cortical lesions consist of vacuolation and cavitation of the ground substance with neuronal shrinkage and necrosis. The capillary endothelium is prominent appearing swollen and hyperchromic and there is dilatation of the perivascular spaces. Advanced lesions contain plump macrophages. Microscopic lesions of the same type but usually less advanced are present also in the basal ganglia, geniculate bodies, corpora quadrigemina and mesencephalic nuclei.[4] The hippocampus is also frequently involved and in this site perivascular haemorrhages are very characteristic.

Evidence from ultrastructural studies has shown that the earliest changes of PE are cytopathic swelling of the processes of astrocytes and satellite cells with accumulation of glycogen granules in their cytoplasm.[5] This oedema of the intracellular compartment can be accounted for in terms of the known functions of thiamine as described under epidemiology. The overall brain swelling may well reduce cerebral circulation to a critical level such that the terminal territories of the cerebral arteries are liable to ischaemic necrosis. This proposal would explain the symmetrical distribution of the gross cortical lesions.

Diagnosis

Under field conditions cases of PE in sheep may not be seen alive and PE should always be considered in the differential diagnosis of sheep found dead. In the live animal evidence of scour coupled with blindness, circling movements, muscular tremors, opisthotonos, and nystagmus are very suggestive of PE but listeriosis, louping-ill, gid *(Coenurus cerebralis* infestation) brain abscess or meningoencephalitis should be considered in the differential diagnosis. Biochemical support for the diagnosis of PE can be obtained from the presence of thiaminase 1 in the rumen contents or faeces, raised levels of blood pyruvate and lactate, considered together with reduced thiamine pyrophosphate and transketolase activity.

Epidemiology

Thiamine pyrophosphate (TPP) is the active form of thiamine and is required for the decarboxylation of pyruvate during its conversion to acetyl CO-A. For energy, brain tissue is critically dependent on glucose which is used via the glycolytic sequence and the tricarboxylic acid cycle with a small additional contribution from the transketolase-dependent pentose phosphate pathway. About two-thirds of the energy produced is used to maintain membrane potentials via the Na-K-ATPase ionic pump. Thus in thiamine deficiency membrane repolarization is impaired, water and electrolyte balance upset, whilst TPP and transketolase activity are reduced and pyruvate and lactate accumulate in the tissues.[6]

Ruminants are not dependent on dietary sources of the vitamin B complex as these vitamins are synthesized during normal ruminal fermentation processes. However, in PE it has been shown that the rumen liquor contains thiaminase type 1 an enzyme which not only rapidly destroys thiamine but may lead to the production of active antagonists which would further deplete the tissues of thiamine.[7]

Some forages, e.g. bracken rhizomes, contain thiaminase type 1 and may sometimes play a role in the development of PE, but more commonly it is believed that thiaminase type 1 is produced in the rumen by bacteria such as *Clostridium sporogenes* and *Bacillus thiaminolyticus*. Such organisms proliferate in the rumen under conditions in which the production of volatile fatty acids is reduced. In a closely related molasses toxicity disorder of cattle

appropriate conditions may occur when the fibre content of the diet is low.[8] However, other causes of altered ruminal fermentation cannot be excluded, and there is evidence that treatment with certain anthelmintics may predispose sheep to PE.

Control, prevention and treatment

If observed early in the course of the disease affected animals often respond favourably to large doses of thiamine or multi-B vitamin preparations given intravenously.[2] Since incidents of the disease are mainly sporadic and insufficient is known concerning the development of high levels of thiaminase 1 in the alimentary tract control procedures are difficult to advocate. Where outbreaks occur however, it would be wise to change the diet to one which contains adequate roughage and would stimulate normal rumen fermentation with production of volatile fatty acids.

REFERENCES

1 Jensen R., Griner L.A. & Adams O.R. (1956) Polioencephalomalacia of cattle and sheep. *Journal of the American Veterinary Medical Association,* **129**, 311–21.
2 Davies E.T., Pill A.H., Collings D.F., Venn J.A.J. & Bridges G.D. (1965) Cerebrocortical necrosis in calves. *Veterinary Record,* **77**, 290.
3 Little P.B. (1978) Identity of fluorescence in polioencephalomalacia. *Veterinary Record,* **103**, 76.
4 Palmer A.C. & Walker R.G. (1970) The neuropathological effects of cardiac arrest in animals: a study of five cases. *Journal of Small Animal Practice,* **11**, 779–90.
5 Morgan K.T.(1973) An ultrastructural study of ovine polio-encephalomalacia. *Journal of Pathology,* **110**, 123–30.
6 Edwin E.E. (1970) Plasma enzyme and metabolite concentrations in cerebrocortical necrosis. *Veterinary Record,* **87**, 396–8.
7 Edwin E.E. & Jackman R. (1970) Thiaminase 1 in the development of cerebrocortical necrosis in sheep and cattle. *Nature, Lond.,* **228**, 772–4.
8 Losada H., Dixon F. & Preston T.R. (1971) Thiamine and molasses toxicity. 1. Effects with roughage free diets. *Revista Cubana de Giencia Agricola (Eng. ed.),* **5**, 369–78.

18 'DAFT' LAMB DISEASE (DLD)

R.M. BARLOW

Synonyms 'Helpless', 'mad', or 'silly' lambs, dafties, Wyn Gweiniaid, inherited cerebellar cortical atrophy.

Knowledge concerning this neurological disease is in an unsatisfactory state. The disease has a familial pattern with a recessive mode of inheritance.[1, 2] The clinical signs are present at birth or shortly thereafter and are fairly characteristic, but morphological changes are inconstant and varied and the pathophysiology is poorly understood.

Cause

DLD has been recognized in several breeds within each of which it appears to be inherited.[1, 2] In the Canadian Corriedale it appears to be conditioned, but not caused directly by a simple recessive gene and there is some evidence that mineral malnutrition may also be involved.[3]

Clinical signs

The clinical signs are evident at birth or develop within the first 2–3 days of life. They vary considerably in intensity; the most severely affected lambs are unable to rise and collapse if placed in the standing position. The head sways continuously from side to side. Often the neck is held in extreme dorsiflexion, so that the crown is adjacent to the withers. Less severely affected lambs may stand unaided but have difficulty in maintaining their balance, possibly due to the abnormal head carriage (Fig. 18.1). Affected lambs have difficulty in sucking their dams but if bottle fed they suck vigorously and may show some improvement in a few days, though the gait remains stiff and waddling with all four legs widely spread to maintain equilibrium. The abnormal head movements and head carriage not infrequently lead to traumatic conjunctivitis and corneal ulceration; a crusty blepharitis develops and the cheeks become 'scalded' with tears. In survivors, the clinical signs gradually remit but growth rate is impaired. Ophthalmoscopy and electroencephalography have failed to

Fig. 18.1 'Daft Lamb' showing characteristic 'star-gazing' head carriage.

reveal any significant abnormalities. Radiographs showed a low bone density in the majority of lambs examined.[2]

Pathology

Macroscopic pathological changes are non-specific and include purulent otitis media and rib fractures with extensive callus formation. It would be tempting to implicate the former in the pathophysiology of DLD; however, otitis media is a not uncommon consequence of bottle feeding.[4]

In the original description of the pathology of DLD[5] degenerative changes, swelling, pallor or hyperchromasia, vacuolation and necrosis were observed among the Purkinje cells and Type II Golgi cells of the cerebellar cortex, the nerve cells of the dentate nucleus, the central cerebellar nuclei and occasionally also in the olives. In cortical regions with severe change the subjacent granular layer also showed a reduced cellular density. Argyrophilic axonal swellings (torpedoes) were present in the granular layer and there was fibrillary astrocytic gliosis which was most marked around blood vessels.

However, several investigations of apparently identical disease have failed to demonstrate the above-mentioned changes in the cerebellar cortex and it has been postulated that the clinical syndrome of DLD comprises more than one pathological entity.[2]

No histological or histochemical changes have been observed in the peripheral nervous system proximal to the intramuscular nerves. Neuromuscular spindles appear normal. In one study[6] variations in argyrophilia of the motor end plates and breaks in their preterminal axons were observed in affected lambs of more than 3 months of age. These changes were associated with atrophy of type II and some type I muscle fibres and with marked hypertrophy of other type I fibres. These dystrophic changes affected many muscle groups but were most severe in the muscles of the neck. It seems probable that such muscle changes are of a secondary nature since they are not evident at birth.

Diagnosis

In the present state of knowledge, the diagnosis of DLD is dependent upon the recognition of the clinical signs of which the abnormal head carriage and swaying movement are the most characteristic. DLD must be differentiated from those other congenital neurological diseases of lambs in which there is locomotor dysfunction, e.g. swayback, border disease, spina bifida, idiopathic hydrocephalus. Histological examination of nervous and muscular tissues may provide confirmation of diagnosis.

Epidemiology

DLD was first described in the Welsh Mountain sheep in North Wales. It has since been recognized in other parts of Britain, in Canada and apparently also in Australia and New Zealand[1], affecting Border Leicester, Scottish Halfbreds, Scottish Blackface, and Corriedale breeds of sheep. It is a hereditary disease with an autosomal recessive mode of transmission, though nutritional factors may be involved in the manifestation of the phenotype.

Control

Since the disease has a genetic basis, control can be achieved by removal of affected lambs and their parents from the breeding flock. No laboratory methods of heterozygote recognition are available.

REFERENCES

1 Innes J.R.M. & Saunders L.Z. (1962) *Comparative Neuropathology.* pp.297–301. Academic Press, New York & London.
2 Terlecki S., Richardson C., Bradley R., Buntain D., Young G.B. & Pampiglione G. (1978) A congenital disease of lambs clinically similar to 'inherited cerebellar cortical atrophy' (daft lamb disease). *British Veterinary Journal,* **134**, 299–307.
3 Stewart W.L. (1950) Cerebellar atrophy in lambs (Daft Lambs). *Veterinary Record,* **62**, 299–303.
4 Macleod N.S.M., Wiener G. & Barlow R.M. (1972) Factors involved in middle ear infection (otitis media) in lambs. *Veterinary Record,* **91**, 360–2.
5 Innes J.R.M., Rowlands W.T. & Parry H.B. (1949) An inherited form of cortical cerebellar atrophy in 'daft' lambs in Great Britain. *Veterinary Record,* **61**, 225–8.
6 Bradley R. & Terlecki S. (1977) Muscle lesions in hereditary 'daft lamb' disease of Border Leicester Sheep. *Journal of Pathology,* **123**, 225–36.

19 NON-SPECIFIC BACTERIAL INFECTIONS OF THE CENTRAL NERVOUS SYSTEM (CNS)

R.M. BARLOW

Under this heading are grouped those afflictions of the nervous system which are not readily specified clinically or aetiologically. They may be manifested by generalized disease or a plethora of localizing signs and one or more of a range of bacterial species may be implicated in the cause. These infections are most common in newborn and young lambs and in older sheep which are debilitated or suffering from intercurrent disease. Conditions such as neonatal enteritis, navel ill, joint ill, tick pyaemia, brain abscess and spinal abscess are discussed here as variations of a single theme but listeriosis,since it represents a more distinct clinical and pathological entity, is presented elsewhere.

Causes

Bacterial infections of the CNS are mostly caused by pyogenic bacteria: staphylococci, streptococci, *Corynebacterium pyogenes* and also by *Escherichia coli*.[1] *Pasteurella multocida*[2], *P. haemolytica*, *Pseudomonas pyocyanea* and *C. ovis* may also be implicated from time to time.

These organisms reach the CNS in three ways: directly as a result of trauma, by local extension from a lesion in adjacent tissues or through the bloodstream from a parenteral site. The last is the most common route. Initial localization and proliferation of bacteria within the cranium or vertebral column is within the meninges or, more rarely, in the choroid plexuses of the ventricles or the capillary bed of the neural tissue itself and this is followed by an acute inflammatory reaction.

Clinical signs

These vary according to the site and extent of the lesions and the functions of the structures involved. In the early stages of diffuse meningitis there is increased irritability and excitement but this is usually quickly replaced by depression, head pressing, apparent blindness and grinding of the teeth. The animal is febrile and there is rigidity of the neck, forced passive movements of which may result in collapse. There is hyperreflexia and hyperaesthesia but as the disease progresses, recumbancy, stupor and coma supervene. Ophthalmoscopy may reveal papilloedema as a sign of raised intracranial pressure and rarely nystagmus may be observed. Diffuse meningitis is most commonly seen in young lambs as a sequel to navel infections, or coliform septicaemia arising from neonatal enteritis.[1]

The clinical signs of focal meningitis are caused by local interference with neural functions and may be due to ischaemia, toxic absorption from the infective focus, or compression of adjacent nervous tissue. Focal meningitis may lead to abscess formation. The clinical signs vary with the site of the lesion, the speed with which it develops and the degree of disruption of nervous tissue which results. Precise localization of the lesion on the basis of the signs is a challenge for the clinical neurologist. Depending on the structures involved brain abscess may result in circling movements, in deviation or rotation of the head, drooping of an ear, ptosis, retention of cud, inequality in size of pupils and alterations in their reactivity to light stimuli. In rams there may be infertility with testicular atrophy if the pituitary is involved. Adult sheep, however, can sometimes tolerate quite large cerebral abscesses with seemingly little distress or functional disability. Spinal lesions affect sensory and/or motor function and depending upon the age of the lesion they may also interfere with reflex activity in regions caudal to the lesion. Focal meningitis with abscess formation is usually caused by staphylococci or *C. pyogenes* and in the former instance, may result from the pyaemia which can follow tick borne fever in young lambs. Abscesses of long duration often become sterile.

Pathology and pathogenesis

Intracranial infection with pyogenic bacteria almost invariably results in meningitis. The subarachnoid CSF provides an ideal vehicle for the dissemination of the infection over the surface of the CNS and into the deepest parts of the Virchow-Robin spaces which may give the impression of encephalitis. However, the pia mater and the glia limitans together with the blood–brain barrier function to prevent or delay invasion of the brain. Where there are contusions, lacerations or abscess formation, or intercurrent disease such as coenurosis or swayback, these defences may be compromised and a true encephalitis supervene.

Meningitis may be inapparent grossly although in most cases there is congestion of meningeal vessels and the leptomeninges have a glairy appearance. A fibrino-purulent exudate may collect in the convolutions of the brain, beneath the medulla and at the ponto-cerebellar angle. Microscopically the exudate is usually rich in neutrophils and macrophages with a variable admixture of lymphoid cells.

Abscess formation in the meninges causes compression of subjacent nervous tissue and may also cause rarefaction and osteomyelitis in overlying bone. The compressed nervous tissue becomes ischaemic with necrosis of neural elements, and swelling and proliferation of astrocytes. A glial capsule may develop or the abscess may rupture into nervous tissue to produce encephalitis or myelitis.

Diagnosis

The acute onset of generalized or localized neurological dysfunction together with fever is very suggestive of bacterial meningitis particularly if combined with stiffness of the neck. In the differential diagnosis, viral meningoencephalitis, listerosis, polioencephalomalacia, swayback, coeneurosis or even aberrant migratory liver fluke may need to be considered. Cultural and cytological examination of CSF tapped from the cisterna magna or lumbo-sacral subarachnoid space may be a useful aid to diagnosis in the live animal. In many instances, however, the diagnosis is made post-mortem.

Epidemiology

In sheep, there is no specific bacterial meningitis comparable to human meningococcal meningitis and most ovine meningitides arise as complications of pre-existing disease states. Thus the epidemiology of ovine meningitis is predominantly that of the various predisposing diseases. These include, septicaemias resulting from coliform enteritis, infection of the neonatal navel, or from pasteurellosis all of which have been dealt with elsewhere. Meningitis or brain abscess may also arise from dosing gun injuries to the nasopharynx, extension of infection in otitis media [3]

(see 'daft' lamb disease) suppurative foci in the paranasal sinuses due to *Oestrus ovis*, or from infection of coenurial cysts. Head trauma sustained by rams during fighting may cause fractures of the skull and contusion of the brain and provide conditions suitable for bacterial colonization. Low grade diffuse pachymeningitis with abscess formation at the base of the brain involving the rete mirabile, sella tursica and hypophysis may sometimes occur.

Spinal meningitis or abscess formation may result from contiguous or haematogenous spread of infection from docking wounds. In tick-borne fever in which there is a marked leucopenia, cutaneous staphylococci, injected during tickbite can rapidly proliferate and produce abscesses in several organs. In the vertebral column, favoured sites for abscess formation are the intervertebral foramina. In tick-free areas, solitary spinal abscesses not infrequently arise from an osteomyelitis which seems to have a predilection for the bodies of the last cervical and first two thoracic vertebrae. Perforation of the dorsal spinal ligament by the abscess results in the sudden onset of posterior paralysis in an otherwise apparently healthy lamb; the pathogenesis of this disorder is not understood.

Control, prevention and treatment

With such a variety of predisposing causes, control and prevention of bacterial meningitis is principally a matter of good husbandry. Treatment if carried out very early in the course of the disease may control the infection but does little to repair any neurological defect. In valuable animals treatment may be considered worthwhile. Parenterally administered antibiotics permeate the damaged blood–brain barrier at the site of the lesion. With chloramphenicol, CSF levels one third to one half of those in the blood can be achieved even in uncompromised subjects.[4] However, clinical application of chloramphenicol in animals should be restricted to cases in which no other antibiotic would be effective. If intrathecal antibiotic medication is contemplated it should be borne in mind that some antibiotics are epileptogenic; however, kanamycin and gentamycin have been used satisfactorily by this route in human subjects.[5]

REFERENCES

1 Terlecki S. & Shaw W.B. (1959) *Escherichia coli* infection in lambs. *Veterinary Record*, **71**, 181–2.

2 Biester H.E., Schwarte L.H. & Packer R.A. (1942) Studies on sheep with the demonstration of Pasteurella localised in the central nervous system. *American Journal of Veterinary Research*, **3**, 268–73.

3 Hull F.E., Taylor E.L. (1941) Abscesses affecting the central nervous system of sheep. *American Journal of Veterinary Research*, **2**, 356–7.

4 Rahal J.J.Jr. (1972) Treatment of Gram-negative bacilliary meningitis in adults. *Annals of Internal Medicine*, **77**, 295–302.

5 Kaiser A.B. & McGee Z.A. (1975) Aminoglycoside therapy of Gram-negative bacilliary meningitis. *New England Journal of Medicine*, **29**, 1215– 20.

20

TETANUS

D. BUXTON

Synonym Lockjaw

Tetanus (derived from the Greek, 'I stretch') had been recognized as a disease entity with a high mortality long before the elucidation of its aetiology. Youatt, writing in 1856, described excellently the clinical signs in the sheep but blamed 'heavy rain' for its cause, although he did allude to its increased incidence following castration. Its infectious nature was first demonstrated in 1884 when Carle and Rattone reproduced the disease in rabbits with the contents of an acne pustle from a man who had died of tetanus. In the same year Nicolair reproduced the disease in rabbits, guinea pigs and mice by inoculation of garden soil, and reported the discovery of a spore-containing bacillus. The bacillus was first obtained in pure culture by Kitasato in 1887, who also reproduced the disease with bacteria-free inoculum. Since that time tetanus has been shown to be caused by a toxin and an effective vaccine has been produced.

Cause

Tetanus in man and animals is caused by the toxin tetanospasmin released from the spore-forming bacillus *Clostridium tetani*.[1] *Clostridium tetani*, a thin rod shaped organism 2–4 μm by 0.5 μm, produces highly resistant spores which when fully developed are spherical, measure more than twice the diameter of the bacillus and are positioned terminally, giving the bacterium the characteristic drumstick appearance.

It is a strict anaerobe which stains Gram-positive in young cultures but can become Gram-negative after about 48 hours, particularly in nutrient broth. The great majority of strains are motile having peritrichous flagella.

Disease occurs more commonly in lambs than adults after spores gain entry to a wound. Deep puncture wounds are most frequently implicated but infected superficial scratches can also allow germination of spores. Germination of the spores into bacilli only occur if the microenvironment is anaerobic. Hence the conditions produced after localized trauma are ideal, with reduced blood supply, necrosis and multiplication of other bacteria, all of which reduce the local oxygen tension.

Toxin production, spread and mode of action

After germination of the spores within a wound *Cl. tetani* bacilli proliferate and produce toxin. Toxin is only released when the bacilli become autolysed and tetanospasmin is therefore not strictly a true exotoxin as it is not secreted by the live organism. However, it is extremely potent, 1 mg of pure toxin (mean mol. wt. of 150 000) being enough to kill about 30 million mice.[2]

Experiments have shown that tetanus toxin diffuses out from the site of infection, is taken up in the nerve endings of alpha axons, and migrates up to the alpha motor neurones in the ventral horn of the spinal cord. However, despite many investigations the precise mechanism of action of tetanus toxin remains unknown.[3] Current research suggests that tetanus toxin acts at the presynaptic level causing spinal disinhibition by blocking synaptic inhibition. This has a similar clinical effect to that of strychnine poisoning, a similarity which was recognized by Simpson as long ago as 1854, although strychnine acts at the post-synaptic level, blocking both synaptic inhibition and the inhibitory effect of glycine.

Clinical signs

The incubation period can be very variable with cases occurring as early as 3 days but more usually about 10 days after injury, such as wounds produced by shearing, docking or castration. Occasionally the incubation period can be several months, if local conditions do not initially favour germination of spores, by which time healing of the primary injury has usually occurred. At first the animal appears slightly stiff, becomes unwilling to move and develops a fine muscle tremor. After 12–24 hours a more severe general stiffness of the limbs, head, neck and tail sets in together with hyperaesthesia and repeated spasms. Mastication becomes extremely difficult due to tetany of the masseter muscles (thus the name lockjaw), the food is chewed clumsily, remains in the mouth and the animal drools saliva. There is retention of urine and constipation. At this stage if the animal falls it is unable to rise, opisthotonus is seen, convulsions ensue and death due to respiratory failure supervenes after a clinical course of about 3–4 days. A few milder cases which develop more slowly can recover over a period of weeks and even months.[4]

Diagnosis

An initial diagnosis is made from clinical signs when these are observed and attempts can be made to isolate Cl. tetani from wounds. Necrotic material from a wound is cultured at 37°C for 2–3 days on blood agar and meat broth. Any non-sporing organisms in the latter are then killed by heating to 65°C for 30 minutes and the culture is then inoculated on to one half of a blood agar plate or on to the water of condensation on an agar slope and incubated anaerobically for 48 hours. Motile strains of Cl. tetani spread over the surface of the medium as a filamentous growth and sub-culture of organisms from the edge of this growth can give a pure culture of Cl. tetani. Recognition of the morphological characteristics of the organism helps to establish a diagnosis. Direct examination of material from wounds is seldom successful as contaminating bacteria usually outnumber Cl. tetani. The most reliable technique for laboratory diagnosis is by the inoculation of mice. A pair of mice is used; one mouse is protected by the subcutaneous injection of 750 units of tetanus antitoxin 2 hours before both mice are inoculated intramuscularly in a hind leg with 0.25 ml of the supernatant from a 48 hour cooked meat culture of the suspect organism. If the culture contains toxin clinical tetanus develops rapidly in the unprotected mouse. No pathological changes characteristic of tetanus have been recognized.

The differential diagnosis should seek to exclude other conditions which also produce symptoms of tetany, such as hypomagnesaemia, hypocalcaemia and even louping-ill.

Epidemiology and transmission

The spores of Cl. tetani are found throughout the world in faeces, dust and soil, and in the latter case more commonly in intensely cultivated soil. Young lambs are more susceptible than adult sheep and this is thought to be because the adult develops some immunity by being exposed to small amounts of toxin produced in the rumen.

Control, prevention and treatment

As the disease follows the contamination of a wound with spores, all surgery such as castration and docking should be carried out with attention to cleanliness.[4] In addition adequate vaccination should be undertaken, especially in enzootic areas where the use each year of temporary pens on clean ground for holding lambs for surgical procedures can also reduce the risk of infection.

Passive protection lasting 10–14 days can be achieved by the injection of 100–150 SI units of antitoxin just after surgery. This form of prophylaxis can augment maternally-derived passive protection produced by the immunization of ewes in the last 2–3 weeks of pregnancy with tetanus toxoid, an alum-precipitated formol-treated preparation of tetanus toxin. In this case if the lamb sucks normally within the first 12–15 hours of birth it will be passively protected during the first 12–16 weeks of life, after which it should be actively immunized with tetanus toxoid.

The primary course of immunization in adults consists of 2 injections of tetanus toxoid at an interval of not less than 6 weeks followed by annual revaccination.

Treatment of cases of tetanus is difficult as their response is usually poor. In an attempt to eliminate the organism parenteral penicillin can be given along with tetanus antitoxin and if an infected wound is found it should be debrided and treated with hydrogen peroxide. Sedation and reduction of muscle tetany, to avoid asphyxia, can also be attempted.

REFERENCES

1 Buxton A. & Fraser G. (1977) *Animal Microbiology*, Vol.1. pp.205–28. Blackwell Scientific Publications, Oxford.
2 van Heyningen W.E. & Mellanby J. (1971) Tetanus toxin. In *Microbial Toxins*. (Eds. Ajl S.J., Kadis S. & Montie T.C.) Vol.II. pp.69–108. Academic Press, London.
3 Bizzini B. (1979) Tetanus toxin, *Microbiological Reviews*, **43**, 224–40.
4 Blood D.C., Henderson J.A. & Radostits O.M. (1979) *Veterinary Medicine*, 5e, p.438–58. Bailliere Tindall, London.

21 COENUROSIS

B.M. WILLIAMS AND T. BOUNDY

Synonyms Bendro, gid, giddy dunt, goggle-turn, staggers, sturdy water brain

Coenurosis is a disease of sheep and other ruminants, which is well recognized in most countries. It develops when the brain and spinal cord is invaded by *Coenurus cerebralis* the larval stage of the tapeworm *Taenia multiceps*.

Coenurosis attracted a great deal of attention during the 19th century, particularly in Europe, where a number of workers attempted to identify its aetiology.[1] It was Kuchenmiester however, who demonstrated the true nature of the disease. He fed a *Coenurus cerebralis* from a sheep to a dog, which developed into a tapeworm that he named *Taenia coenurus*. Gravid proglottids of the tapeworm were then fed to a sheep which developed acute coenurosis.

The disease has been reported from all sheep-rearing countries of the world, but some workers allege that the disease is much less common than it used to be and relatively few losses now occur.[2] The reasons for the apparent reduced prevalence are not always obvious. While a reduction in prevalence can be expected in those countries which have introduced control programmes for hydatidosis, a reduction has also been recorded in other countries without the introduction of such control measures. In the USA it is suggested that modern sheep management and predator control practices have probably made the parasite extinct.[3]

In Britain where new systems of sheep husbandry and management comparable to those in the USA have been developed, there is no indication of a significant reduction in prevalence. Coenurosis is regularly reported from all the sheep-rearing areas, with serious losses recorded from time to time in some of these. In the Dyfed area of south-west Wales, it was estimated from a survey of 600 flocks, that an average 3 per cent of sheep under 12 months of age in each flock, developed gid each year.[4] In the same area, a slaughter house survey revealed that the brains of 5.8 per cent of fat lambs showed evidence of gid.[4] In contrast to this only 0.5 per cent of the brains of lambs sent for slaughter from the Snowdonia area of north Wales showed evidence of gid. [5]

Cause

Coenurus cerebralis is the larval stage of *Taenia multiceps*, the tapeworm of the dog, fox, wolf and jackal. The tapeworm was fully described by Leske in 1780.[1] There has over the years been a great deal of confusion and debate over the identity and taxonomy of the various species of the genus *Taenia*. It has been postulated by a number of authors, that *T. multiceps* and *T. serialis* are different strains of the same tapeworm and that the differing morphological characteristics of the coenuri and their location within the host, are largely due to the species of the host.[6] Attempts to infect sheep with gravid segments of *T. serialis* and rabbits with gravid segments of *T. multiceps* failed.[4] Moreover a comprehensive revision of the taxonomy of the genus *Taenia*, identifies *T. multiceps* and *T. serialis* as species in their own right.[7]

The tapeworm becomes gravid in the dog about 42 days after infection, following which gravid segments are passed in the faeces almost daily. There is a rapid disintegration of the proglottids, but the eggs however remain viable on pasture for periods of up to 8 weeks in Great Britain[4] and up to 4 months in colder climates.[8] The viability of scolices in the coenurus in contrast is much less and under ambient temperatures in Britain it is unlikely to be longer than 3 days in the dead animal.[4]

Infection of the final or definitive host follows the ingestion of embryonated eggs. The eggs hatch in the small intestine and the hexacanth embryos are released. They penetrate the intestinal mucosa, enter the submucosal blood vessels and are subsequently disseminated throughout the body. Only the embryos that reach the brain and spinal cord develop into a coenurus, while those that reach other sites rapidly die. Occasionally however, a cyst may develop in the subcutaneous tissues.

The embryos reach the brain between 7 and 12 days after ingestion, actively migrate through the brain tissue for a few days before finally coming to rest. If the number of embryos reaching the brain is large, then the animal may die from an acute meningoencephalitis at this stage. In the vast majority of cases, the number of embryos reaching the brain is small and rarely more than one develops into a coenurus, over a period of 4–6 months. The mature coenurus may reach 6 cm in diameter, but its size is determined largely by its location in the brain, and consists of a thin wall cyst with between 80 and 320 scolices on its inner surface.

Clinical signs

The clinical signs of coenurosis have been well-described, but most of the descriptions are those of the chronic phase. This is understandable as only a very small proportion of those animals which develop the classical or chronic phase exhibit clinical signs of the acute phase and only a few die of the acute disease.

The acute form is usually observed in young lambs, 6–8 weeks of age, 10–14 days after the ingestion of the embryonated eggs, when the embryos are migrating in the brain. Some lambs may show only mild clinical signs in the form of deviation of the head laterally or they adopt a 'star gazing' posture for 2–3 days. Others show a progressive depression, followed by aimless wandering, blindness, incoordination recumbency and convulsions, with death occurring in 4 or 5 days.

The clinical signs seen in older sheep during the chronic phase are largely governed by the site at which the coenurus develops. These appear 2–4 months after infection and include incoordination, blindness, high stepping gait, deviation of the head and circling. Occasionally when the coenurus occludes the cerebral aqueduct, a sheep may be found recumbent without having shown any previous clinical signs. As the disease progresses there is a loss of condition, recumbency and death. In the majority of cases when the coenurus is located in the cerebrum, there is a softening of the overlying cranial bones.

In about 5 per cent of cases, the coenurus is located in the spinal canal. The clinical signs in such cases are weakness, incoordination recumbency and paresis of the hind legs. On two occasions the authors have encountered congenital infection in lambs with coenuri of 3 mm in the cerebral hemispheres. Neither of the lambs had shown nervous signs and their deaths were attributable to other conditions.

Pathology

Acute phase On macroscopic examination the brain surface exhibits a number of yellowish-white tortuous tracks and there are accumulations of pus around the brain stem and optic chiasma. When the brain is sectioned, more tracks become visible, particularly in the depths of the cerebral convolutions, the lateral and third ventricle, the brain stem and occasionally the cerebellum. Minute cysts can sometimes be detected at the end of a track.

Microscopically the tracks consist of a central core

of necrotic tissue, surrounded by a haemorrhagic zone with a small number of leucocytes. Outside this zone, there is another of more intense cellular reaction, the predominant cells being eosinophils and giant cells. In areas remote from the tracks, blood vessels show a cuffing reaction in which the majority of cells are eosinophils. It appears that the damage produced is not solely due to the direct action of the parasite, but also to the amount of vascular damage produced.

Chronic phase The macroscopic appearance of the brain depends on the number of coenuri present and their location. Frequently the coenurus is located on the surface of the brain, covered only by the meninges and a thin layer of brain parenchyma (Fig. 21.1). The affected portion of the cerebrum is tense with flattened gyri, some of which may even be completely obliterated. If the coenurus is large, the cerebellum may be compressed and the vermis herniated through the foramen magnum. The presence of a cyst in one of the ventricles or the aqueduct system may lead to a very obvious hydrocephalus. The cranium overlying the coenurus is often rarefied, presumably the result of pressure atrophy.

Histologically, the lesions are not as spectacular as those seen during the acute phase. As the coenurus develops it exerts pressure on the surrounding brain tissue causing atrophy. Around the cyst, a network of collagen fibres is formed, in which eosinophils are common. Secondary lesions may develop away from the cyst such as displacement of some parts and compression of others.

When the spinal cord is involved, the pathological picture is similar to that seen in the brain during the chronic phase. The coenurus is smaller because of its location within the spinal canal but is sufficiently large to cause compression of the cord.

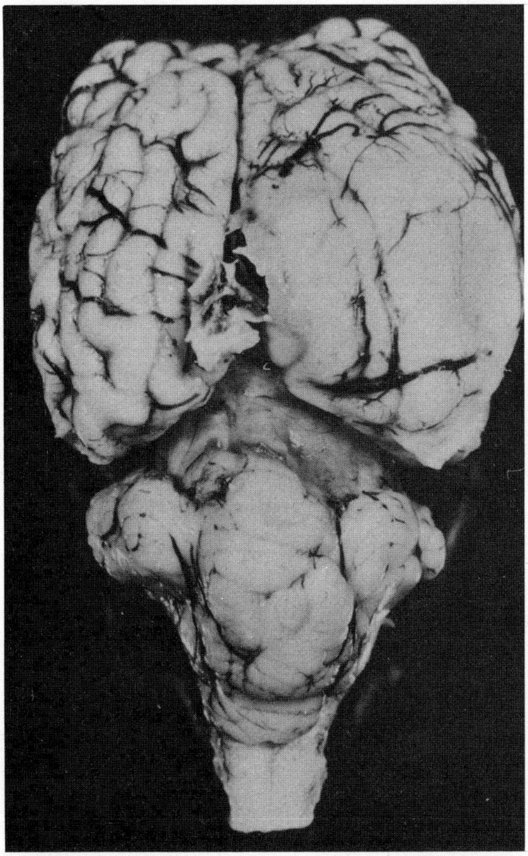

Diagnosis

A tentative diagnosis of chronic coenurosis can be arrived at after careful consideration of the history and clinical manifestations. Young sheep are usually affected with more than 80 per cent of field cases occurring in animals under 12 months of age. The slow onset of clinical signs, the characteristic circling movements, rarefaction of the cranial bones etc., are suggestive of gid. In a proportion of cases there is an obvious retinal papilloedema which can be detected with an ophthalmoscope. Diagnosis of the acute form is more difficult and nervous signs particularly in young lambs may suggest a bacterial meningoencephalitis, louping ill or cerebrocortical necrosis as well as acute coenurosis. The examination of blood samples is not helpful, but an increase in the cell count of the CSF, predominantly eosinophils, is a characteristic feature of the acute phase.[4]

Serological tests: CF, IHA, and ELISA tests are not particularly useful [4, 9, 10] because *T. multiceps* shares antigens with other cestode parasites such as *T. hydatigena* and *Echinococcus granulosus*. Intradermal tests with coenurus fluid have a limited value in confirming a diagnosis for the same reason.[4, 9] Negative serological and intradermal tests however would indicate that an affected animal was unlikely to have gid. Other methods, X-ray, angiography have also been used, but these too are of little value.

The only certain method of confirming a diagnosis is the demonstration of a coenurus, either by surgery or post-mortem examination.

Fig. 21.1 Typical chronic coenurosis of the right cerebral hemisphere.

Epidemiology

Comparatively few studies have been carried out on the epidemiology of coenurosis, especially the role of definitive hosts other than the dog. Observations in Russia [11] demonstrate clearly that although carnivorous species other than the dog can act as definitive hosts, the major source of infection for farm livestock is sheepdogs in daily contact with the flock. In Britain sheepdogs and fox hounds are the main definitive hosts [12, 13, 14] and although some foxes carry *T. multiceps*, they are not considered to be important in the epidemiology of the disease.[14, 15] Dogs and hounds which do not receive regular anthelmintic treatment and those that have access to sheep carcasses are most frequently found to be harbouring *T. multiceps* [12, 13]

Because *Coenurus cerebralis* develops only within the CNS of sheep, cattle and other ruminants and is, therefore, not readily accessible, it has been postulated that other species, particularly wild rodents may act as intermediate hosts. Extensive investigations in a number of countries including Russia and Britain, failed to demonstrate that these species act as intermediate hosts.

Control and treatment

The surgical treatment of coenurosis has been advocated and practised for over 200 years in Great Britain and varying degrees of success have been claimed. However the success of such an approach is clearly dependent on the location of the coenurus, its accessibility, and the degree of brain damage produced by the coenurus and by surgical interference. Thus the removal of a coenurus from the cerebrum is more likely to be successful than its removal from the cerebellum. Some animals make a complete clinical recovery even when there is considerable brain damage.(Fig. 21.2).

Affected sheep must be carefully observed prior to the surgical operation, so that the clinical signs may be interpreted and the probable location of the coenurus assessed. If the animal is blind in one eye, then the coenurus is usually located in the cerebral hemisphere of the opposite side. Animals with cerebellar cysts are ataxic, often failing to maintain balance, become recumbent, 'paddle' and show a pronounced opisthotonus.

The operation is performed under sedation and local anaesthesia. A general anaesthetic is not recommended. The sheep is restrained in lateral recumbency and the operation site, posterior to the base of

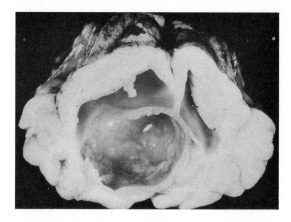

Fig. 21.2 Cross-section of cerebral hemispheres of brain from a ewe killed 4 months after removal of the coenurus. Note the cavity remaining although the ewe made a complete clinical recovery following the operation.

the horn or horn bud area, is prepared in the usual way. If there is evidence of bone rarefaction, then a 3 mm (1/8 in) orthopaedic drill is used to penetrate the bone and a 38 mm (1.5 in) 18 gauge needle is introduced into the hole and directed towards the midline. The appearance of clear fluid under pressure indicates that the coenurus has been penetrated and the needle should be removed. A short length of 1.5 mm (1/16 in) soft polythene tubing is then introduced and attached to a 50 ml hypodermic syringe. The plunger is withdrawn to create a negative pressure and the tube is gradually withdrawn through the aperture in the bone. This procedure usually results in a portion of the cyst being drawn into the tube and enables it to be removed from the brain.

When there is no bone rarefaction, then a more radical approach must be adopted. A triangular flap of skin is folded back over the base of the ear and a circular portion of bone 1–2 cm diameter removed with a trephine. The dura mater is then penetrated with a needle and a circular portion removed with blunt scissors. The same procedures are then adopted as before (Fig. 21.3). After removal of the cyst the skin is sutured back into position.

Post-operative treatment includes sedation for a few days and maintaining the animal in a quiet environment. Treatment with corticosteroids to reduce brain swelling and antibiotics to control secondary infection is also advocated As soon as possible a companion animal is introduced, which often produces a dramatic improvement.

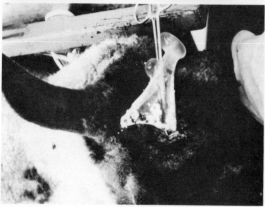

Fig. 21.3 (a) A Kerry Hill yearling that required trephining. The circular piece of bone removed is shown as a flap and the dura mater can be seen bulging upward consequent on the pressure of the cyst.
(b) A large cyst containing some 60 ml of fluid being removed through a small bone hole in the skull of a Suffolk ram lamb.
Both of these sheep recovered completely.

The control measures recommended for coenurosis are similar to those which have been adopted for the control of hydatidosis and are based on the regular anthelmintic treatment of dogs and hounds, and denying them access to sheep and cattle carcasses and heads. All farm dogs are treated at intervals of 6–8 weeks with an effective taenicide and hounds which range over large areas of the countryside are similarly treated. Pet dogs in rural areas are not considered important in the epidemiology of the disease, but these too should be treated if possible.

It is not always possible for farmers to locate and arrange for the rapid disposal of sheep carcasses, especially on mountain and upland grazings. Thus dogs that wander from farmsteads may gain access to

animals which die from gid. It is therefore recommended that all farm dogs should be confined when they are not being used for sheep gathering or when they cannot be supervized.

Action should be taken against stray dogs as these animals do regularly feed on unburied carcasses. No action is necessary against foxes as they are not considered to be important in the epidemiology of the disease.

The artificial immunization of sheep against gid has been considered from time to time. A compound vaccine produced from the strobila of *T. multiceps* and from *Coenurus cerebralis* produced a good immunity in experimental sheep [16], but such a vaccine has little practical application.

REFERENCES

1 Williams B.M. (1979) Coenurosis: a historical review. *State Veterinary Journal*, **32**, 235–9.
2 Blood D.C., Henderson J.A. & Radostits O.M. (1979) *Veterinary Medicine*, 5e, pp.317–18. Bailliere Tindall, London.
3 Becklund W.W. (1970) Current knowledge of the gid bladder worm *Coenurus cerebralis (Taenia multiceps)* in North American sheep. Proceedings of the Helminthological Society of Washington, **37**, 200–03.
4 Williams B.M. (1982) In press
5 Edwards G.T., Hackett F. & Herbert I.V. (1979) *Taenia hydatigena* and *multiceps* infections in Snowdonia II. The role of hunting dogs and foxes as definitive hosts and sheep as intermediate hosts. *British Veterinary Journal*, **135**, 433–9.
6 Esch G.W. & Self J.T. (1965) A critical study of the taxonomy of *Taenia pisiformis, Multiceps multiceps,* and *Hydatigera taeniformis. Journal of Parasitology,* **51**, 932–7.
7 Verster A. (1969) A taxonomic revision of the genus *Taenia Limerus* 1758, S. Str. *Onderstepoort Journal of Veterinary Research,* **36**, 3–58.
8 Abasov M.T. (1965) Survival outside the host of *Multiceps multiceps, Echinococcus granulosus* and *Taenia hydatigena* oncospheres and larvocysts. *Veterinariya,* **42**, No.9, 46–7
9 Dyson D.A. & Linklater K.A. (1979) Problems in the diagnosis of acute coenurosis in sheep. *Veterinary Record,* **104**, 528–9.
10 Hackett F., Willis J.M., Herbert I.V. & Edwards G.T. (1981) Micro ELISA and Indirect Haemagglutination tests in the diagnosis of *Taenia hydatigena* metacestode infections in lambs. *Veterinary Parasitology,* **8**, 137–42.
11 Erchov V.S. (1961) IV. Present conceptions on the epizootiology of echinococcosis and coenurosis. *Bulletin Off. International Epizootology,* **56**, 977–92.

12 Williams B.M. (1976) The epidemiology of adult and larval cestodes in Dyfed 1. The cestodes of farm dogs. *Veterinary Parasitology*, 271-276.

13 Williams B.M. (1976) The epidemiology of adult and larval cestodes in Dyfed. 2. The cestodes of fox hounds. *Veterinary Parasitology*, 277- 280.

14 Edwards G.T., Hackett F. & Herbert I.V. (1979) *Taenia hydatigena* and multiceps infections in Snowdonia. I.Farm dogs as definitive hosts. *British Veterinary Journal*, **135**, 426–32.

15 Williams B.M. (1976) The intestinal parasites of the red fox in south west Wales. *British Veterinary Journal*, **132**, 309–12.

16 Pukhov V.I., Zinichenko I.I. & Chernobaev (1966) Experimental immunisation of lambs against Coenurosis. Contributions to Helminthology published to commemorate the 75th birthday of K.I. Skrjabin. Israel Program for Scientific Translations Jerusalem. pp.567–71. Oldbourne Press.

FOOT ROT AND FOOT CONDITIONS

22

T. BOUNDY

FOOT ROT

Synonyms Contagious foot rot (CFR); hoof rot, pietin (Fr); pedero (Sp)

Foot rot is a highly contagious disease that can affect every sheep in the flock at one time. Some sheep may become carriers for life. It can cause severe loss of production and on occasion death and can also affect cattle, goats and deer. Foot rot is widespread throughout the sheep producing areas of the world. In Australia, sheep of the predominant Merino breed appear to be particularly susceptible to the disease and losses in meat and wool amount to millions of pounds per year.

The lameness associated with foot rot may be so severe that affected animals graze on their knees, lie down for unduly long periods and rapidly lose condition. It is well to remember that lameness in sheep is probably the commonest disease from which they will suffer in their lifetime. Neglected lameness can cause misery and pain and can have far reaching serious consequences even affecting the fertility of both ewe and ram and may result in abortion. Herein lies a welfare aspect that must not be overlooked. In

so many cases early and simple treatment produces rapid recovery and a contented sheep.

Cause

Bacteroides nodosus is the transmitting agent which acts in synergism with the environmental agent, *Fusobacterium necrophorum (Sphaeroides necrophorus)*. Given the correct conditions of wetness and humidity *F. necrophorum* invades the interdigital skin and induces a dermatitis. The affected interdigital skin is then susceptible to the action of *B. nodosus*.

The disease is always introduced to a clean flock by an infected animal. The infection can persist from year to year by the contamination of pastures from carrier animals but outbreaks of disease may not occur each year. An infected sheep may show no signs of lameness for 2–3 years after purchase then, in the correct environmental conditions, lesions recur and the sheep becomes a virulent carrier of the disease.

Sheep are the predominant carriers but *B. nodosus*

from cattle can cause infection in sheep although usually only an interdigital dermatitis results.[1] Goat isolates have been shown to be invasive and damaging to the feet of sheep.

B. nodosus persists in the environment for a maximum of 4 days and there is no useful evidence to indicate that this organism is transmitted by vehicle wheels, boots or surface water. The disease thrives in sheep in areas with a high rainfall when the temperature is above 10°C [2], spreading rapidly among flocks on lush lowland pastures. Flocks grazing mountain and sandy soil areas remain naturally free.

Clinical signs

Depending on the strain of *B. nodosus*, two clinical syndromes have been described.

1. Virulent foot rot, where 10–14 days after contact with *B. nodosus* spread occurs from the initial dermatitis in the cleft between the claws to the hoof. This causes marked separation of the horn beginning at the heel and extending to the sole and toe. The interdigital skin lesion persists and is characterized by inflammation in the form of redness, moistness, erosion and loss of hair (Fig. 22.1). The inflamed tissue beneath the separated horn is covered by a thin layer of white exudate with a characteristic odour. The sole may develop several flaps of separated horn. In more chronic cases the underlying tissue has a moist white-yellow appearance with pockets of white foul-smelling exudate.

2. Benign foot rot, is non-progressive and usually referred to as 'scald'. It is caused by less invasive strains of *B. nodosus* and is confined to the skin between the claws and the skin–horn junction, with occasional simple separation of the soft horn at the heel bulbs and posterior sole regions. There is little or no underrunning or accummulation of necrotic exudate.

Diagnosis

Films of necrotic skin material stained by Gram's method show *B. nodosus* as a large Gram-negative rod with terminal swellings. It may not be the most prevalent organism present and *F. necrophorum* may or may not be present in such smears. A specific fluorescent antibody staining method for *B. nodosus* has been developed.[6]

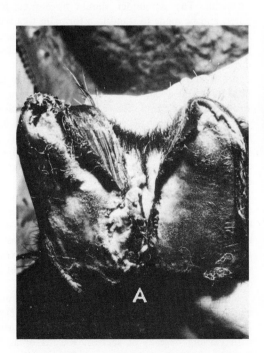
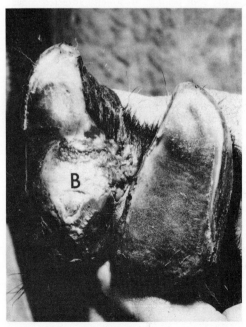

Fig. 22.1 Virulent foot rot showing (A) separation spreading from the interdigital space, the extent being revealed (B) after removal of the underrun sole.

B. nodosus is readily cultivated from lesions especially if the isolation medium contains 4 per cent agar. Plates inoculated with lesion material should be examined 3–4 days after anaerobic incubation.

In any complete eradication programme it is necessary to differentiate between benign and virulent strains of *B. nodosus*.

A proteolytic index test developed by Egerton and Parsonson [3] appears to correlate with the virulence of the organism. Depiazzi and Richards [4] have also developed a test called the degrading proteinase test which distinguished between benign and virulent strains of *B. nodosus*. In addition a highly specific diagnostic test, based on the production of elastase has been described by Stewart [5] who divided *B. nodosus* field isolates into two distinct populations. His results showed that there was a remarkable correlation between the degrading proteinase test, elastase production and the virulence of wild isolates. Ovine virulent isolates invariably produced elastase whereas ovine benign isolates were elastase negative.

Treatment

Topical or parenteral treatment can be highly successful in the majority of cases. It is essential to prepare the foot prior to treatment. Infected areas should be opened up with a sharp knife or secateurs. It is necessary to expose to the air as far as possible any deep pockets of infection. Care should be taken to avoid unnecessary cutting into the sensitive tissue of the hoof. This type of mutilation can only result in the production of severe painful lesions, the formation of granulation tissue, continued lameness and increased pain on exposure to topical irritants such as formaldehyde with a resulting slowing of recovery.

Topical application can be applied by foot bathing, e.g. 5 per cent formalin or by treating lesions with an antibiotic spray or paint, e.g. tetracycline. The ideal topical application should penetrate deeply into affected tissues and agents capable of penetrating both skin and horn are being developed. Such preparations could immensely improve topical methods of treatment.[7]

Parenteral treatment with antibiotics can be highly successful and the injection of a mixed penicillin and streptomycin mixture is indicated. If this method is used, results are improved by some attention to the feet as well.

Both methods give better results if the sheep are able to stand on dry ground for a few hours after treatment and not returned immediately to wet pastures.

Control

Where small flocks are heavily infected with chronic foot rot it may be advisable to slaughter them and to rest the ground before new, clean sheep are introduced. In less severely affected flocks it is necessary to identify and eliminate all cases of infection with *B. nodosus* and take full precautions to avoid reintroduction of this organism.

The ability to differentiate between virulent and benign foot rot must be considered in any severely-infected flock. Benign foot rot or 'scald' so often responds to simple measures, e.g. change to dry pasture plus simple topical treatment.

A typical eradication programme would entail drafting 'clean' and 'dirty' (i.e. infected) sheep into two groups. Each clean sheep should have its feet pared and be made to stand in a foot bath of 5 per cent formalin for 5 minutes after which the sheep should be moved on to clean pasture. This group should then be walked through 5 per cent formalin once per week for 3 weeks.[8] The dirty sheep should be examined carefully and those worst affected culled forthwith. The remainder should have their feet pared and be made to stand in a formalin foot bath for 5 minutes then returned to the contaminated pasture. Subsequently this group should be walked through a formalin foot bath each 5 days for approximately 5 treatments. Two weeks after the final treatment a careful inspection should be made and if they are clear of disease they can join the clean group. If not, repeat the treatment 3 times at 5 day intervals. After a further 3 weeks the extent of infection should be carefully reappraised and the owner advised that sheep still affected should be culled forthwith.

The need to avoid contact with infection once a flock is clean cannot be over-emphasized. It is also essential during any foot-paring operation that parings and filth associated with the work be swept up and burned as *F. necrophorum* may persist for several weeks in such debris. All equipment used should be disinfected and oiled before being put away.

B. nodosus vaccine can prevent the disease and also help in the treatment of affected flocks. In New Zealand and Europe such vaccines have been used with varying degrees of success. Results obtained from vaccination can be improved by careful attention to foot care. On no account must the use of vaccines reduce other efforts in the normally accepted methods of foot care. Vaccination with a vaccine containing only some serological types of *B. nodosus* could enhance the action of other serotypes.

Vaccines do have a place in controlling and treating foot rot and it is important that sheep are vaccinated to make them more resistant at times when transmission is likely to occur. For this it is essential to have a knowledge of the epidemiology of the disease in the area concerned.

WHITE LINE DISEASE

White line disease is probably the commonest cause of simple lameness in sheep today and is most often seen in white feet with a shelly, brittle type of horn. Separation of the insensitive horn from the sensitive laminae occurs at the white line (Fig. 22.2). The pocket that is produced becomes packed with soil and dirt and presses on to the sensitive laminae resulting in pain and the development of infection (Fig. 22.3). Unless this area is opened to eliminate the pocket, infection bursts out at the coronary band with subsequent abscess formation which is a common sequel. If such lesions are present concurrently with virulent foot rot, they may act as a pocket in which *B. nodosus* is protected from topical treatment.

Regular attention to these cases by opening out the affected area, which is usually the outside wall near the toe, and dressing with a topical application invariably results in success.

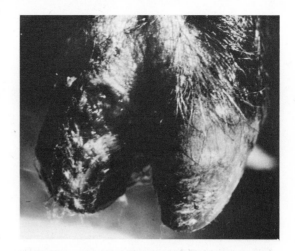

Fig. 22.3 Foot abscess that has burst at the coronet.

FOREIGN BODIES

Injuries to the sole of the foot from foreign bodies, e.g. thorn, glass, stones occur sporadically. These are most common in sheep with soft feet such as lambs, and may result in bruising, which is usually transient, abscess of the sole or, more seriously, in an abscess within the hoof. The latter causes marked lameness which abates only after the pus has drained through a sinus discharging at the hoof head (Fig. 22.4) or

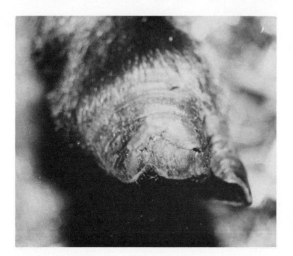

Fig. 22.2 White line disease with separation of the outer wall [of the hoof (claw: clay)] from the sensitive laminae.

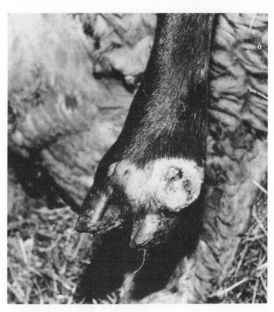

Fig. 22.4. Foot abscess breaking out at the coronet with discharging sinuses involving the joint.

through a crack in the adjacent solar or laminar horn. The joints and tendon sheaths may become involved in the infection resulting in a severe and painful foot condition.

The foot should be cleaned in warm antiseptic solution and subsequent treatment should be directed at thinning the sole to identify in the clean horn the sinus track through which the foreign body has penetrated. The track when extended leads to a cavity containing pus all of which must be exposed and treated with antibiotics. The exposed surface should then be protected by a moist dressing such as paraffin gauze and bandages for 7–10 days. A covering of aluminium foil helps to keep the foot dry. Finally, waterproof adhesive tape should be wrapped around the foot.

When soft tissues are infected the foot should be desensitized by ring-block anaesthesia achieved by the subcutaneous infiltration of local anaesthetic just above the fetlock.

Complete drainage is helped by placing a tampon in the opening for 2–3 days, after which the foot can be placed in a plaster cast up to the fetlock for 2–3 weeks. Parenteral antibiotics should be given during this period.

Severe infections which invade the joints or tendon sheaths may necessitate amputation of a claw at the level of either the first of second phalanx.

GRANULOMAS

Granulomatous tissue may develop in untreated cases of foot rot or where excessive paring of a foot has taken place. After applying a tourniquet above the knee and infiltrating the foot with local anaesthetic the horn adjacent to the granuloma should be carefully trimmed down to the base of the granuloma and the latter removed completely. Removal may be undertaken using a line-firing instrument, a scalpel or a hoof knife. Subsequent treatment is as for a foreign body in the foot, except that the dressings should be changed after 3 days and then weekly for 3 weeks. Penicillin and phenylbutazone should be given for 3 days.

FIBROMAS

'Fibromas' in the interdigital cleft are common in some breeds and there may be a hereditary factor associated with these growths. Lambs under one year old are often affected but most problems are created when growths develop in the hind feet of rams. Rams should be inspected before purchase for such growths and each spring the breeding rams should be examined similarly.

These may be single and large extending throughout the entire interdigital space or smaller bilateral growths developing from either side of the interdigital cleft (Fig. 22.5). Bruising, ulceration and infection

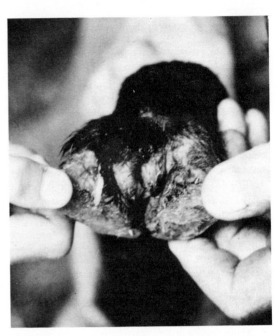

Fig. 22.5 Bilateral interdigital growths showing sites of origin in the interdigital space at the medial junction of the skin and horn on either side.

can occur which results in a painful lameness. Topical antibiotic therapy sometimes cures the infection and removes the associated oedema so that the lameness improves. However, large growths should be removed. Before attempting to remove a growth, a tourniquet should be applied to the leg and 2–3 ml of local anaesthetic injected into the interdigital space below the grease gland to the depth of 3/4 in (18 gauge) needle and also around the heel. Post-surgery dressings and parenteral treatment are applied as described for foreign bodies of the foot.

ORF AND STRAWBERRY FOOT ROT

Orf lesions can occur around the coronary band and may cause severe lameness. Individual sheep should have their feet dressed with an antibiotic spray or ointment. If the flock is extensively infected with orf then vaccination should be considered.

Lesions of strawberry foot rot respond to antibiotic treatment topically or systemically. For further information on orf see Chapter 38.

ACUTE LAMINITIS

Sheep affected with metritis or being fed a high content of grain in their diet or those grazing lush pastures can develop acute laminitis. This usually affects all four feet so that affected sheep are reluctant to move and remain lying down for long periods. The cause of the condition should be eliminated. If the laminitis is associated with metritis the ewe should be treated with antibiotics and if the condition is due to excessive feeding of grain or lush grass then a poorer quality ration should be fed or the sheep allowed access to other poorer pastures. It is essential to give maximum dosage of phenylbutazone to reduce acute pain and inflammation.

CONGENITAL AXIAL ROTATION

Congenital axial rotation or 'corkscrew claw' is not an uncommon condition which is usually found affecting an outer claw. If persistant lameness is to be avoided such claws require frequent paring to keep the whole of the sole of the foot in contact with the ground. This type of claw if unattended rapidly becomes under-run and subsequently infected.

ERYSIPELOTHRIX RHUSIOPATHIAE INFECTION

Lameness caused by this organism is often associated with sheep that have been dipped, for example in the control of sheep scab and has been reported by a number of workers.[9] Outbreaks appear to occur sporadically and to be associated with misuse of dip. Further information on this condition is contained in Chapter 23.

REFERENCES

1 Wilkinson F.C., Egerton J.R. & Dickson J. (1970) Transmission of *Fusiformis nodosus* infection from cattle to sheep. *Australian Veterinary Journal,* 46, 382–4.

2 Graham N.P.H. & Egerton J.R. (1968) Pathogenesis of ovine foot-rot: the role of some environmental factors. *Australian Veterinary Journal,* 44 , 235–40.

3 Egerton J.R. & Parsonson I.M. (1969) Benign foot-rot - a specific interdigital dermatitis of sheep associated with infection by less proteolytic strains of *Fusiformis nodosus. Australian Veterinary Journal,* 45, 345–9.

4 Depiazzi L.J. & Richards R.B. (1979) A degrading proteinase test to distinguish benign and virulent ovine isolates of *Bacteroides nodosus. Australian Veterinary Journal,* 55, 25–8.

5 Stewart D.J. (1979) The role of elastase in the differentiation of *Bacteroides nodosus* infections in sheep and cattle. *Research in Veterinary Science,* 27, 99–105.

6 Roberts D.S. & Walker P.D. (1973) Fluorescin-labelled antibody for the diagnosis of foot-rot. *Veterinary Record,* 92, 70–71.

7 Malecki J.C. & McCausland I.P. (1981) Personal Communication.

8 Scanlan W. (1981) Personal Communication.

9 Hungerford T.G. (1975) *Diseases of Livestock.* 8e, McGraw-Hill, Sydney.

ARTHRITIS

23

M.H. LAMONT

NEONATAL POLYARTHRITIS

Synonyms Joint-ill, navel-ill, joint-evil, big joint, schooley

Neonatal polyarthritis is normally a suppurative arthritis occurring in lambs up to one month of age.

Cause

The commonest organisms causing this condition are streptococci of Lancefield's Group C [1] but a number of other organisms including *E. coli*, Group D streptococci and *Corynebacterium pyogenes* may be isolated from affected joints. *Erysipelothrix rhusiopathiae* and *Fusobacterium necrophorum* have also been isolated from the arthritic joints of young lambs but these organisms do not cause suppuration.

Clinical signs

Severe lameness and depression are the most obvious clinical signs. Closer inspection reveals swellings of one or more joints. The larger joints such as the carpus, stifle and hock, are most often affected but any joint may become diseased. The swellings over the joints are hot and painful and, if the animal survives, may rupture spontaneously. Affected lambs are normally pyrectic and anorectic and may be underweight in more chronic cases.

Pathology

Incision of affected joints releases large quantities of yellow or grey fluid pus. The infection may be localized in the joint alone or may involve adjacent tendon sheaths. This is particularly likely to occur in infection of the carpal joints when the common extensor tendons of the digits become affected.

The surface of the articular cartilage is pitted and ulcerated but the soft structures in the joint show fewer signs of reaction to the naked eye.

Suppuration is not a feature of *Erysipelothrix* infection. Joints infected with this organism contain sheets or floccules of fibrin in varying volumes of cloudy synovial fluid. *Fusobacterium* infection, which is relatively rare, causes necrosis rather than suppuration and might be found in association with necrotic lesions elsewhere in the body, particularly in the liver.

Diagnosis

Joint-ill is readily diagnosed on the basis of a history of lameness of sudden onset, in young lambs which have swollen joints and pyrexia. It is not necessary to resort to laboratory methods for confirmation unless it is desired to ascertain the causal organism. Joint fluid should then be aspirated from diseased joints, taking care to avoid external contamination, and cultured using routine bacteriological techniques.

Epidemiology and transmission

The pathogenesis of neonatal polyarthritis is closely related to that of other neonatal infections. Most of the organisms which cause neonatal disease are commonly found in the environment of the neonate. The development of disease in the young lamb is due less to the virulence of the causal organism than to the susceptibility of the neonate.

Neonatal lambs are far more susceptible to disease than older lambs or adults, due to their dependence on colostrum as a source of circulating antibody. Anything which interferes with the intake of an adequate amount of colostrum leaves the young lamb less able to combat infection. Hence chilling, mismothering and competition from siblings all increase the likelihood of neonatal infection.

Insanitary conditions in lambing sheds or paddocks may present the young neonate with such an overwhelming challenge that it will be infected before it has acquired colostral antibody. Under extreme circumstances the infective dose of pathogens may be so heavy that, even though the lamb has obtained colostrum, its passive immunity may be insufficient to withstand infection.

Control

The general rules for the control of neonatal infections, including polyarthritis, are simply stated:

1. Ensure that neonates are born into a clean environment to avoid infection before colostrum has been taken.
2. Colostrum intake must be sufficient for each lamb. This may entail bottle feeding lambs removed from multiple litters.
3. Increase the specific immunity of the lambs by vaccination of the ewes where appropriate. Of the neonatal infections only *E. rhuisopathiae* vaccine is available at present.

TICK PYAEMIA

Synonym Cripples

Tick pyaemia is a disease of young lambs in flocks grazing tick-infested pastures and is considered to be a serious cause of loss of production in many hill flocks. The disease is characterized by the formation of abscesses in the joints, spinal column and internal organs and is generally accepted as being closely associated with tick-borne fever.[2,3,4]

Cause

Tick-borne fever is transmitted to young lambs by the tick *Ixodes ricinus* and is characterized by pyrexia and leucopoenia. The leucopenia is derived initially from a lymphocytopenia which is followed by a neutropenia and a thrombocytopenia. The lowered resistance, which is a sequel to tick-borne fever in lambs, appears to allow *Staphylococcus aureus*, normally present on the skin and mucous membranes, to invade and produce a bacteraemia with a consequential tendency for abscesses to develop in the joints and internal organs.

Clinical signs

The clinical signs of tick pyaemia cannot be distinguished from those present in cases of joint-ill or navel-ill. Lambs of 2–6 weeks of age become dull and lame. The lameness, which is caused by abscesses in one or more joints, particularly the carpal or tarsal joints, is obvious. Soft fluctuating swellings are present in affected joints which later become indu-

rated and ankylosed. When abscesses of the internal organs occur, unaccompanied by joint lesions, then unthriftiness, dullness and a 'tucked-up' appearance may be all that can be appreciated. Abscess formation in the spine causes clinical signs which vary from mild ataxia to complete paraplegia, depending on the extent the abscess impinges on the spinal cord. Frequently the ataxia progresses to unilateral or bilateral paralysis of the hind limbs. Occasionally sudden deaths result from multiple internal abscesses without producing any obvious premonitory signs.

Pathology

The lesions in the joints and tendon sheaths are those of a suppurative polyarthritis. Pus is present in the joint capsule, sometimes accompanied by erosion of the articular surfaces, which proceeds to necrosis and osteomyelitis. Where spinal abscesses occur, osteomyelitis of the body of the associated vertebra with compression of the spinal cord may be seen. In the internal organs, particularly the liver, abscesses of varying diameter can be found.

Diagnosis

The presence of suppurative polyarthritis, seen as swollen joints, ataxia or paraplegia, in young lambs grazing tick-infested pastures in the spring, generally indicates the presence of tick pyaemia. Cultures demonstrate the presence of *Staphylococcus aureus*. It is possible that joint-ill or navel-ill can be present without tick pyaemia and the finding of other bacteria such as *Corynebacterium pyogenes* or *Fusobacterium necrophorum* indicate this. Ataxia or paraplegia caused by spinal abscesses can be confused with delayed swayback, and histopathological examination of the brain and spinal cord may be necessary to distinguish the cause.

Epidemiology

Because of the relationship of tick-borne fever and tick pyaemia with areas populated by *Ixodes ricinus* the prevalence and incidence of these diseases are related to the biology of this tick. In years climatically favouring a high tick rise, when, over a prolonged period in spring, young susceptible lambs are grazing indigenous herbage, the incidence can be quite high. Nearly all known tick areas support the presence of tick-borne fever, in contrast to louping-ill and babesiosis, which are absent from some tick areas. Management practices such as extensive grazing by cattle, or pasture improvement schemes, which destroy the basal herbage mass, necessary as an environment for tick populations, may substantially reduce these populations and the overall incidence of tick-borne diseases.

Control, prevention and treatment

While dipping ewes *pre partum* in known tick areas, augmented by dipping lambs prior to their removal from in-bye lambing fields to hill pasture, does reduce the level of tick infestation, there is little evidence to suggest that this will control tick pyaemia to a marked degree. The use of acaricidal creams smeared on lambs was advocated in the past but does not appear to be practised widely in the present day. Exposure to tick-borne fever by injecting lambs with infected blood prior to their entry to tick areas has been tried, with varying results. It is difficult to assess the efficacy of this approach because of year to year variations in the rise of ticks in the spring. Because strains of tick-borne fever exist, with no cross immunity it is essential to use the strain specific to the farm in question. Strains also vary in virulence and this may be another complicating factor.

Prophylaxis by the use of the ultra long-acting penicillin [5] benzathine penicillin, at the time of risk did appear to be useful, although this preparation is not now available. When the indications of a high risk year are present, the use of long-acting procaine penicillin or oxytetracycline prophylactically should be considered or administered to individual clinically- affected lambs. However, the use of the latter preparation may inhibit the development of immunity to tick-borne fever. Joint lesions can be drained and dressed with a moderate degree of success, although a frequent sequel is ankylosis of the joint.

In the long term, the reduction of tick populations by management practices such as the burning of heather, pasture improvement by liming and reseeding to remove the indigenous herbage mass, or the use of cattle to destroy bracken and heather by trampling, are worthy of consideration. Where practicable, intensive and frequent dipping of sheep flocks may be of benefit but it would appear that the traditional single tick-dip has not done much to influence the prevalence of tick-borne disease in years of high tick rise.

ERYSIPELOTHRIX POLYARTHRITIS

Synonyms Stiff lamb disease, erysipelas

Erysipelothrix usually causes a non-suppurative polyarthritis affecting lambs between 2–6 months of age. Less common manifestations of *Erysipelothrix* infection in sheep are neonatal septicaemia and polyarthritis, post-dipping lameness and endocarditis. Endocarditis is a rare condition in sheep which is seldom diagnosed. A review of *Erysipelothrix* infections in sheep is available.[6]

Cause

The causal agent is *Erysipelothrix rhusiopathiae* which has also been called *E. insidiosa*. At the time of writing over 20 serotypes have been identified. These serotypes are not host specific and serotypes 1 and 2 have been identified as the most frequent cause of disease in both pigs and sheep.

Clinical signs

In any outbreak a variable proportion of the lambs may be affected. In a severe outbreak the morbidity rate can reach 40 per cent but it is usually much lower. The disease is characterized by lameness or stiffness and pyrexia ($> 40.5°C: > 105°F$) in the acute disease but depression is not evident unless septicaemia is present. There is little obvious swelling of affected joints and careful palpation is necessary to identify those affected. Lambs in the acute stage of the disease may therefore not be noticed by the shepherd or flock-master. Spontaneous recovery can take place but the untreated acute disease frequently proceeds to the chronic stage. Lameness is then more pronounced and a variable degree of muscle wasting will have occurred. Chronically diseased joints are more readily palpated because of the loss of bodily condition and periarticular swelling.

Diagnosis

In the UK a tentative diagnosis may be reached on the basis of the clinical signs and history of the outbreak. In the acute disease the organism can be isolated from synovial fluid aspirated from diseased joints. If the disease is of long standing the possibility of isolating *E. rhusiopathiae* is much reduced. Serological tests, particularly the agglutination test, are useful in confirming a diagnosis of chronic *Erysipelothrix* polyarthritis. Such tests can also be applied to synovial fluid samples.

Pathology

In the acute disease lesions are restricted to the joints and drainage lymph nodes which become enlarged and slightly oedematous. The joint capsule may be oedematous and distended by synovial fluid which is turbid or flocculent, due to an increased cellular and fibrin content, and has a green or brown colour. Larger sheets of fibrin cover the articular surfaces which otherwise appear normal. Histological examination of the synovial membrane confirms the marked exudation of fibrin and neutrophil polymorphs in the joint space and tissues. These inflammatory lesions may be diffuse but they are often localized and ulcerated.

When the disease is chronic, the prolonged lameness results in wasting which may even tend to emaciation. Diseased joints are rarely distended by fluid but they are enlarged because of periarticular fibrosis and oedema. The synovial fluid is turbid and greenish but is seldom present in excess, which partly explains the difficulty of aspirating a sample from the live animal. The synovial membrane undergoes proliferative changes and the villi develop into branching fronds which consist of highly vascularized connective tissue, throughout which are scattered numerous granulomata. These granulomata consist of well-defined aggregates of lymphocytes and plasma cells which resemble lymphoid follicles but have no germinal centres. In the sheep, pannus formation is restricted to the edges of the articular cartilage and is not as extensive as similar lesions in the pig. Articular cartilage loses ground substance and takes on a fibrillated appearance and in long-standing cases may be replaced by granulation tissue.

Epidemiology and transmission

There are two important epidemiological features associated with this infection. The first is the ability of the organism to survive for long periods away from animal hosts even though it does not form spores. The second is its wide host range.

Earlier this century the belief that *E. rhusiopathiae* was a soil saprophyte became accepted as an explanation of its epidemiology. The organism is capable of surviving for long periods in the external environment, especially at low temperatures, but laboratory studies [7] have failed to prove that the organism is capable of multiplying in soil. Recent observations [8] suggest that *E. rhusiopathiae* is capable of existing as a saprophyte only in the presence of organic matter such as faeces.

Direct transmission of infection from host to host is not likely to be a frequent means of spread. Diseased animals or healthy carriers excrete the organism and contaminate the immediate environment. Given favourable conditions the bacteria multiply and may infect other susceptible hosts. Sheep are the obvious source of infection for sheep, but field observations suggest that other hosts such as pigs, turkeys or other birds may occasionally be the source of infection.

Treatment

Best results are obtained if the disease is treated in its early stages by parenteral administration of penicillin. It is important that treated animals are given a full course of treatment since inadequate treatment may allow the disease to recur some weeks later.

In lambs or sheep which have been affected for a long time, treatment is seldom successful. In a commercial flock, such animals should be slaughtered, since they are unlikely to recover. In the case of an especially valuable sheep, immunosuppressive therapy or other anti-arthritic treatment may be justified. This author has no experience of such techniques nor are there any reports in the literature.

Prevention and control

The most effective means of preventing disease due to *E. rhusiopathiae* is to ensure that the flock is kept in clean, dry conditions. This is particularly important if the flock is lambed indoors or housed for long periods. During wet weather, feeding troughs should be moved at frequent intervals, to avoid the area becoming heavily soiled with faeces. The maintenance of good hygiene is especially important during docking or open castrating as contamination of open wounds readily leads to the establishment of *E. rhusiopathiae* within the body.

Outbreaks of disease may be alleviated to some extent by sharply reducing stocking density or by moving the flock to a fresh pasture.

Vaccination has been shown to be beneficial in

preventing outbreaks of polyarthritis. A licensed bacterin is available in the UK for use in sheep. Its use is not justified in flocks which do not have a known disease problem but it can be useful in controlling or preventing disease in problem flocks. Two, or preferably three doses of vaccine should be given to the pregnant ewes, with the last dose being administered about 3 weeks before the expected date of lambing. This should protect the lambs until about 8 weeks of age. Lambs can be actively immunized with two doses of vaccine, 3 weeks apart, from 6–8 weeks of age.

The protection provided by vaccination is wide and protects animals against disease caused by serotypes not included in the vaccine preparation. There have been reports that serotype 10 may cause disease in pigs vaccinated with heterologous serotypes [9] but this has not been reported in sheep.

POST-DIPPING LAMENESS

This condition was first described in Australia in 1948 [10] but has since been seen in many parts of the world including the UK [11] and the Americas.

Cause

The causal organism is *Erysipelothrix rhusiopathiae.*

Clinical signs

The condition normally presents as an explosive outbreak of lameness in a flock of sheep dipped within the preceding 2–7 days. Affected animals will be pyrectic and severely lame. A variable proportion up to 70–80 per cent of the flock may be affected. In cases not complicated by septicaemia or polyarthritis the lesions are restricted to the coronary band, the hoof and the skin and subcutaneous tissues up to the fetlock. The diseased foot feels hot to touch and the skin is reddened and swollen. Septicaemia or polyarthritis may occur, causing the clinical signs already described.

Pathology

The most common lesion is a non-suppurative cellulitis of the subcutaneous tissue of the coronary band extending into the sensitive laminae. The pathology of polyarthritis has already been described.

Diagnosis

The history and clinical signs are sufficient for the clinician to reach a tentative diagnosis. In the absence of polyarthritis, which is an occasional sequel, serology is of limited value. In early cases of the disease the organism can be isolated from the lesion [12] by simply incising the skin of the coronary band and culturing the blood and exudate collected.

Epidemiology

E. rhusiopathiae is capable of rapid multiplication in heavily soiled sheep dip solution. Thus the re-use of a sheep dip, a day or so after it was first used, may expose the flock to a severe challenge with the organism. The infection is presumed to enter the small abrasions around the coronary band although such abrasions are seldom visible to the naked eye. In the past decade, the disease has become more common because of the use of highly concentrated dip preparations which no longer contain antibacterial substances, such as phenol and copper sulphate.

Control, prevention and treatment

Post-dipping lameness is unlikely to occur in a flock dipped in a clean bath containing freshly-prepared dip solution. If it is necessary to re-use a dip on a subsequent day then a suitable anti-bacterial substance should be added to the bath. It is preferable to add this to the bath on the first day it is used, while it is still clean, to prevent multiplication of *E. rhusiopathiae*. It is not possible to recommend here a particular anti-bacterial because the substance chosen must be compatible with the dip formulation. Dip manufacturers recommend substances which are known to be compatible and their advice should be followed.

Post-dipping lameness may recur on a farm and it may be worth routinely vaccinating the flock before dipping each year. Two annual vaccine doses should suffice.

Penicillin is the antibiotic of choice for treatment, but care must be exercised to ensure that each animal receives a full course of antibiotics; a single dose of a long-acting preparation may not be sufficient. Polyarthritis may occur some weeks later in a flock which has been inadequately treated.

CHLAMYDIAL POLYARTHRITIS

Synonyms Ovine chlamydial polyarthritis, stiff-lamb disease

This disease has been reported from the UK but most reports describe outbreaks in feedlot lambs in the USA [13] and a similar disease is recognized in Australia and Scandinavia. An extensive review of the subject of chlamydial polyarthritis is availible. [14]

Cause

The disease is caused by a strain or strains of the species *Chlamydia psittaci*, which is well-known in the UK. Members of the same species also cause enzootic abortion of ewes (EAE), psittacosis and ornithosis in birds and a pneumonic disease in man. Although all of these syndromes are caused by *Chlamydia psittaci* there are a considerable number of strains within the species. There is no evidence that the agent causing polyarthritis causes any of these other conditions. The subject of chlamydial infections in mammals has been concisely reviewed by Foggie. [15]

Clinical signs

These are similar to *Erysipelothrix* polyarthritis already described. There may also be the additional sign of conjunctivitis. The disease is also slightly different in that spontaneous recovery is more frequent than in *Erysipelothrix* polyarthritis.

Pathology

The early lesions and their subsequent development are superficially similar to those of *Erysipelothrix* polyarthritis.

Diagnosis

Joint fluid smears stained by the modified Ziehl-Neelsen method may allow the intracellular organisms to be seen. Isolation requires culture in either chick embryos or cell cultures.

Serology may be a useful aid to diagnosis. The test commonly in use is the CFT. Because this test relies on the reaction of antibody with an antigen, common to the species *Chlamydia psittaci*, suspect sera may be tested by the EAE CFT which is in common use.

Obviously, interpretation of CF titres are difficult in sheep flocks which have concurrent infection with the enzootic abortion chlamydia. However, if the arthritic lambs are not old enough to have been vaccinated or to have picked up active EAE infection, then the existence of rising CF titres in cases of polyarthritis provides evidence of infection with the polyarthritis agent.

Epidemiology and transmission

This is not fully understood. Following ingestion or inhalation of infective particles which have been shed in faeces, urine or ocular discharges, infection is carried to the joints. Biting arthropods may transmit the disease but their role is not clear.

Treatment, control and prevention

Therapy must be given in the early stages to have most effect. A wide range of injectable antibiotics has been used, including tylosin and penicillin. Tylosin appears to be the most effective.

There is no vaccine available which can be used against this disease. Control depends on isolation and prompt treatment of diseased lambs, reducing overcrowding and reducing stocking density.

OSTEOARTHRITIS

This is a sporadic condition which occurs in older sheep. Osteoarthritis or osteoarthrosis is insidious in onset and is most often seen as a complication of a primary condition such as infection or structural abnormality. Structural abnormalities may be congenital or acquired following injury or skeletal fracture.

Clinical signs are variable and range from stiffness to frank lameness. These are common symptoms in older sheep and are frequently ascribed to foot rot or similar hoof defects. As a result, osteoarthritis is rarely diagnosed in the live animal and may only be recognized in the abattoir during meat inspection.

A limited abattoir survey [8] has revealed a previously unreported syndrome of bilateral osteoarthritis in the elbow joints of older sheep. The condition is remarkable because the extensive degeneration of the articular surfaces and resulting bony changes gave rise to few recognizable clinical signs. The most constant finding was degeneration and calcification of the lateral collateral ligaments of the joint with associated changes on the articular surfaces. The aetiology of the disease has not been determined but there is no evidence that it is a sequel to an infectious arthritis.

The diagnosis of osteoarthritis is based mainly on the history and clinical findings. Radiographic examination can confirm a diagnosis but this diagnostic approach would seldom be justified in a commercial flock. Similarly, treatment of osteoarthritis in a commercial ewe could not be recommended, although the symptoms of arthritis would be alleviated by treatment with anti-inflammatory drugs.

UNUSUAL INFECTIOUS ARTHRITIDES

None of these causes of arthritis has been reported in the UK.

Actinobacillus seminis: which also causes posthitis in rams and mastitis in ewes has been reported as causing septicaemia and polyarthritis in 6 week old lambs.[16]

Corynebacterium ovis has been isolated from lambs with a non-suppurative polyarthritis which clinically resembled disease caused by *E. rhusiopathiae* or *Chlamydia psittaci.* (Although polyarthritis due to this cause has not been reported from the UK the organism is occasionally isolated from sheep).[17]

Haemophilus agni causes septicaemia in older lambs (6–7 months) but survivors may have polyarthritis.[18]

Mycoplasma capricolum was reported from Zimbabwe as the cause of chronic arthritis in lambs.[19]

Maedi-visna The virus responsible for causing maedi-visna has also been reported as a cause of chronic arthritis in adult ewes suffering from progressive pneumonia caused by this virus.[20]

ACKNOWLEDGMENTS

The author gratefully acknowledges the help of Mr A. Whitelaw, Hill Farming Research Organisation, Edinburgh in compiling the information on tick pyaemia.

REFERENCES

1 Cornell R.L. & Glover R.E. (1925) Joint ill in lambs. *Veterinary Record,* 5, 833–9.
2 Foggie A. (1962) Studies on tick pyaemia and tick-borne fever. Symposia of the Zoological Society of London, 6, 51–60.
3 Foster W.N.M. & Cameron A.E. (1968) Aetiology of enzootic staphylococcal infection (tick pyaemia) in lambs. Field investigations into the relationship between tick-borne fever and tick pyaemia. *Journal of Comparative Pathology,* 78, 243–50.
4 Watt J.A. (1971) *The Shepherd's Guide,* 1971 p.27. HMSO.
5 Watt J.A. & Foster W.N.M. (1968) Benzathine penicillin as a prophylactic in tick pyaemia. *Veterinary Record,* 83, 507–8.
6 Lamont M.H. (1979) *Erysipelothrix rhusiopathiae:* epidemiology and infection in sheep. *Veterinary Bulletin,* 49, 479–95.
7 Wood R.L. (1973) Survival of *Erysipelothrix rhusiopathiae* in soil under various environmental conditions. *Cornell Veterinarian,* 63, 390–410.
8 Lamont M.H. (1978) An introductory epidemiological and immunological study of arthritis in sheep of slaughter age. PhD. thesis, University of Edinburgh.
9 Wood R.L. (1979) Specificity in response of vaccinated swine and mice to challenge exposure with strains of *Erysipelothrix rhusiopathiae* of various serotypes. *American Journal of Veterinary Research,* 40, 795–801.
10 Whitten L.K., Harbour H.E. & Allan W.S. (1948) Cutaneous erysipelothrix infection in sheep. An etiological factor in post-dipping lameness. *Australian Veterinary Journal,* 24, 157–63.
11 Harbour H.E. & Kershaw G.F. (1949) *Erysipelothrix rhusiopathiae* infection in sheep. Two outbreaks of post-dipping lameness. *Veterinary Record,* 61, 37–8.
12 Vaz A.K. (1981) The minimum number of *Erysipelothrix rhusiopathiae* necessary to cause post-dipping lameness in sheep and some considerations about its diagnosis. MSc. Thesis, University of London.
13 Mendlowski B. & Segre D. (1960) Polyarthritis in sheep. 1. Description of the disease and experimental transmission. *American Journal of Veterinary Research,* 21, 68–73.
14 Cutlip R.C., Smith P.C. & Page L.A. (1972) Chlamydial polyarthritis of lambs: a review. *Journal of the American Veterinary Medical Association,* 161, 1213–16.

15 Foggie A. (1977) Chlamydial infections in mammals. *Veterinary Record*, **100**, 315–17.

16 Watt D.A., Bamford V. & Nairn M.E. (1970) *Actinobacillus seminis* as a cause of polyarthritis and posthitis in sheep. *Australian Veterinary Journal*, **46**, 515.

17 Marsh H. (1947) Corynebacterium ovis associated with an arthritis in lambs. *American Journal of Veterinary Research*, **8** 294–8.

18 Kennedy P.C., Frazier L.M., Theilen G.H. & Biberstein E.L. (1958) A septicemic disease of lambs caused by Hemophilus agni (new species). *American Journal of Veterinary Research*, **19**, 645–54.

19 Swanepoel R., Efstratiou S. & Blackburn N.K. (1977) Mycoplasma capricolum associated with arthritis in sheep. *Veterinary Record*, **101**, 446–7.

20 Oliver R.E., Gorham J.R., Parish S.F., Hadlow W.J. & Narayan O. (1981) Ovine progressive pneumonia: pathologic and virologic studies on the naturally occurring disease. *American Journal of Veterinary Research*, **42** 1554–9.

24 OSTEODYSTROPHIC DISEASES

J.A. SPENCE

The sheep skeleton is prone to most of the bone diseases seen in other domestic animals (fractures, local abscesses, osteomyelitis and tumours for example) but most are seen only sporadically or rarely. However, the group of conditions called the osteodystrophies occur more regularly and, when they do, may become flock problems of economic significance. The osteodystrophies are defined as diseases of bone where abnormalities of development or, in the adult, defects in skeletal metabolism predominate.

Several specific osteodystrophic syndromes have been described. 'Cappi', or 'double scalp', occurs in the autumn and winter in some parts of Britain. It used to be most prevalent on the poor hill pastures of the north of England and the south and north of Scotland [1] but is now only reported from the far north. 'Generalized rickets' [2] and 'Bent Leg' [3] are generic terms applied world-wide to a number of osteodystrophic conditions. However, the presenting signs of these and other similar conditions, bone weakness, lameness, skeletal distortion and secondary fractures are non-specific and their aetiology often remains uncertain or oversimplified. Hence, an appreciation of how the skeleton can respond to metabolic/physical/chemical insults and the circumstances under which such responses occur is the most profitable approach to the diagnosis and treatment of osteodystrophies rather than attempts to relate skeletal disease to specific named syndromes.[4]

The pathology of osteodystrophies

The skeleton can respond to 'insult' in a limited manner which differs in the growing and adult skeleton. The final pathological conditions attained by these responses are as follows:

Osteoporosis (matrix osteoporosis, bone atrophy). The amount of proteinaceous bone matrix per unit volume of the skeleton is reduced but that present is well mineralized (Fig. 24.1). Osteoporosis is found in both the growing and adult skeleton (Fig. 24.1). It may be induced by reduced matrix formation, excess bone resorption or a mixture of the two. Anatomically, it is recognized by reduction of trabecular bone and increased porosity of cortical bone. Whatever the cause, the bone is more fragile than normal.

Rickets This is limited to the growing skeleton where, as well as the changes of osteomalacia, the cartilage matrix of active growth plates show mineralization defects and arrest of endosteal ossification (Fig. 24.2).

Osteomalacia (mineral osteoporosis, bone softening). In this state, the adult form of rickets, the amount of matrix per unit volume of bone is normal but mineralization of matrix is either slowed or reduced. Large osteoid seams at the site of mineralization and increased bone resorption delineate the

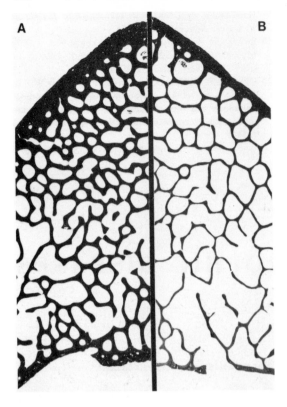

Fig. 24.1 Microradiographs of transverse sections taken from a lumbar vertebral body. A = normal animal; B = osteoporotic animal with thinner cortex, finer trabeculae and less trabecular bone per unit area.

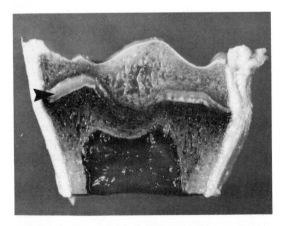

Fig. 24.2 Cross-section through a tibial head showing increased thickness of growth plate (arrowed) in rickets due to arrested endochrondral ossification and build-up of unmineralized cartilage.

condition (Fig. 24.3). Such bone may be more fragile or flexible than normal.

Stunted growth   Though not a disease state *per se*, suboptimal bone growth resulting in small bones with mineral and matrix in the correct proportions is commonly seen in flocks with subclinical osteodystrophy and, as such, may be of significance in the diagnosis of osteodystrophies.

Fig. 24.3 Thin section (6 μm) from same case as Fig. 24.2 showing mineralized bony areas (white arrow) and enlarged unmineralized osteoid seams (dark arrow). In the normal animals osteoid seams would be almost totally absent. (x 170 Silver stain)

Whilst each of these pathological states are theoretically separable, in practice their clinical expression depends upon the exact age of the sheep, the cause of the condition and its duration. It is common to find two or more of these pathological states evident within the same bone and their expression differing between bones, of varying maturity, within the same animal.

The pathogenesis and aetiology of primary matrix and mineral osteoporosis differ. Matrix osteoporosis is usually due to a factor, or factors, that inhibit normal production of proteinaceous matrix while rickets and osteomalacia (or mineral osteoporosis) are associated with defects in mineralization of apparently normal matrix and, as such, are most often due to defects of mineral metabolism. Where either mineral or matrix osteoporosis are present alone they are of great diagnostic value. However, some conditions affect both parts of normal bone formation and others, that start as one or other, progress to present features of both osteoporoses. Although diagnosis may then be difficult, examination of a number of animals from an outbreak can indicate the primary or earlier forms of the disease and aid diagnosis.

Clinical signs

All osteodystrophies lead to brittle or flexible bones and, though their clinical presentation may differ in detail, most clinical signs can be related to such bone weakness. Shifting lameness due to pain is an early sign of disease that appears in both growing and adult sheep. It may become severe enough to put the animal off its legs. Bowing or bending of the forelegs due to a distortion of the axial relationships between the long bones is a common sign of osteodystrophia in growing sheep [3] and is seen world-wide (Fig. 24.4). Swelling of the distal ends of the long bone and, less commonly, ribs close to the growth plates may accompany the bowing or appear alone; the metacarpal and metatarsal joints are the commonest sites to be affected. In some osteodystrophic conditions involving sheep less than a year old an imbalance of mandible growth can, over a period of time, give severe malocclusion and an inability to close the mouth.[5]

A generalized loss of condition, varying with the severity and distribution of the bone lesions, is common, probably resulting from an inability or lack of desire to move and feed. The thin or soft bones of the skull, especially the frontal bone, can be depressed into the frontal sinus, thus the local term of 'cappi' or 'double scalp'. Severe and multiple fractures of the long bones and spine, vertebral collapse and subsequent hemiplegia are all secondary sequelae to osteodystrophy. Such fractures, rachitic growth plates and muscle insertions may all show compensatory palpable exostoses in chronic disease.

Causes and epidemiology

Nutritional problems are the commonest causes of osteodystrophic disease in sheep in Britain. The metabolism of calcium, phosphorus and vitamin D is central to bone development and skeletal health and absolute deficiencies of any of the three or imbalances of calcium and phosphorus can precipitate

Fig. 24.4 Ewes showing typical carpal/metacarpal distortions of the forelegs in 'bent leg'.

disease. The significance and role of these nutritional factors are discussed in Chapter 36 and should be read in conjunction with this section.

Deficiencies in phosphorus or low vitamin D availability are most often implicated in nutritional osteodystrophy, resulting in an osteomalacia or rickets. Thus, amongst the named osteodystrophies, 'rickets', 'dental malocclusion', 'cruban' or 'osteomalacia' and many of the 'bent leg' syndromes have been associated with either hypophosphataemias, hypovitaminosis D or both together. Sheep are not so liable to primary phosphorus deficiency from low soil levels as cattle but it can occur, especially when animals are growing fast on lush pasture or added concentrates. Thus hoggs often become rachitic in early spring when pasture growth commences. Phytates in herbage have been reported to reduce phosphorus availability to the sheep where the diet contains high levels of calcium but there is no evidence for this suggestion. High levels of calcium, alone, do not induce a phosphorus deficiency.

In Britain most phosphorus deficiencies are

complicated by a hypovitaminosis D. This situation is well recognized in Britain and New Zealand where solar radiation is considered inadequate for normal winter requirements of the vitamin (see section on vitamin D). Thus indoor rearing of lambs, often further aggravated by fast growth rates under these conditions, may predispose to rachitic syndromes.[4] Lush green feeds, particularly cereals, when fed in winter, reduce vitamin D intake and may give rachitic syndromes in hoggs in late winter.

The role of calcium in clinical osteodystrophies is unclear. While fluctuating calcium requirements throughout the year are reflected in bone quality experimentally, there are no reports of clinical disease due to straight calcium deficiency. A number of the gastrointestinal helminths, such as *Trichostrongylus vitrinus,* inhibit calcium and phosphorus uptake from the gut and induce a subclinical mineral osteoporosis.

More recent work has shown the significance of adequate dietary protein for normal bone growth.[6] Reduced protein availability leads to stunting and osteoporosis and has been suggested as a complicating factor in some osteodystrophies where both osteomalacia and osteoporosis appear. Since some gastrointestinal helminths reduce net protein uptake and induce an osteoporosis even with subclinical burdens [7] (see Chapter 11) they too may be implicated in the aetiology of some osteodystrophic syndromes. It is probable that many of the conditions seen in Britain have a multifactorial aetiology.

A lack, or excess, of certain trace elements also induces sporadic cases or outbreaks of osteodystrophic syndromes. Thus, excess lead in old lead mining areas has occasionally produced lameness associated with a matrix osteoporosis (see Chapter 47).[8] So-called secondary copper deficiency from high molybdenum pasture levels with resulting osteoporosis has been seen in Great Britain while in New Zealand higher levels of molybdenum used as a pasture dressing has induced osteoporosis and growth plate degenerative lesions distinct from those of rickets.[9] Experimental primary copper deficiency gives osteoporosis but in the field skeletal lesions in sheep have not been reported.

Distinctive osteodystrophic lesions are also seen in industrial fluorosis, mottling and excess wear of teeth being the most visible signs. Lastly, overdosing with vitamin D, doses in the region of 1 000 000 iu, may give vitamin D toxicosis in growing lambs characterized by scoliosis of the cervical spine, hypercalcaemia, osteoporosis and inhibited mineralization of developing bone.

Diagnosis and control

In the absence of neurological signs, clinical signs of stiff gait, lameness, recumbency and loss of condition are usually suggestive of musculo-skeletal disease. Where skeletal deformity is present, an osteodystrophy is indicated but otherwise the possibility of joint or muscle disease must be ruled out. Since the commonest causes of these conditions are dietary, analysis of the diet and pasture, its protein, calcium, phosphorus and vitamin D status, may allow a presumptive diagnosis confirmed by the replacement of the deficient nutrients. Predisposing environmental factors such as fast growth, parasitism and indoor rearing should also be taken into account.

Where such an empirical approach does not resolve the problem recourse to laboratory diagnosis is necessary. While serum calcium and phosphorus estimations may help, levels of these minerals are often within physiological limits until late in the disease and are not considered good diagnostic features. This situation is further complicated by normal fluctuations in the levels through the year associated with the demands for calcium in pregnancy and lactation and the physiological osteoporosis that may result. Plasma copper concentrations may be used in the diagnosis of 'so-called' secondary copper deficiency but there are limitations to its use; these are detailed in the chapter on trace elements.

Radiographic examinations of limb bones may aid in diagnosis of rachitic conditions, irregularities and thickening of the radiolucent line in the growth plates, swelling of the bone epiphysis and malalignment of the bones having diagnostic value. However, less specific osteomalacic and osteoporotic changes can be missed on radiographs; a variable proportion of bone mineral (that may be as much as 25 per cent of total) must be lost before it can be detected, even with sophisticated radiographic techniques. Exostoses of fluorosis and molybdenosis and multiple fractures are further features that may aid in differential diagnosis.

Histopathological examinations of bone may be necessary to come to a firm diagnosis of some of the osteodystrophies. Osteoporosis, rickets, and osteomalacia can all be differentiated histologically and specific lesions are seen in molybdenosis and fluorosis. The mottled osteons of fluorosis are pathognomonic while the most likely cause of an epiphysiolysis and growth plate dyschondroplasia is molybdenosis. High fluorine levels in bone (over 2000–3000 ppm in cortical bone) will confirm fluorosis. Ashing techniques to determine mineral to protein ratios, amount of ash per unit volume and matrix per unit volume are commonly used in diagnosis of pathological states[6]: such procedures are not as sensitive as histological bone quantitation.

Treatment of the simple nutritional osteodystrophies (calcium, phosphorus, vitamin D and protein) by replacement of the suboptimal nutrient leads to apparently complete recovery. Further details of methods of treatment and dietary intakes may be found in Chapter 35. However, there is a suggestion that the skeleton may not be able to compensate for retarded skeletal growth and that permanent stunting of the skeleton and production loss may result. Distortion of the skeleton and fractures may also leave permanent skeletal defects. Treatments of primary or induced hypocuprosis are discussed in the section dealing with trace elements. Withdrawal of animals from fluorotic pasture ameliorates the lesions but recovery is slow and slaughter may be advised.

REFERENCES

1 Nisbet D.I., Butler E.J., Bannatyne C.C. & Robertson J.M. (1962) Osteodystrophic diseases of sheep. I. An osteoporotic condition of hoggs known as Double Scalp or Cappi. *Journal of Comparative Pathology,* **72**, 270– 80.

2 Nisbet D.I., Butler E.J., Smith B.S.W., Robertson J.M. & Bannatyne C.C. (1966) Osteodystrophic diseases of sheep. II. Rickets in young sheep. *Journal of Comparative Pathology,* **76**, 159–69.

3 Hidiroglou M., Dukes T.W., Ho S.K. & Heaney D.P. (1978) Bent-limb syndrome in lambs raised in total confinement. *Journal of the American Veterinary Medical Association,* **173**, 1571–4.

4 Benzie D., Boyne A.W., Dalgarno A.C., Duckworth J., Hill R. & Walker D.M. (1960) Studies of the skeleton of the sheep. IV. The effects and interactions of dietary supplements of calcium, phosphorus, cod-liver oil and energy, as starch, on the skeleton of growing Blackface wethers. *Journal of Agricultural Science,* **54**, 202–21.

5 Duckworth J., Benzie D., Cresswell E., Hill R., Dalgarno A.C., Robinson J.F. & Robson H.W. (1961) Dental mal-occlusion and rickets in sheep. *Research in Veterinary Science,* **2**, 375–80.

6 Sykes A.R., Nisbet D.I. & Field A.C. (1973) Effects of dietary deficiencies of energy, protein and calcium on the pregnant ewe. V. Chemical analyses and histological examination of some individual bones. *Journal of Agricultural Science (Cambridge),* **81**, 433–40.

7 Coop R.L., Sykes A.R., Spence J.A. & Aitchison G.U. (1981) *Ostertagia circumcincta* infection of lambs. The effect of different intakes of larvae on skeletal development. *Journal of Comparative Pathology,* **91**, 521– 30.

8 Butler E.J., Nisbet D.I. & Robertson J.M. (1957) Osteoporosis in lambs in a lead mining area. I. A study of the naturally occuring disease. *Journal of Comparative Pathology,* **67**, 378–96.

9 Pitt M., Fraser J. & Thurley D.C. (1980) Molybdenum toxicity in sheep: epiphysiolysis, exotoses and biochemical changes. *Journal of Comparative Pathology,* **90**, 567–76.

Plate 5 Sheep scab: early lesion in the fleece (Crown copyright).

Plate 6 Fleece held apart to demonstrate dried exudate at several levels on wool staples indicating recurrent attacks of *D. congolensis* infection.
By courtesy of Mr. E.A. McPherson

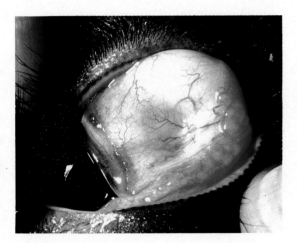

Plate 7 Follicle development in the conjunctival sac of a gnotobiotic lamb infected with *Mycoplasma conjunctivae*. Similar follicle development is observed in lambs infected by the ocular route with *Chalamydia psittaci ovis*.

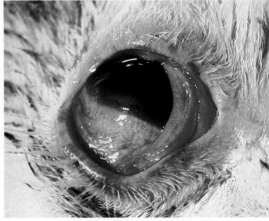

Plate 8 Naturally-occurring keratoconjunctivitis in a lamb. The blood vessels of the conjunctiva are congested, and can be seen spreading centripetally across the limbus towards the centre of the cornea. An opaque band precedes the invading blood vessels. Epiphora, and the caking of periorbital hair with dried ocular discharges are apparent.

SECTION 4

REPRODUCTION

25

ENZOOTIC (CHLAMYDIAL) ABORTION

I.D. AITKEN

Synonyms Enzootic abortion of ewes (EAE), ovine enzootic abortion, ovine viral abortion, kebbing.

Enzootic abortion is the major diagnosed, infectious cause of sheep abortion in Britain[1], occurring predominantly in lowground flocks which are intensively managed at lambing time. The disease is characterized by a chlamydia-induced necrotic placentitis which results in abortion in late pregnancy and premature lambing. EAE occurs in many European countries, in the Middle East and in North America but it is not a problem in Australia or New Zealand.

In parts of central and southern Scotland the disease had been recognized since the early 1900s but it was not until 1950 that the cause was discovered. The work of Stamp and his colleagues[2] clearly incriminated a small unique micro-organism of the psittacosis lymphogranuloma group. Originally considered to be large viruses these organisms are now known as chlamydiae and are amongst the most widespread and commonly occurring microbial parasites within the animal kingdom.

Once the infective agent had been identified experimental vaccines based on formalin-inactivated organisms were developed and gave good results in both laboratory and field trials.[3] Infection of sheep was not completely prevented but the incidence of abortions due to EAE was significantly reduced. Subsequent development and general application of commercial vaccine effectively brought EAE under control and while some cases continued to be encountered the disease was no longer of major concern to sheep farmers. However, since about 1978 there has been a resurgence of EAE in both vaccinated and non-vaccinated flocks[4] and the disease has appeared in parts of Britain where previously it had not been recognized. The situation has not yet been resolved but epidemiological and research studies point to a changing pattern of disease incidence and the possible emergence of serotype variants of the causative organism for which a vaccine of wider antigenicity may be needed.

Cause

Like other members of the group the chlamydiae which cause EAE are intermediate between bacteria and viruses. They resemble bacteria in that they have a cell wall, possess both DNA and RNA, multiply by binary fission and are susceptible to antibiotics like penicillin and tetracyclines. However, like viruses, chlamydiae can multiply only within living host cells and exhibit a unique development cycle with alternating intracellular and extracellular phases associated with two different types of particle. The extracellular infective form, known as an elementary body (EB) is small (300 nm) and round with a thick wall and dense cytoplasmic core. In contrast, the intracellular replicative form or reticulate body (RB) is larger (1000 nm), thin-walled and has a homogeneous cytoplasm. It does not survive outside the host cell.

The chlamydial infection cycle is initiated when the EB adheres to the surface of a susceptible cell and is drawn into the cytoplasm by a phagocytic-like process. In the cell the EB transforms to an RB which divides many times within a membrane-bounded vesicle before the progeny particles revert to EB forms. These are released by cell rupture to renew the infection cycle.

In the laboratory, chlamydiae can be grown readily in the yolk sac of the developing chick embyro and in cell cultures. Infectivity is rapidly destroyed by heat (10 minutes at 60°C) and by treatment with formalin or ether. However, infective particles can survive for several days at low environmental temperatures and for years at $-70°C$.

Only two species of chlamydiae are recognized. One, *Chlamydia trachomatis*, is a human pathogen with affinity for ocular and urogenital tissue. The other, *C. psittaci*, has a much wider host range and is able to infect many wild and domestic birds and mammals, including man.[5, 6, 7] The gut appears to be the natural habitat of *C. psittaci* and in most cases there is a well adapted and innocuous enteric parasitism. However, under certain circumstances, of which pregnancy is one, chlamydiae can spread to and localize in specific tissues and cause a variety of disease syndromes, including placentitis and abortion. EAE is one of the best characterized chlamydial diseases of animals.

Chlamydiae are antigenically complex. A heat-stable complement-fixing antigen is common to all isolates and antigens specific for each species are detectable by direct immunofluorescence and by immunodiffusion of disrupted organisms. Within each species are several poorly defined 'serotypic' antigens that are implicated in the infective process and which seem to be important in the development of protective immunity. At least 15 distinct serotypes of *C. trachomatis* are recognized but little is known of serotypic variants of *C. psittaci*. Isolates of ovine and bovine origin can be differentiated from avian chlamydiae and fall into two groups distinguishable by serology. Group 1 includes isolates from intestinal infections and cases of abortion and group 2 contains chlamydiae isolated from cases of polyarthritis, encephalomyelitis and conjunctivitis.

Clinical signs

In EAE there are rarely any premonitory signs of impending abortion. Behavioural changes and a vulval discharge may be present shortly before the fetus is expelled but under farming conditions these signs are seldom noticed. Individual ewes may abort as early as 10–12 weeks of pregnancy with the expulsion of an immature and sometimes mummified fetus but such early cases are likely to be missed unless a discharge is noted on the ewe's tail. Most commonly it is the discovery of dead lambs 2–3 weeks before lambing is due that provides the first indication that something is wrong.

Despite an associated necrosis of the fetal membranes the majority of lambs aborted at this late stage in pregnancy are well-developed and quite fresh, the absence of autolytic change implying that death *in utero* has been fairly recent. A number of these aborted lambs may appear 'pot bellied' due to accumulation of blood-tinged fluid in serous cavities and in some the fleece may be partially covered with or discoloured by flecks of pink-brown material originating from placental exudate. However, true degenerative changes such as corneal opacity and easily detachable wool, indicative of death some days or weeks before abortion, are seen in only a minority of cases. An occasional undersized, mummified fetus may also be encountered.

As well as frank abortion, premature or full-term delivery of stillborn, moribund or weak lambs is also a feature of the disease. Even with nursing many of the weak lambs fail to survive and their deaths contribute to overall losses from EAE. In contrast, some infected ewes give birth to normal live lambs which they rear successfully and in multiple pregnan-

cies it is not uncommon to have one dead lamb and one or more live weak or healthy lambs produced by the same ewe.

For several days after abortion ewes pass varying amounts of a discoloured uterine discharge but are otherwise clinically normal. The discharges eventually dry up and future breeding potential is not impaired. However, if the placenta is retained and an associated metritis develops there can be loss of condition and death as a result of secondary bacterial infection.

Pathology

From examination of field material [2] and chronological study of experimentally-infected pregnant ewes [8] it is clear that placentitis is the major and typical pathological component of EAE. The fetal membranes of aborted, premature and weak full-term lambs reveal gross changes of variable degree and extent. The principal features are discolouration and necrosis of some of the cotyledons, oedema or a rough thickening of adjacent intercotyledonary tissue and a dirty pink exudate containing flakes of similarly coloured material (Plate 4). Similar but milder and more restricted lesions may be detected also in the placentae of infected ewes which give birth to normal live lambs.

The crucial lesion occurs in the vascular placentome, the intimate apposition of endometrial caruncle with chorionic cotyledon that is the anatomical and physiological unit for maternal-fetal transfer of oxygen and nutrients. Chlamydiae reaching the placentome in the maternal circulation infect and replicate within the trophoblastic epithelial cells of the chorionic villi in which they produce readily visible cytoplasmic chlamydial inclusions.[2, 8] Placentomal infection also induces local infiltration of macrophages, neutrophils and lymphocytes. Loss of epithelial cells and progressive necrosis of the underlying villi and tips of the caruncular septa are accompanied by accumulation of necrotic cellular debris and organisms in the placentomal haematomas. With time, the reaction spreads to the periplacentome and then to intercotyledonary regions of the chorion. In these areas the principal pathological features are infection and destruction of chorionic epithelial cells, oedema and cellular infiltration of the underlying stroma and accumulation of exudate between the chorion and endometrium.

In contrast to these changes the fetal side of the chorion remains unaffected. Similarly, in those areas of the endometrium opposed to infected and diseased chorion there is only limited infection and loss of epithelial cells and the associated cellular reaction is mild and confined to the immediately underlying tissue.

Not all placentomes of a placenta become infected and in those that do, the degree and extent of inflammatory and destructive change is variable. However, loss or reduction of the functional integrity of even a proportion of placentomes impairs maternal–fetal exchange and is probably a major contributory factor in fetal deaths. In the latter stages of gestation the chorionic epithelial cells constitute the major source of progesterone, the hormone responsible for maintaining pregnancy. A fall in the level of progesterone consequent upon destruction of these cells, rather than fetal death, may be the trigger for abortion and premature births.

Pathological changes in the fetus are confined to the development of inflammatory or necrotic foci within parenchymatous and lymphatic organs and in lungs, skin and brain though seldom are the changes so severe or extensive as to be grossly visible. Congestion and slight swelling of the liver and pinpoint white foci have been reported but are not consistent features.

Diagnosis

Abortion of well preserved lambs in the last 2–3 weeks of pregnancy and an associated necrotic placentitis provide reasonable grounds for a provisional diagnosis but it should be appreciated that some cases of toxoplasmosis may present similarly and that more than one micro-organism may be involved in an outbreak of abortion. Necrotic placentitis is not a feature of the less common or sporadic causes of infectious ovine abortion some of which, e.g. vibriosis and border disease are dealt with elsewhere in this section. Rarely, abortion occurs as a consequence of infection of pregnant ewes with the rickettsia, *Coxiella burnetii*, which is highly infectious for man, causing an acute, febrile influenza-like illness known as Q-fever. Vast numbers of the organism are shed in infected placentae and discharges. In stained smears *C. burnetii* resembles *C. psittaci* but the two micro-organisms are antigenically unrelated and may be distinguished serologically.

Diagnosis of EAE is most easily confirmed by the demonstration of large numbers of chlamydial EB's in smears made from affected cotyledons or chorion and stained by a modified Ziehl-Neelsen procedure to reveal red bodies, singly and in clusters, against a blue background. When examined by dark-ground illumination the stained bodies stand out clearly as bright, green, coccoid structures. In the absence of suitable placental material vaginal swabs from ewes which have aborted may be used to prepare smears or smears may be made directly from the wet fleece of a recently aborted or stillborn lamb. However, chlamydiae may be less abundant in smears prepared from these sources, and they do not occur in fetal stomach contents.

If laboratory facilities for staining and examination of smears are not readily accessible, a small piece of affected placental tissue, free from gross contamination, should be placed in a suitable transport medium containing streptomycin (not penicillin) for despatch by post. Under these conditions chlamydiae survive for several days and will be evident in tissue smears. If necessary, isolation of chlamydiae may be attempted by inoculation of tissue extracts into chick embryo yolk sacs or on to cell culture monolayers.

Serology may be used for confirming or refuting diagnosis. Infected and vaccinated ewes generally have low or moderate levels of complement-fixing antibodies but aborting ewes experience an episode of chlamydaemia which often results in a significant post-abortion rise in antibody titre. Ideally, paired blood samples should be collected, one at, or soon after abortion and the second about 2 weeks later. Alternatively, single samples from representative aborting and non-aborting ewes may reveal in the former a significant elevation of antibody titre.

Accurate diagnosis early in the course of an abortion outbreak allows adoption of appropriate control measures. However, it is advisable to monitor diagnosis as long as abortions continue so that mixed infections such as EAE and toxoplasmosis or salmonellosis can be detected.

Epidemiology and transmission

Characteristically, EAE is a disease of lowground flocks which practise intensive management over the lambing period. The disease is rarely encountered in flocks maintained under the sort of extensive management systems used for hill sheep. In lowground flocks the practice of passing successive groups of pregnant ewes through the same lambing field or sets of lambing pens is conducive to a high level of environmental contamination by the agent of EAE. Exposure of susceptible females to heavy infection in this way is a crucial component in transmission of the disease.

The principal sources of infection are the ewes which abort or deliver stillborn or weak lambs as a result of placental infection. Vast numbers of chlamydiae shed in the diseased placentae and uterine discharges can, under cold weather conditions, remain viable for several days and so afford opportunity for spread of infection. Susceptible companion ewes become infected by consuming food or water contaminated by chlamydiae but infection generally remains inapparent and does not endanger the current pregnancy. Not until the subsequent pregnancy does the infection progress and so cause abortion in the year following infection. However, recent evidence indicates that this is not always the case.[9] Experiments have shown that within a few days of oral infection at mid to late pregnancy chlamydiae are established in the placenta though normally a period of at least 5–6 weeks elapses between infection and abortion. Thus, if abortions occur in the earliest lambing group of a flock with a prolonged lambing period the risk of infection and abortion occurring in the same pregnancy should be considered. How frequently such a single pregnancy infection cycle occurs in the field is unknown but it could be a feature of abortion storms in flocks experiencing EAE for the first time.

Most ewes which have aborted or undergone placental infection remain intestinal carriers of chlamydiae for an undefined period. Variable numbers of chlamydiae excreted intermittently in the faeces have the potential to cause intestinal infection in other sheep but the significance of faecal excretors in maintaining and transmitting EAE has not been established. Pregnant ewes carrying intestinal chlamydiae are not immune to EAE. Rams also may acquire intestinal infection and although chlamydiae are able to spread to genital tissue and to infect semen there is no evidence that the ram plays any role in the transmission of EAE.

Live lambs born to mothers with active placental infections and lambs fostered by ewes which have aborted or produced dead lambs are very likely to be infected as a result of close contact with their mothers though no clinical evidence of infection may be apparent. Field and experimental evidence suggests that up to one third of neonatally-infected ewe lambs may develop placental infection during their first pregnancy and that some may abort.

A clean flock usually becomes infected following the introduction of infected replacement females. When these infected replacements abort they disseminate infection to ewes of all ages in the home flock which, in the following year, experiences a serious outbreak of EAE with up to 30 per cent of the ewes being involved. In succeeding years it is mainly the younger ewes which are likely to be affected and the annual incidence will continue at about 5–10 per cent unless control measures are introduced.

Clean replacements added to an already infected ewe flock run a high risk of picking up EAE at their first lambing and are likely to abort in the following year.

Control and prevention

When an outbreak occurs the primary aim is to prevent spread of infection. Ewes which abort or deliver a dead or weak lamb should be clearly marked for later identification and isolated from other sheep until their uterine discharges have dried up (about 7–10 days). Aborted fetuses, dead lambs, placentae and any contaminated bedding and food must be removed and destroyed. Lambing pens in which abortions have occurred should be cleaned and disinfected and, if possible, should not be reused. The shepherd should also wash his hands thoroughly after dealing with affected ewes or abortion material. As lambing progresses, continued vigilance is needed to ensure detection and swift isolation of all affected ewes particularly those that prematurely deliver live rather than stillborn lambs. It must be recognized that the strict segregation of aborting ewes for control of EAE is contrary to the policy recommended in other infectious abortion diseases such as toxoplasmosis and vibriosis in which spread of infection to non-pregnant females is encouraged in order to achieve early development of immunity. Where more than one infection is contributing simultaneously to an abortion problem, control of EAE should take precedence.

Treatment with a long acting oxytetracycline may moderate the severity of incipient EAE. In experimentally infected ewes treatment extended the duration of their threatened pregnancies and resulted in the delivery of more liveborn lambs than were born to infected but untreated controls.[10, 11] In addition, in a field trial in south east Scotland treatment initiated after EAE had been diagnosed significantly reduced the incidence of abortion.[12] As the main effect of treatment is to suppress chlamydial multiplication antibiotic cover must be sustained and two intramuscular doses (20 mg/kg) at a two week interval are recommended. While this type of therapy may diminish the number of organisms ultimately shed it does not eliminate the infection nor can it reverse the pathological damage already inflicted on a heavily infected placenta. Accordingly, some abortions and stillbirths occur despite treatment.

Although ewes which have experienced EAE are able to breed normally again they do constitute a continuing source of infection and their fate must be considered. Where only a few members of a flock have been involved it is advisable that they be culled. However, if infection has been widespread a reasonable level of flock immunity will have been established and for this reason it may be worthwhile retaining ewes which have aborted. Provided no fresh source of chlamydial infection is introduced to the flock few if any of the ewes previously affected abort in a subsequent pregnancy.

The most effective way to avoid the introduction of infection to a clean ewe flock is to keep it closed but that is seldom feasible. Replacement stock should be obtained from sources known to be free of EAE and it is preferable to select females which have not yet been bred and so have not been exposed to the risk of infection by passing through lambing pens. Bought in and home-bred replacements, whether entering a clean or previously-infected ewe flock, should be vaccinated before tupping and again after not more than three years.

Properly used, the inactivated vaccine currently available mitigates the worst effects of an EAE outbreak but does not completely eliminate infection. Moreover, available evidence indicates that some of the 'strains' of chlamydia currently circulating are able to overcome vaccinal immunity though whether this is due to enhanced virulence or serotypic variation has yet to be established. While the nature of vaccinal immunity in EAE is not fully understood it is likely that the principal targets are the chlamydial EB's in their brief extracellular phases. Therefore, stimulation of immunity before conception

occurs affords the best defence against colonization of the developing fetal membranes.

REFERENCES

1 Linklater K.A. (1979) Abortion in sheep. *In Practice*, **1**, 30– 33.

2 Stamp J.T., McEwen A.D., Watt J.A. & Nisbet D.I. (1950) Enzootic abortion in ewes. 1. Transmission of the disease. *Veterinary Record*, **62**, 251– 4.

3 Foggie A. (1973) Preparation of vaccines against enzootic abortion of ewes. A review of the research work at the Moredun Institute. *Veterinary Bulletin*, **43**, 587–90.

4 Linklater K.A. & Dyson D.A. (1979) Field studies on enzootic abortion of ewes in south east Scotland. *Veterinary Record*, **105**, 387–9.

5 Storz J. (1971) *Chlamydia and Chlamydia-induced Diseases*. Charles C. Thomas, Springfield, Illinois.

6 Foggie A. (1977) Chlamydial infections in mammals. *Veterinary Record*, **100**, 315–17.

7 Shewen P.E. (1980) Chlamydial infection in animals: a review. *Canadian Veterinary Journal*, **21**, 2–11.

8 Novilla N.M. & Jensen R. (1970) Placental pathology of experimentally induced enzootic abortion in ewes. *American Journal of Veterinary Research*, **31**, 1983–2000.

9 Blewett D.A., Gisemba F., Miller J.K., Johnson, F.W.A. & Clarkson M.J. (1982). Ovine enzootic abortion: the acquisition of infection and consequent abortion within a single lambing season. *Veterinary Record*, **111**, 499–501.

10 Rodolakis A., Souriau A., Raynaud J-P. & Brunault G. (1980) Efficacy of a long acting oxytetracycline against chlamydial ovine abortion. *Annales de Recherches Veterinaires*, **11**, 437–44.

11 Aitken I.D., Robinson G.W. & Anderson I.E. (1982) Long-acting oxytetracycline in the treatment of enzootic abortion of ewes. *Veterinary Record*, **111**, 446.

12 Greig A., Linklater K.A. & Dyson D.A. (1982) Long-acting oxytetracycline in the treatment of enzootic abortion of ewes. *Veterinary Record*, **111**, 445.

TOXOPLASMOSIS

26

D. BUXTON

Toxoplasmosis is caused by the protozoon parasite *Toxoplasma gondii*.[1] The organism is an intestinal coccidium of felids with a wide range of warm-blooded intermediate hosts including the sheep, in which it can cause considerable losses at lambing time.

The name *Toxoplasma* comes from the Greek and refers to its crescentic shape (*toxon* = arc, *plasma* = form). It was first discovered in 1908 in a rodent, *Ctenodactylus gundi*, by Nicolle and Manceaux in Tunis and about the same time Splendore independently described *Toxoplasma* in a laboratory rabbit in Brazil.

However the considerable economic importance of *T. gondii* infection to the sheep industry was not recognized until the 1950s when Hartley and colleagues working in New Zealand described a previously unrecognized form of abortion.[2, 3]. Soon after the condition was reported in the UK by Beverley and co-workers. More recent work has indicated that *T. gondii* infection can immunosuppress the host making it more susceptible to some unrelated infections and less responsive to some vaccines.[4, 5]

Cause

Life cycle *Toxoplasma gondii* is a world-wide zoonosis with any warm-blooded animal, including birds, being susceptible to infection. The life cycle of *Toxoplasma* can be divided into two parts; an asexual cycle with little host specificity and a sexual cycle, confined to the enteroepithelial cells of cats, which results in the production of oocysts.[1] Two developmental stages are involved in the asexual cycle, the tachyzoite and the bradyzoite. The crescent-shaped tachyzoite measures 5 μm by 1.5–2 μm, has one end more pointed than the other and a nucleus usually situated nearer the rounder end. The tachyzoite actively penetrates the host cell becomes surrounded by a parasitophorous vacuole and multiplies by endodyogeny. Multiplication continues until the cell ruptures when the organisms are released to parasitize further cells. This process continues until the host dies or more usually it develops immunity to the infection. In the latter case chronic infection is

established, extracellular parasites are thus eliminated, intracellular multiplication slows and tissue cysts develop. A small cyst contains only a few bradyzoites (the second stage of the asexual cycle) but a large one may contain thousands. Intact cysts are found most frequently in brain and skeletal muscle and represent the quiescent stage of the parasite within the host. When a cyst ruptures the bradyzoites are released to enter other cells and so the asexual cycle is complete.

Initiation of the sexual cycle occurs when a non-immune cat ingests food containing tissue cysts. The cyst wall is dissolved by proteolytic enzymes in the stomach and small intestine and the released bradyzoites penetrate the epithelial cells of the small intestine. The toxoplasms then pass through several stages before gametogeny begins. Gametocytes develop in epithelial cells in the small intestine but most commonly in the ileum 3–15 days after infection. Microgametes develop, are released and 'swim' to and penetrate a mature macrogamete after which formation of the oocyst wall begins around the fertilized gamete. The oocysts which are then discharged into the intestinal lumen are subspherical to spherical, 10 x 12 μm in diameter and each is almost filled by the sporont. The oocysts pass out of the cat in the faeces and sporulation occurs within 1–5 days depending on aeration and temperature. The sporulated oocyst is subspherical to ellipsoidal, 11 x 13 μm in diameter and each contains two ellipsoidal sporocysts. Each sporocyst contains 4 sporozoites each with a subterminal nucleus and a few PAS-positive granules in the cytoplasm.

Simultaneously with the sexual enteroepithelial cycle the ingested bradyzoites can penetrate the lamina propria of the feline intestine, multiply as tachyzoites and spread to extra-intestinal tissues within a few hours.

Thus 3–10 days after ingesting tissue cysts the cat is capable of shedding hundreds of thousands and sometimes millions of oocysts in its faeces for a period of about 8 days, after which it will probably not excrete oocysts again. However, experiments have shown that recrudescence of infection with re-excretion of oocysts can occur if the cat becomes stressed through unrelated illness. Cats can also become infected by ingesting oocysts and tachyzoites but in this case reduced numbers of oocysts are produced after 19 or 20 days and even then by only about half of the animals.

Resistance to physical and chemical reagents
Toxoplasma tachyzoites, bradyzoites and oocysts show considerable variability in their ability to remain viable outside the host. Tachyzoites can survive in normal saline for up to a day, in human colostrum for up to 5 days and in normal rabbit serum for as long as 6 weeks. However, like bradyzoites within tissue cysts, they rapidly lose viability in distilled water or after freezing/thawing or drying. Tissue cysts in infected tissues can survive for as long as 68 days at 4°C but for only three hours in peptic juice. Tachyzoites survive only a minute in peptic juice. Oocysts on the other hand can be highly resistant and survive for long periods (548 days) at room temperature in water. Native cat faeces have also remained infective outdoors for a similar period. Oocysts are destroyed by 5 per cent ammonium solution within minutes, 95 per cent ethanol within an hour and 10 per cent formalin within 24 hours.

Antigens Serological tests have indicated that *T. gondii* contains several antigenic determinants, but as it is an obligate intracellular parasite which has not, as yet, been cultured in a cell-free medium, great difficulty has been experienced in producing preparations of the parasite uncontaminated by host antigen. However, recent studies with monoclonal antibodies to *Toxoplasma* have confirmed the presence of cytoplasmic antigen and at least four major and several lesser membrane antigens.

Clinical signs

In a typical outbreak, disease first becomes apparent when ewes lamb a few days early with a significant proportion of lambs, although outwardly quite normal, being stillborn and some accompanied by a 'mummified' fetus. In addition a proportion of lambs born alive are weak and die within the first few days of life despite careful nursing. The ewes remain clinically normal. However, the real effect of infection in susceptible pregnant ewes can be more widespread and depends upon the stage of gestation at which infection occurs. Infection in early pregnancy produces fetal resorption with the ewes ending up barren, infection between about 70 and 120 days

presents the clinical picture described while ewes which become infected in late pregnancy produce infected but clinically normal lambs.

Pathology

The most characteristic macroscopic changes are found in the fetal membranes.[2,3,6] The cotyledons are usually bright to dark red and speckled with white foci 1–2 mm in diameter. These foci may be sparse or so numerous that they can become confluent and on cut section they can be seen to occur on any plane. Cotyledons with macroscopic changes can be found alongside apparently normal ones in the same placenta. The inter-cotyledonary allanto-chorion appears normal.

Visible changes in lambs can vary, the most obvious being the mummified fetus, a small chocolate brown miniature of a lamb, often with its own small grey-brown placenta (Fig. 26.1). Lambs dying later in gestation are born in various stages of decomposition often with clear to bloody subcutaneous oedema and a variable amount of clear to bloodstained fluid, sometimes flecked with strands of fibrin, in body cavities. However, while these latter changes indicate an intrauterine infection they are not specific to *Toxoplasma* infection. Lambs which die just before or after birth appear quite normal.

Fetal membranes are very often too autolytic to allow useful histopathological interpretation. However, examination of well-preserved material can reveal moderate oedema of the mesenchyme of the fetal villi with a diffuse hypercellularity due to the presence of large mononuclear cells. In addition there can be foci of swollen trophoblastic epithelium in fetal villi which can progress to necrosis and desquamation. The larger affected areas give rise to necrotic nodules initially showing a loss of cellularity but with caseous and granular calcium-containing material eventually being laid down. Small numbers of intracellular and extracellular toxoplasms are sometimes visible, usually near a necrotic area or in a villus which is in the early stages of reaction. The organisms appear ovoid, 1–4 μm long, with cytoplasm which is poorly eosinophilic and nuclei which are moderately basophilic and located centrally or towards the blunt end.[2,6] Histopathological changes are often found in the fetus[3,7] and are most readily recognized in lambs that died at or soon after birth. The most consistent lesions are found in the brain and consist of mild lymphoid perivascular cuffing and distinctive discrete foci of glial cells. In

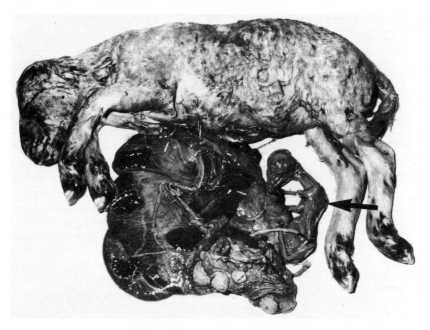

Fig. 26.1 Stillborn lamb together with small mummified fetus (arrowed) and diseased placenta arising from a *Toxoplasma* infection in mid-pregnancy.

addition a focal leucomalacia, thought to arise from anoxia induced by cotyledonary damage, is also seen. These foci occur most commonly in cerebral white matter cores, corpus striatum and cerebellar white matter. Less commonly, spongy change is present and usually confined to cerebellar peduncles. The distinctive glial foci are quite characteristic of *Toxoplasma* infection whereas focal leucomalacia on its own, while common in *Toxoplasma* abortions, has been observed, albeit infrequently, in cases of ovine chlamydial abortion.

Changes in other organs include, in the liver, accumulations of reticular cells in the portal triads and focal granulomata, sometimes encapsulated, in parenchyma. In the lung small foci of mononuclear inflammatory cells can sometimes be seen in alveoli while in the spleen there is hyperplasia of white pulp.

Diagnosis

A diagnosis of toxoplasmosis as a cause of abortion and neonatal mortality depends not only upon the clinical and post-mortem picture but also upon histological and serological findings and upon the outcome of any attempts to isolate the parasite from tissues.

Isolation Attempts to isolate *T. gondii* from dead lambs and fetal membranes are best made by inoculation of laboratory mice as they are more sensitive to infection than chick embryos or cell cultures. The choice of tissue for inoculation is dictated by the quality of the material available, the best being the cotyledon and fetal brain. About 0.5 ml of homogenate should be injected per mouse by the intraperitoneal (i/p) route, or the subcutaneous route for specimens which are contaminated with bacteria. Oral dosing is best used for preparations of unpreserved cat faeces thought to contain *Toxoplasma* oocysts, in which case extra care should be taken to prevent spillage in view of the ability of oocysts to survive for long periods.

Ascites might develop in 7–14 days after i/p injection,when tachyzoites may be demonstrated in smears of this fluid or impression smears of mesenteric lymph node. After 14 days tachyzoites are less easily demonstrated but tissue cysts are more readily found. However, tissue cysts are best sought in mice 6–8 weeks after inoculation by which time they will have reached an easily detectable size. Brains from mice are homogenized with saline either in a pestle and mortar or by passing them through a 16 gauge needle 10 times by means of a syringe. A drop of the suspension is then dried on a slide, stained with

Giemsa, mounted under a coverslip and examined with a microscope. Failure to demonstrate cysts does not rule out a positive diagnosis unless serum from the same mouse is negative when tested for the presence of antibody to *Toxoplasma*.

Serology Of the many tests available the most reliable is the dye test of Sabin and Feldman [8] which uses live virulent *Toxoplasma* tachyzoites, a complement-like 'accessory-factor' and test serum. When specific antibody acts on the tachyzoite it is altered so that it does not stain uniformly with alkaline methylene blue. The test is expensive to operate and is not free from hazard. The IFAT [8] gives titres comparable with the dye test and is safer as it uses killed tachyzoites. Other tests include the IHA, [9] the agglutination test, [10] the CF test, [11] the radioimmuno-assay (RIA) [12] and the ELISA.[13] The ELISA test can also be modified to detect *Toxoplasma* antigen in body fluids.[14]

The IHA test is easy to perform; generally titres rise later than dye test titres but are sustained longer. Titres of 1/80 or less are generally regarded as being negative and 1/160 and above as positive. Positive CF titres, which rarely exceed 1/256, point to recent acute infection. The ELISA and RIA are readily automated and therefore suitable for handling large numbers of test sera and they can also distinguish between IgG and IgM antibody.

As a single positive serum sample only indicates infection of the host sometime in the past, serological diagnosis of recently acquired toxoplasmosis in the ovine depends upon the demonstration of a rising titre, thus paired sera 14 days apart should be tested. However, if a single dye test or IFAT titre of 1/1000 or higher is recorded it can be assumed to indicate an acute infection.

Serological diagnosis of toxoplasmosis in newborn lambs is best done with precolostral serum or CSF samples when a single positive titre indicates infection as does detection of antigen in body fluids by means of the ELISA test. In post-colostral samples it is necessary to demonstrate IgM antibody against *Toxoplasma* as IgG antibody could represent absorbed maternal antibody.

Differential diagnosis The differential diagnosis should seek to eliminate other causes of abortion and neonatal mortality such as enzootic (chlamydial) abortion of ewes, vibriosis, listeriosis, border disease and salmonellosis.

Epidemiology and transmission

Epidemiological studies have produced strong evidence which indicates that the major source of *Toxoplasma* infection for susceptible animals is *Toxoplasma* oocysts excreted in cat faeces [15] (Fig. 26.2).

Susceptible cats becoming infected with *T. gondii* for the first time after infection by tissue cysts, excrete large numbers of oocysts which then become infective within a few days and can remain so for many

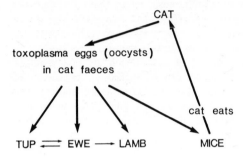

Fig. 26.2 The transmission of toxoplasmosis

months. During this time they can become disseminated over a wide area not only in dust but by passive transmission by flies, cockroaches and earthworms. In addition birds can spread infection in two ways. After ingestion of oocysts they can both excrete a proportion of the same oocysts unchanged in their faeces and at the same time become infected themselves. In the latter instance after an initial acute phase of infection they become chronically infected, with tissue cysts in brain and muscle, and act as a source of infection for cats.

However, the most important source of infection for cats is wild rodents as it has been shown with mice that infection can be passed vertically from generation to generation although the offspring appear normal. Further experiments with laboratory mice have indicated that they have a greater chance of being caught as they are less responsive than normal mice to novel stimuli.

The importance of the cat as a disseminator of infection hinges on its role as a carnivore and hence its potential for ingestion of tissue cysts which give rise to so many *Toxoplasma* oocysts. Ingestion of oocysts or tachyzoites, while being equally efficient at infecting a susceptible cat, results in far fewer oocysts being produced and hence much less contamination of the environment.

On a farm the environment of a cat tends to be centred on farm buildings although individual male cats can command a territory of up to 80 hectares (200 acres). Cats acquire infection when they start hunting so that over 60 per cent will have seroconverted by adulthood. As about 1 per cent of a population are excreting oocysts at any time, infection of the environment, which includes hay, bedding and concentrated feeds, is readily maintained. This reservoir of potential infection is therefore a danger to susceptible pregnant sheep and studies on outbreaks of toxoplasmosis suggest that infection occurs during the winter months when sheep are in mid-pregnancy when supplementary feeding with hay and concentrates can be common.

Other modes of transmission are of less importance. Infection can be picked up by susceptible animals from diseased placentae at lambing time. Vertical transmission can take place when ewes become infected during late pregnancy and give birth to live infected lambs. It has also been demonstrated that chronically-infected ewes can produce infected lambs but to what extent this occurs is not clear. Toxoplasms can also be passed in milk and semen during acute infections but there is little evidence to suggest that a great deal of infection is transmitted in this way. There is similarly very little evidence to suggest that contact transmission, outside lambing time, occurs.

Control, prevention and treatment

Control and prevention of toxoplasmosis is extremely difficult as no vaccine exists and therefore the kind of measures attempted depend upon the extent of *Toxoplasma* infection in the flock.

Once an outbreak of toxoplasmosis has started in lambing sheep there is little that can be done other than to prevent sheep in earlier stages of pregnancy contacting diseased lambs and placentae.

Practical control measures aimed at preventing stillbirths and perinatal mortality due to *T. gondii* infection depend upon the degree of penetration of the infection through the flock. Thus a 'clear' flock should not be allowed access to an environment, especially food and bedding, inhabited or contaminated by cats. In addition replacement sheep should be free from recent infection. However, where a flock is already substantially infected there is a good case to be made for spreading infection to young replacement stock and uninfected bought-in animals, 2–3 months before tupping, by exposure to a contami-

nated environment. Identification of a contaminated environment is not easy but it is likely to be any area in and around farm buildings where cats live.

In general, vermin should be kept to a minimum as should the number of cats which, if kept in a healthy state, are unlikely to suffer a recrudescence of *Toxoplasma* infection which could result in the re-excretion of *Toxoplasma* oocysts.

Treating and preventing an outbreak of toxoplasmosis in a flock of sheep with pharmaceuticals is not a practical proposition. However, if treatment is to be employed on an individual basis it is only useful during the acute phase of disease as it is only effective against tachyzoites.

Sulphonamides such as sulphadiazine, sulphamezathine and sulfamerazine are effective against toxoplasms and if given with pyrimethamine the two drugs act synergistically by blocking the metabolic pathway involving para-aminobenzoic acid and the folic-folinic acid cycle respectively. Daily doses for some weeks are often necessary.

Trimethoprim together with sulphadoxine gives a sequential double blockade of folinic acid synthesis and has been used in the treatment of toxoplasmosis.

REFERENCES

1 Frenkel J.K. (1973) Toxoplasmosis: parasite life cycle, pathology and immunology. In *The Coccidia, Eimeria, Isospora, Toxoplasma and Related Genera*. (Eds. Hammond D.M. & Long P.L.) pp.343–410. Butterworth, London.

2 Hartley W.J., Jebson J.L. & McFarlane D. (1954) New Zealand type II abortion in ewes. *Australian Veterinary Journal*, **30**, 216–18.

3 Hartley W.J. & Marshall S.C. (1957) Toxoplasmosis as a cause of ovine perinatal mortality. *New Zealand Veterinary Journal*, **5**, 119–24.

4 Reid H.W., Buxton D., Gardiner A.C., Pow I., Finlayson J. & Maclean M.J. (1981) Immunosuppression in toxoplasmosis: studies in lambs and sheep infected with Louping-ill virus. *Journal of Comparative Pathology*, **92**, 181–90.

5 Buxton D., Reid H.W., Finlayson J., Pow I. & Anderson (1981) Immunosuppression in toxoplasmosis: studies in sheep with vaccines for chlamydial abortion and louping-ill virus. *Veterinary Record*, **109**, 559–61.

6 Beverley J.K.A., Watson W.A. & Payne J.M. (1971) The pathology of the placenta in ovine abortion due to toxoplasmosis. *Veterinary Record*, **88**, 124–8.

7 Buxton D., Gilmour J.S., Angus K.W., Blewett D.A. & Miller J.K. (1981) Perinatal changes in toxoplasma infected lambs. *Research in Veterinary Science*, **32**, 170–6.

8 Frenkel J.K. (1971) Toxoplasmosis: mechanisms of infection, laboratory diagnosis and management. *Current Topics in Pathology,* **54**, 28–75.

9 Thorburn H. & Williams H. (1972) A stable haemagglutinating antigen for detecting toxoplasma antibodies. *Journal of Clinical Pathology,* **25**, 762–7.

10 Fulton J.D. & Turk J.L. (1959) Direct agglutination test for *Toxoplasma gondii. Lancet,* **ii**, 1068–9.

11 Fulton J.D. & Fulton F. (1965) Complement fixation tests in toxoplasmosis with purified antigen. *Nature, Lond.,* **205**, 776–8.

12 Finlayson J. (1980) A microtitre radio-immunoassay for *Toxoplasma gondii* antibody. *Journal of Comparative Pathology,* **90**, 491–3.

13 Araujo F.G. & Remington J.S. (1980) Antigenemia in recently acquired acute toxoplamosis. *Journal of Infectious Diseases,* **141**, 144–50.

14 Denmark J.R. & Chessum B.S. (1978) Standardization of enzyme-linked immunosorbent assay (ELISA) and the detection of Toxoplasma antibody. *Medical Laboratory Science,* **35**, 227–32.

15 Frenkel J.K. & Ruiz A. (1981) Endemicity to toxoplasmosis in Costa Rica. Transmission between cats, soil, intermediate hosts and humans. *American Journal of Epidemiology,* **113**, 254–69.

27   BORDER DISEASE

R.M. BARLOW AND A.C. GARDINER

Synonyms 'B' Disease, 'hairy-shaker' disease, congenital trembles, fuzzy lambs.

Although lambs with similar clinical signs were observed at the turn of the century, it was not until the 1950s that border disease (BD) was recognized as a distinct clinicopathological entity.[1] On a flock basis the main diagnostic feature is the birth of weakly lambs of poor conformation, the majority of which thrive badly and die before weaning. In normally fine-woolled breeds of sheep, affected lambs exhibit a coarse birth coat with a fuzzy halo of longer fine hairs which is most prominent over the rump and nape in which sites the fleece may also be abnormally pigmented. Some lambs are affected by tremor which varies from violent rhythmic contractions of the muscles of the hind legs and back to fine fibrillations of the ears and tail which are only evident on close inspection. The tremor is associated with a deficiency of stainable myelin in the white matter of the central nervous system.[2] As ewes usually develop immunity following infection the disease in endemic areas predominantly affects the progeny of primiparous ewes.

In recent years BD has been widely recognized in the UK and in most of the other major sheep rearing countries of the world. A virus aetiology closely related to that of bovine virus diarrhoea/mucosal disease (BVD/MD) has been established[3] and it is now known that the infection can be associated with significant wastage in sheep flocks in addition to that attributable to hairy-shaker lambs.

Cause

BD is the result of infection during fetal life with a pestivirus which has antigenic relationships to the viruses of BDV/MD and European swine fever or hog cholera (HC). Both cytopathic and non-cytopathic strains have been recognized and some strains appear to differ in their pathogenicity and immunogenicity in certain breeds of sheep.[4]

Clinical signs

The mature sheep exposed to BD virus experiences only a transitory infection associated with slight fever and a mild leucopenia occurring a few days after infection. Neutralizing antibodies are commonly detectable in serum some 3 weeks after infection. In pregnant ewes however the fetus becomes infected and may either be aborted or born with various malformations. Abortion may occur at any stage of pregnancy but is most common around day 90 of gestation at which stage a brown, mummified or swollen anasarcous fetus often voided. Resorption of the conceptus can also occur. Since the conceptus is small, the uterine discharge is non-purulent and the placenta rarely retained, abortions at these earlier stages of pregnancy may go unnoticed, the shepherd attributing the brown 'greasy' staining of the perineum to a 'touch of scour'. Aborting ewes show no signs of discomfort and continue to feed well. Thus the first clear evidence of disease may be at lambing time when an excessive number of ewes are found to be barren, and hairy-shaker or otherwise defective lambs of low viability begin to appear.

Diseased live lambs may be born 2–3 days early and are usually small and weakly. The birth coat abnormalities become evident as soon as the lamb has been licked dry. Many are quite limp until they are helped to stand whereupon the tremor becomes obvious. Those that can stand usually have some degree of teat-seeking ability but may be unable to prehend the teat; if held to the udder many of them will suck vigorously. In smooth-coated breeds the general hairiness of the fleece and the presence of halo hairs can be best appreciated at this time. Abnormal, tan to brown-black pigmentation, especially at the nape of the neck is often present. In the coarse-woolled breeds of sheep however the birth coat of affected lambs does not appear abnormal.

In addition to being small, hairy-shaker lambs frequently have altered conformation.[5] The frontal bones are domed, the back short and arched and the limbs short and fine with incomplete extension of the carpal joints. There may be marked shortening of the mandible so that the incisors occlude behind the dental pad.

In some outbreaks of BD, weakly lambs of normal size or moderate oversize may occur and may show little or no hairiness or obvious tremor. Some of these may have excessively long legs with long hyperextended pasterns 'camel-legged' a flattened cranium and poorly sprung rib cage. Such lambs may stand unaided but wander aimlessly and show little sucking drive. They usually die within 1–2 days.

With careful nursing a proportion of hairy-shaker lambs can be reared, though deaths attributable to enteritis or pneumonia may occur at any time. Halo hairs are soon lost and the birth coat gives way to a coarse kempy fleece. The tremor also gradually becomes less severe and by 4–5 months of age it has usually disappeared, except under stress when a fine tremor of ears and tail may be appreciated. Affected lambs often continue to thrive badly and weaning usually precipitates a further number of deaths.

Pathology

No characteristic pathological changes have been found in acutely infected non-pregnant mature animals, though there may be some lymphoreticular proliferation, small nodules or mild diffuse infiltrations of lymphocytes occurring in organs such as kidney, pancreas, liver and heart.

In pregnant ewes however, infection causes an acute necrotizing placentitis which develops in about 10 days.[6] The lesions may become diffuse resulting in fetal death or remain as multiple small foci of necrosis which heal in about 25 days and are compatible with sustained fetal life. There may be a variable degree of hydrops of the fetal membranes and the arcade haematomata of the placentomes may be extended and discoloured. Histologically the lesions commence as mural necrosis of capillaries in the basal third of the caruncular crypts, extending to involve the crypt wall. Capillary haemorrhage occurs with dissection of the union between crypt lining cells and the trophoblastic epithelium of the cotyledonary villi. Macrophages and a very few maternally-derived lymphocytes participate in the reaction. Local villus atrophy may occur. If the pregnancy is sustained the necrotic debris is phagocytosed by macrophages, possibly aided by trophoblastic epithelial cells and the caruncular stroma proliferates. The healed stroma has a more fibrous quality than usual and villus branches are less profuse at the site of the lesion.

In hairy-shaker lambs the most characteristic pathological changes are in the CNS and skin.[2] In the CNS, at all levels of the brain and spinal cord there is a deficiency of myelin which is associated with an increased density of interfasicular glia. The lesion is usually diffuse and non-inflammatory. However, the hypomyelinogenesis may be intensified in foci around blood vessels. The lesions are best appreciated in newborn lambs, in sections stained for myelin, whereupon it can be seen that the nerve fibres especially the large diameter axons, are naked or invested with disproportionately thin myelin sheaths. Three types of interfasicular glial cells are seen; astroglia, oligodendroglia and the so-called

Type III cell. Astroglia, either normal or slightly swollen, are present in fairly normal numbers while oligodendroglia are reduced in number compared to control lambs. The Type III cell on the other hand, which is only occasionally seen in normal lambs, constitutes the major proportion of glial cells in the affected lamb. In appropriately stained sections from new-born hairy-shaker lambs, Type III cells can be seen to contain numerous fine lipoidal droplets.

In older BD lambs with remitting clinical signs, myelin defects are less obvious and the lesions have usually resolved by about 20 weeks of age. The density of interfasicular glia remains high however and cells with swollen nuclei persist for up to 3 years.

The abnormalities in the fleece of BD lambs are attributable to 2 changes in the follicular structure of the skin viz. increased size of primary wool follicles and decreased numbers of secondary wool follicles.[7] There is thus a greater than normal proportion of large fibres many of which are medullated and a reduced number of fine fibres. In the recovered BD case, at certain times of the year, the fleece abnormality may not be very obvious but serial skin biopsies have shown that the follicular alterations are permanent.

It has been shown experimentally that hairy-shaker lambs with the above characteristics result only from infections acquired before 80 days gestation. Such lambs normally produce no cellular or humoral immune response to infection and are tolerant, persistent excretors of virus. However, under certain poorly understood circumstances infections acquired in the same early stages of gestation may produce a violent necrotizing and inflammatory process within the fetal CNS. Focal glial nodules form within the spinal cord, and the germinative layers of the brain may undergo extensive destruction, which if the fetus survives, results in varying degrees of cerebellar hypoplasia and dysplasia, hydranencephaly and porencephaly.[3, 8] In one investigation 18 out of 25 such lambs had specific virus neutralizing antibody in their precolostral serum. Typically these lambs are not hairy and have little tremor or spinal hypomyelinogenesis. However, not infrequently they have skeletal abnormalities giving them a 'camel-legged' appearance.

BD virus infections acquired after 80 days gestation do not produce fleece abnormalities and tremor is rarely observed. Gross destructive lesions of the CNS have not been seen and hypomyelinogenesis is absent or confined to the more rostral, later myelinating, CNS structures, e.g. cerebellar foliar cores, corpus callosum and pyriform lobe. In these lambs the main pathological feature is microscopic and consists of a disseminated nodular periarteritis affecting small and medium sized arterioles mainly in the CNS and meninges but sometimes also in the viscera. The inflammatory exudate consists of an infiltration of the media and adventitia by lymphocytes and macrophages. These lesions appear to be relatively stable and can persist with little change for at least the first year of postnatal life.

The sporadic deaths from enteritis and pneumonia which occur among recovered hairy-shaker lambs have often been attributed respectively to helminthiasis or bacterial infections. In recent work in which tolerant, recovered hairy-shaker lambs were 'superinfected' with the homologous strain of virus the experimental animals developed depression, wasting, respiratory distress and persistent diarrhoea. They all died over a period of 2–3 weeks. Though none seroconverted, at necropsy all were found to have focal lymphoid cell infiltrations in the periventricular brain substance, choroid plexuses and a variety of visceral organs including lung and large intestine. These findings suggest that an ineffective immune response had occurred as a result of changes in the host/virus relationship. Similar lymphoproliferative disease has been observed following field outbreaks of BD indicating that such altered host/virus relationships may contribute to the sporadic deaths which occur in recovered hairy-shakers, and hitherto have been attributed to other causes.

Diagnosis

Diagnosis on clinical grounds alone presents little difficulty in outbreaks in which several hairy-shaker lambs are born. Confirmation of diagnosis is based upon the histological examination of the CNS of affected lambs supported by virus isolation from spleen, blood clot or lymph node and by demonstration of viral antigen in the tissues using specific immunofluorescent staining. The presence of neutralizing antibody in maternal or post-colostral lamb sera cannot be considered as diagnostic as antibodies in maternal serum may persist from one year to the next.

Diagnosis is more difficult in outbreaks of BD in which abortion, barrenness or the birth of weakly or 'camel-legged' lambs predominate. It must be based on virus isolation coupled with an assessment of serum neutralizing antibody levels in mothers of affected lambs, considered in relation to those in the flock as a whole and the general status of sheep in the region. In differential diagnosis it is important to consider bacterial abortions, *Salmonella abortus ovis*, *Brucella ovis*, *Listeria monocytogenes* and *Campylobacter fetus*; chlamydial abortion; toxoplasmosis; bacterial meningoencephalitis due to *E. coli* or pyogenic bacteria. Nutritional disorders such as swayback (enzootic ataxia) in newborn lambs may also be confused with BD.

Epidemiology and transmission

The epidemiology of BD is poorly understood. Serological survey data from several countries indicate that about 70 per cent of mature cattle have experienced infection with BVD/MD viruses whereas in the same regions the prevalence of antibody in sheep is much lower, varying from less than 5 per cent to about 16 per cent. It has been shown that some cattle strains of virus will infect sheep and vice versa and will cause abortion and/or malformations. However, it is rare for outbreaks of BD to be preceded by the occurrence of clinical BVD/MD in the cattle population. Probably the most frequent form of introduction of infection is by the addition to a breeding flock of a recovered case of BD. Such animals are persistently infected and spread virus in their secretions and excretions over long periods. Susceptible pregnant ewes in contact with excretors become infected and have affected progeny but infections in non-pregnant animals are subclinical. Persistently infected females can conceive and may abort or produce affected lambs over a period of years. Affected ram lambs which reach maturity usually have small soft testicles and are subfertile. They can however transmit the infection in their semen either to their own progeny if conception occurs, or to other progeny if the ewe returns to service with another ram, at the first or second subsequent oestrus.[9] Early abortions provide potent sources of infection for inquisitive pregnant flock mates. The virus is susceptible to desiccation and ultraviolet light and in favourable conditions will not persist long outside the host but may be protected from these effects when in aborted fetal tissue in moist conditions with little sunlight.

Though several species of deer have been shown to be susceptible to BVD/MD virus infection it is not known to what extent these or other wild animals may contribute to the spread of infection.

Control

The control of BD depends on the extent of infection in the flock. In outbreaks where circumstances indicate a low level of infection, segregation of affected lambs and their dams with strict attention to hygiene and with disposal of surviving lambs prior to the next breeding season may be all that is required.

However, by the time the presence of the disease is recognized it is probable that the infection will have been widely disseminated through the flock and much of the damage already done. In this event it is wise to segregate in-lamb ewes from those with affected lambs at foot (to minimize the extent of the periarteritic form of the disease) until lambing time is past. Since infection outwith pregnancy results in an immunity to reinfection with the same strain of virus, susceptible animals retained for breeding should be thoroughly mixed with surviving hairy-shakers through the grazing period to maximize their chances of becoming immune. However, all lambs from a flock in which BD has occurred should be sent for slaughter before the breeding season comes round again.

Vaccination against challenge with a homologous strain of virus has been shown to be effective experimentally.[10] However, immunization did not protect against challenge with a heterologous strain of virus.[11] The heterologous challenge also resulted in the birth of some lambs with massive intracranial malformations such as hydranencephaly. As these pathological features are more life-limiting than hypomyelinogenesis, vaccination is not a practical procedure until polyvalent vaccines containing a wide spectrum of antigenic variants have been developed.

REFERENCES

1 Hughes L.E., Kershaw G.F. & Shaw I.G. (1959) 'B' or Border disease. An undescribed disease of sheep. *Veterinary Record*, **71**, 313–17.

2 Barlow R.M. & Dickinson A.G. (1965) On the pathology and histochemistry of the central nervous system in Border disease of sheep. *Research in Veterinary Science*, **6**, 230–7.

3 In: Border Disease of Sheep; a Virus Induced Teratogenic Disorder. Eds. R.M. Barlow, D.S.P. Patterson. (Fortschritte der Veterinarmedizin H36 1982) Berlin & Hamburg, Paul Parey.

4 Barlow R.M., Vantsis J.T., Gardiner A.C. & Linklater K.A. (1979) The definition of Border disease: problems for the diagnostician. *Veterinary Record*, **104**, 334–6.

5 Terlecki S., Herbert C.N. & Done J.T. (1973) Morphology of experimental Border disease of lambs. *Research in Veterinary Science*, **15**, 310–17.

6 Barlow R.M. (1972) Experiments in Border disease IV. Pathological changes in ewes. *Journal of Comparative Pathology*, **82**, 151–7.

7 Carter H.B., Terlecki S. & Shaw I.G. (1972) Experimental Border disease of sheep: effect of infection on primary follicle differentiation in the skin of Dorset horn lambs. *British Veterinary Journal*, **128**, 421–7.

8 Barlow R.M. (1980) Morphogenesis of hydranencephaly and other intra-cranial malformations in progeny of pregnant ewes infected with pestiviruses. *Journal of Comparative Pathology*, **90**, 87–98.

9 Gardiner A.C. & Barlow R.M. (1981) Vertical transmission of Border disease infection. *Journal of Comparative Pathology*, **91**, 467–70.

10 Vantsis J.T., Rennie J.C., Gardiner A.C., Wells P.W., Barlow R.M. & Martin W.B. (1980) Immunization against Border disease. *Journal of Comparative Pathology*, **90**, 349–54.

11 Vantsis J.T., Barlow R.M., Gardiner A.C. & Linklater K.A. (1980) The effects of challenge with homologous and heterologous strains of Border disease virus on ewes with previous experience of the disease. *Journal of Comparative Pathology*, **90**, 39–45.

28

VIBRIOSIS

N.J.L. GILMOUR

Synonym Campylobacteriosis

Vibrionic abortion in sheep was first described in 1913 and has since been reported in many countries. The causal organism was orginally assigned to the genus *Vibrio*, hence the name vibriosis, but it is now in the genus *Campylobacter*. However, 'vibriosis' has priority and ease of pronunciation in its favour.

Cause

Vibriosis is caused by infections with *Campylobacter fetus* subspecies *intestinalis* or *C. fetus* subsp. *jejuni*.[1] The bacteriological differentiation of these two closely related organisms is shown in Table 28.1. Five different serotypes of the subsp. of *C. fetus* have been described [2] *C. fetus* subsp. *jejuni* belongs to serotype 1; subsp. *fetus* and *intestinalis* occur in groups II, III and V.[3] Serotyping is not however a procedure which is used routinely. *C. fetus* subsp. *fetus* causes the epidemiologically distinct, venereally transmitted, vibrionic abortion in cattle.

C. fetus is a Gram-negative, non-sporing, motile bacterium which is comma-shaped (1.5–5.0 μm long and 0.3–0.5 μm wide) in 48 h cultures and filamentous in old cultures. It is microaerophilic and grows best in an atmosphere consisting of approximately 6 per cent (v/v) oxygen, 10 per cent (v/v) carbon dioxide and 84 per cent (v/v) hydrogen or nitrogen. On primary isolation, after incubation for 3–4 days on 10 per cent blood agar plates, the colonies are low, flat, smooth, greyish-brown and translucent with a tendency to coalesce along the streak and are non-haemolytic.

Both *C. fetus* subsp. *intestinalis* and *jejuni* have been isolated from the blood and intestines of man, in whom they cause a gastroenteritis, and from the intestines of a wide range of wild-life.

Clinical signs

The clinical sign of an active infection in pregnant sheep is abortion during late pregnancy. Lambs may be carried to full term but are born dead or in a weak condition.

Pathology

The principal lesion is a placentitis. The cotyledons separate easily from the caruncles. There are arteriolitis and thrombosis in the hilar zone of the

Table 28.1 The differentiation of the subspecies of *C. fetus*.

	C. fetus subsp. *fetus*	*C. fetus* subsp. *intestinalis*	*C. fetus* subsp. *jejuni*
Growth at 25°C	+	+	−
Growth at 42–45°C	−	−	+
Growth in 1% glycine	−	+	+
Production of H$_2$S	−	+	+

placentomes. Large numbers of organisms are found in chorionic cells and capillaries. Pale necrotic foci are sometimes present in the enlarged and haemorrhagic livers of aborted fetuses. Blood-stained serous fluid occurs in the body cavities.

Diagnosis

Smears from the cotyledons and stomach contents of a freshly-aborted fetus should be examined microscopically for the presence of small, comma-shaped bacteria. These are often difficult to see on conventional Gram staining and it is recommended that strong (Ziehl-Neelsen's) carbol fuschin be used. For culture from contaminated specimens, 10 per cent blood or chocolate agar incorporating bacitracin (15 iu/ml), novobiocin (10 µg/ml) and polymyxin (2 iu/ml) is of value. Optimum growth conditions can be obtained by the use of gas generating kits specifically for *Campylobacters* (Oxoid Ltd., Basingstoke, Hampshire) in conjunction with standard anaerobic jars. Specimens sent to a laboratory for diagnosis should include cotyledons and fetal stomach contents from a number of abortions. Vaginal swabs from aborted ewes are less useful, especially since other *Campylobacter* sp can be cultured from normal ewes. It is unwise to ascribe abortions to *Campylobacter* infection on the isolation of *C. fetus* from a single specimen since infections with other abortifacient agents, e.g. chlamydia or toxoplasma, can occur simultaneously in a flock.

Epidemiology and transmission

Between 40 and 80 diagnoses of fetopathy due to *C. fetus* are made annually in UK but this is unlikely to be a true reflection of the prevalence of the disease. Vibrionic abortion occurs all over Britain and in one recent year involved 22 different breeds or crossbreeds and all ages of ewes. Within flocks the abortion rate varies from 5–50 per cent. Experimental

evidence suggests that abortions are commoner if infection is acquired beyond the twelfth week of pregnancy and that abortion occurs between 7 and 25 days after infection.

Vibriosis in sheep is transmitted orally. It occurs in many closed flocks and the source of infection is not always clear. Sheep can carry *C. fetus* in their intestines and excrete the organisms in the faeces. It is therefore likely that some outbreaks are due to the contact of pregnant sheep with a faecal excretor which may have been introduced into a flock. The appearance of the disease in closed flocks has been attributed to contamination of pastures by wild-life vectors, as sparrows, carrion crows and magpies have been found to be intestinal carriers. Serotype 1 isolates from naturally-infected carrion crows produced abortion in ewes experimentally infected.[4] After the first abortion infection is further spread as a result of contamination of pastures from the products of this abortion and by direct contact with fetal membranes and fetuses.

Control and prevention

The disease is self-limiting in that is is unusual for vibrionic abortions to occur in a self-contained flock the year after an outbreak of abortion from this cause. Ewes which are in contact with abortions but which do not themselves abort, and non-pregnant females, do not abort at subsequent pregnancies. This development of immunity to abortion can be used to ensure that an outbreak is confined to one lambing season. Ewes which abort should not be removed from the flock and potential additions to the breeding flock should be mixed with the infected flock. However, hygienic measures should be taken to prevent spread of infection from fetal membranes and aborted fetuses to adjacent flocks and to possible wild life vectors. Infection can easily be spread on boots and it should be noted that the organism remains viable in vaginal discharge on pasture or bedding for several days.[4] *C. fetus* causes gastroenteritis in man and is a further reason for proper sanitary precautions during outbreaks.

Treatment with penicillin and streptomycin and concurrent vaccination of in-contact ewes after the first abortion has prevented further abortions [5] but there is no evidence that therapy alone is useful.

Vaccination with disrupted, formalinized cells of *C.fetus* subsp. *intestinalis* in oil adjuvant has been shown to protect against experimental challenge. Also vaccination of ewes after abortion in an experimentally-infected contact ewe resulted in a reduction in abortions.[2] This suggests that vaccination in the face of an outbreak might be of value, although routine vaccination of commercial flocks is less justifiable economically, since in Britain the prevalence of the disease is low, outbreaks are sporadic and an active immunity develops which prevents abortions in the following year. Any vaccine for use here would have to contain the three serotypes (I, II and V) which occur in Britain. In the USA there are a number of commercial *Campylobacter* vaccines for use in sheep but the efficacy of some of them is in doubt.[4] It seems probable that an efficient adjuvant is important in vaccination against vibriosis.

REFERENCES

1 Watson W.A. (1979) *Fertility and Infertility in the Domestic Animals.* (Ed. Laing J.A.) 3e, p.213. Baillière Tindall, London.

2 Gilmour N.J.L., Thompson D.A. & Fraser J. (1975) Vaccination against Vibrio (*Campylobacter*) foetus infection in sheep in late pregnancy. *Veterinary Record,* **96**, 129–31.

3 Smibert R.M. (1974) Genus II Campylobacter. *Bergey's Manual of Determinative Bacteriology.* (Eds. Buchanan R.E. & Gibbons N.E.) 8e, pp.209–10. Williams & Wilkins, Baltimore.

4 Bryner J.H., Foley J.W. Hubbert W.T & Matthews P.J. (1978) Pregnant guinea pig model for testing efficacy of *Campylobacter fetus* vaccines. *American Journal of Veterinary Research,* **39**, 119–21.

5 Ullman V. (1979) Methods in Campylobacter. In *Methods in Microbiology.* (Eds. Norris J.R. & Bergan T.) Vol.2, p.445. Academic Press, New York.

29 SALMONELLOSIS AND SALMONELLA ABORTION

K.A. LINKLATER

Synonym Paratyphoid

Several serotypes of salmonellae have been associated with deaths and abortion in sheep in all parts of the world. The number of flocks affected annually may be small but within those the losses may be catastrophic, as many animals can be affected in any one outbreak.

Cause

The organisms of the *Salmonella* group are short Gram-negative rods which are aerobic and mostly grow well on laboratory media at 37°C. They belong to the family *Enterobacteriaceae* and are differentiated from other members of that family by their biochemical and serological reactions. Salmonellae do not ferment lactose or sucrose but ferment glucose, mannite and dulcite. They produce hydrogen sulphide but not urease or indole. They are further identified by the serological reactions of antigens carried on their cell surfaces (somatic or O antigens) and on their flagellae (H antigens).

The large number of organisms within the *Salmonella* group are also differentiated from each other by these antigens and are classified into groups by means of their O antigens and into individual serotypes by means of their H antigens. For example *Salmonella typhimurium* possesses O antigens 4 and 12 which place it in Group B and H antigen i which distinguishes it from other members of that group.

Only a few members of the *Salmonella* group are host-specific and among these is *S. abortus ovis* which causes abortion in sheep. It used to be of major significance to the sheep industry in S.W. England but is now rarely seen. Of the other serotypes which are potentially pathogenic for most animals and man those which are most commonly associated with disease in sheep are *S. typhimurium, S. dublin* and *S. montevideo*.[1]

The last of these has regularly been associated with abortion in ewes in Scotland. From 1970 to 1981 a total of 67 incidents was reported [2] but in 1982 there was a sudden upsurge mainly in the South East of the country where the infection was confirmed as a cause of abortion in 38 flocks. *S. montevideo* has therefore become a significant cause of loss in sheep flocks in certain parts of Scotland.

Clinical signs

Clinical signs of salmonellosis are varied and include general systemic and enteric symptoms as well as abortion. Signs vary with the serotype, for example with *S. abortus ovis* the main symptom is abortion usually in the last six weeks of gestation. Affected ewes may have a transitory pyrexia, which can go unnoticed, and are rarely seen ill unless post-parturient metritis due to secondary bacterial infection develops. Scouring is not usually a feature of this infection except occasionally in young lambs in contact with aborting ewes.

With *S. montevideo* the clinical picture is very similar to that with *S. abortus ovis*. Apart from aborting, affected ewes usually appear normal. Many ewes in an affected group may excrete the organism in their faeces without aborting and the infection can pass through a group of non-pregnant animals without any symptoms being apparent.[3]

On the other hand with *S. typhimurium* enteric and systemic symptoms predominate.[4] Affected animals are anorectic, have high temperatures (over 41°C/106°F) and scour profusely. Those which do not die of septicaemia may continue to scour and die from dehydration with sunken eyes and tight skins. Pregnant animals may die of septicaemia before aborting and if this serotype is present in a group of in-lamb ewes all symptoms are usually present including some apparently sudden deaths without prior signs of ill-health.

Symptoms with *S. dublin* infection are similar to those produced by *S. typhimurium* with pyrexia, malaise and diarrhoea being common as well as abortion. Outbreaks have been reported where abortion is the predominant symptom, with little clinical disturbance in lambs over one week old.[5]

Pathology

Aborted fetuses and placentae are usually fresh and no typical macroscopic lesions are present. Post-mortem findings in carcasses of animals which die of salmonellosis are variable. Usually signs of septicaemia are seen, i.e. enlargement of spleen and congested organs. In animals which have not eaten for several days, the gall bladder is distended and the liver very friable. Signs of acute abomasitis and severe enteritis are present in animals which live longer with the associated lymph nodes being grossly enlarged.[4] Intestinal contents tend to be very fluid and inflammatory changes may also be detected in the caecum and colon.

Diagnosis

Confirmation of diagnosis depends on the isolation of the causal organisms. In cases of abortion, typical Gram-negative organisms are seen on direct smears made from stomach contents of fetuses and from placentae. These can be confirmed as salmonellae by direct overnight culture on desoxycholate citrate agar (DCA) or MacConkey agar. Typical non-lactose fermenting colonies are present in profuse numbers and it should not be necessary to resort to enrichment media, such as selenite F for primary isolation if salmonellae are the cause of abortion. *S. abortus ovis* grows more slowly than other serotypes and may require up to 72 h for colonies to reach a significant size on solid media.[6]

In enteric and septicaemic cases isolation of the causal organisms should also be possible on direct culture of organs, faeces, intestinal contents and drainage lymph nodes. Initial cultures in a selective liquid medium such as selenite broth are sometimes helpful from faeces, intestinal contents and lymph nodes and allow the salmonellae to multiply preferentially over other members of the family *Enterobacteriaceae*. Typical non-lactose fermenting colonies are then identified by subculturing overnight on DCA or MacConkey.

In chronic enteric infections the predilection sites for the infection are the kidney-shaped, posterior mesenteric lymph nodes which become enlarged. It appears that the infection is carried in these for longer than elsewhere in the body and it is possible to recover organisms from them after faecal excretion has ceased.

A serum agglutination test may be used for detecting ewes infected with *S. abortus ovis* but it is advisable to sample several sheep soon after abortion as the antibody levels usually drop 2–3 months after parturition. Serological tests are not commonly used for other serotypes in sheep.[7]

Epidemiology and transmission

As *Salmonella abortus ovis* is host-specific, it is invariably introduced into a flock by an infected sheep. On the other hand there are many sources of infection for the other serotypes, e.g. food, water, other animals, wild birds and man and frequently it is not possible to determine the origin of an outbreak. Infection probably takes place by mouth and in all probability occurs a long time before the abortions are actually apparent. Experimentally it has been shown that ewes infected with *S. montevideo* at 12 and 14 weeks gestation aborted 2–3 weeks later while ewes infected at 16 and 18 weeks gestation with the same dose of organisms lambed normally at term. This suggests that, when outbreaks of abortion occur in a flock due to *S. montevideo*, infection may have occurred several weeks before clinical abortions are apparent. Hence the difficulties experienced in trying to trace the sources of infections in natural outbreaks of the disease.

This epidemiological pattern also means that *S. montevideo* is likely to be of much greater significance and to cause greater losses if it gets into a flock when the ewes are in the early stages of gestation than if it is introduced near to the expected date of onset of lambing.

Aborted fetuses and placentae are heavily contaminated with organisms and are rich sources of infection for other animals which become infected by licking and eating the material. Organisms are also excreted in faeces which can readily contaminate food and water troughs as well as pasture. Fields containing affected animals therefore become heavily contaminated with organisms which may find their way into water courses. Consequently, animals drinking from rivers or streams downstream from such infected fields may inadvertently become infected.[4]

Food can also be a source of infection, especially inadequately sterilized animal proteins and bone meal. These are most frequently the sources of so-called exotic serotypes of salmonellae, rather than *S. abortus ovis*, *S. typhimurium*, and *S. dublin*, and usually originate in countries outwith the UK.

Wild birds have been incriminated in spreading some serotypes especially *S. montevideo* which has frequently been reported as a cause of ovine abortion in areas frequented by wild geese. The evidence however, remains circumstantial because, as abortions do not occur until several weeks after infection, it is difficult in the field situation to tell whether geese have infected the sheep or vice versa.

It is more likely that foci of infection with *S. montevideo* occur within the sheep in a flock and lateral spread to neighbouring flocks is brought about by mechanical transfer by wild birds. This is supported by work on gulls which has shown that salmonellae do not appear to multiply and establish in these birds but that large numbers of serotypes may be carried by them from rubbish tips, sewage outfalls and other places where they feed.[8] Sheep troughs in winter could be another source of infection for them if *S. montevideo* or any other serotype is present in a flock.

With salmonellae, animals frequently become carriers of infection. This is especially true of *S. dublin* in cattle which can then be introduced into susceptible populations of other animals including sheep by symptomless carrier cattle. With *S. typhimurium*, sheep do not tend to become carriers and usually animals which survive acute episodes of disease caused by that organism quickly throw off the infection and stop excreting the organism within 6 weeks of being infected. With *S. abortus ovis* carriers are difficult to detect. This may be because that organism is difficult to grow or because the carrier state is rare.[6] Nevertheless, it seems that infected sheep are the main route by which *S. abortus ovis* is introduced into a flock.

Control, prevention and treatment

Usually by the time salmonellosis is detected in a flock of sheep the infection has spread throughout the group. Nevertheless, isolation of affected animals is worthwhile to try to limit the weight of infection which may pass to the rest of the flock.

Animals which are septicaemic have to be treated with suitable parenteral antibiotics and those which scour may require supportive therapy with electrolyte solutions. Ewes which abort frequently develop post-parturient metritis and injections of long-acting preparations of antibiotics at the time of abortion help to prevent this. Attempts have also been made to minimize losses by the administration of antibiotics and furazolidone in food as soon as the causal agent is identified. The treatment should be continued for at least 7 days at therapeutic levels but care must be taken in the administration of furazolidone lest toxicity due to overdosing occurs. Results from these treatments are equivocal and evidence for their efficacy are mostly circumstantial.

With *Salmonella abortus ovis* immunity appears to follow natural infection and abortion storms are only experienced in the first year following introduction into a flock. Thereafter in the endemic situation only replacements tend to be affected. Attempts to protect these have been made with apparent success by mixing non-pregnant female replacements with aborting ewes. However, this may produce more carriers but this is of little consequence in flocks which are already infected. It has, however, the disadvantage that should enzootic abortion of ewes (EAE) also be present in a flock it will be spread at the same time unless replacements are vaccinated against EAE before mixing.

It is unusual for other serotypes of salmonellae to become endemic in flocks in the UK and normally, after an episode of disease, the infection disappears from a flock.

A dead vaccine against *Salmonella abortus ovis* is available in some countries but its efficacy is in doubt. The fairly limited distribution of this problem in the UK does not warrant the production of a commercial vaccine in this country. Because they share a common H antigen, a commercially-available live *S. cholera-suis* vaccine has been used in south west England with some apparent success for the control of abortion in sheep due to *S. abortus ovis*. It is possible that the same vaccine might also be useful for controlling abortion due to *S. montevideo* as that serotype and *S. cholera-suis* also have some antigens in common.

Other serotypes of salmonellae are so sporadic in occurrence that, even if efficient ones were available, the use of vaccines cannot be justified unless perhaps in flocks considered to be particularly at risk in certain situations, for example, those in contact with infected cattle.

REFERENCES

1 Animal Salmonellosis Annual Summary (1979). Ministry of Agriculture, Fisheries and Food and Department of Agriculture and Fisheries for Scotland.

2 Sharp J.C.M., Reilly W.J., Linklater K.A., Inglis D.M., Johnston W.S. & Miller J.K. (1983) *Salmonella montevideo* infection in sheep and cattle in Scotland, 1970–81. *Journal of Hygiene, Cambridge,* (In press).

3 Linklater K.A. (1983) Abortion in sheep associated with *Salmonella montevideo* infection. *Veterinary Record,* (In press)

4 Hunter A.G., Corrigall W., Mathieson A.O. & Scott J.A. (1976) An outbreak of *S. typhimurium* in sheep and its consequences. *Veterinary Record,* **98**, 126–30.

5 Baker J.R., Faull W.B. & Rankin J.E.F. (1971) An outbreak of Salmonellosis in sheep. *Veterinary Record,* **88**, 270–7.

6 Jack E.J. (1968) *Salmonella abortus ovis:* an atypical salmonella. *Veterinary Record,* **82**, 558–61.

7 Jack E.J. (1971) Salmonella abortion in sheep. In *The Veterinary Annual.* 12th Issue,(Ed. Grunsell C.S.G.) p.57. John Wright & Sons Ltd., Bristol.

8 Fenlon D.R. (1981) Seagulls (*Larus* spp.) as vectors of salmonellae: an investigation into the range of serotypes and numbers of salmonellae in gull faeces. *Journal of Hygiene, Cambridge,* **86**, 195–202.

30

INFERTILITY

A. GREIG AND D.W. DEAS

RAM INFERTILITY

An accurate figure for the incidence of infertile rams has not been produced but some workers [1,2] have put forward figures of between 4–10 per cent based on the examination of animals with a history of doubtful breeding performance. In recent years in the UK there has been increased interest in the evaluation of semen collected by electro-ejaculation from rams prior to the breeding season and 10–12 per cent of rams examined have been suggested as being infertile or subfertile. This may be an over-estimate of the size of the problem because workers[3] at the Meat and Livestock Commission have shown that semen collected by electro-ejaculation can be inferior and more variable in quality than semen collected into an artificial vagina which must be considered as a more normal ejaculate. However, in a limited study Lees[4] found that about 30 per cent of rams were less than fully satisfactory and 10 per cent were infertile.

Developmental abnormalities as causes of infertility

Cryptorchidism Cryptorchidism is attributed to a hereditary factor transmitted by the male. The testes may be found intra-abdominally or in the inguinal canal either unilateral or bilateral. Hyperthermia of the testes in these abnormal positions causes degenerative changes so that spermatogenesis does not occur. Thus, if it is a bilateral condition the ram will be sterile. If unilateral, normal semen will be produced from the normal descended testicle but such a ram would be expected to serve fewer ewes successfully. However, because of the hereditary etiology of this condition cryptorchids should not be used for breeding replacement stock rams.

Testicular hypoplasia Like the previous condition this too is usually an inherited trait with either or both testicles affected to varying degrees. Histologically there is failure of development of the spermatogenic epithelium but the interstitial tissue and cells of Leydig are normal. If hypoplasia is severe and bilateral the ram will be sterile whereas if unilateral and slight the semen will be dilute and contain an increased percentage of abnormal sperm.

Epididymal hypoplasia or aplasia The epididymis is the site where sperm mature and are stored and therefore any reduction in the size of this organ will adversely affect its function and semen quality. Epididymal hypoplasia and aplasia are considered to be congenital anomalies.

Spermatocoele This condition arises from an obstruction of the epididymis and can occur at puberty from imperfect formation of epididymal tubules or as a sequel to an inflammatory condition in the epididymis. Stasis of sperm ensues and the increased accumulation of spermatozoa results in duct dilation, destruction of duct epithelium and release of sperm into the interstitial tissue which evokes a granulomatous reaction. Part or whole of the epididymis can be involved but the tail is most commonly affected.

On palpation the affected portion of epididymis will feel enlarged and nodular and grossly is comprised of green-yellow milky or caseated material surrounded by fibrous tissue. The testicle is generally unaffected. Cultural examination is frequently unrewarding.

It is generally a progressive condition which, if bilateral, will lead to infertility. Before gross changes are palpable semen quality deteriorates with reduction in sperm numbers and increase in percentage of abnormal forms. If unilateral, the affected testicle can be removed.

Inflammatory causes of infertility

Orchitis Inflammation of the testicle whether caused by an infectious agent, e.g. *Corynebacterium pyogenes* or trauma, is rapidly reflected in subnormal fertility. In the acute stage heat, pain and swelling are evident while chronic orchitis is characterized by reduced testicular mobility and induration of testicular tissue.

When orchitis affects both testicles, infertility is permanent while if it is a unilateral condition the contralateral testicle undergoes degenerative changes during the acute phase but regeneration may follow so that a degree of fertility will return. Semen changes occur early in the process so that ejaculates contain only a few normal sperm and inflammatory exudate.

Treatment with systemic antibiotics can give variable results and, when unilateral, removal of the affected organ early in the disease may prove the best course of action.

Epididymitis This is a condition which has caused serious problems in some countries, e.g. South Africa, Australia and New Zealand where *Brucella ovis* and *Actinobacillus* sp have been incriminated. While these organisms cause endemic problems other organisms, e.g. *Corynebacterium, Pasteurella, staphylococci* and *streptococci* can cause sporadic disease. While infection may be blood-borne there is evidence that organisms present in the sheath can migrate, under the influence of systemic hormonal stimulation, to the deeper-lying organs of the reproductive tract.[5] Thus in a *Br. ovis* or *Actinobacillus* endemic situation infection can be passed from older carrier animals to young stock. In the UK cases of epididymitis are sporadic and frequently coexist with orchitis. Trauma has been put forward as a possible cause since bacteriological examination has often proved unrewarding.

In the acute stage acute pain and swelling is evident while induration is a feature of the chronic disease.

Treatment and prognosis are the same as for orchitis except that in infection with *Brucella ovis* and *Actinobacillus* sp culling is recommended.

Infection of accessory glands These are organs which have received little attention in the ram although Jansen[5] found that infection in these preceded that of testicles and epididymes and this could explain why rams producing infected semen had normal epididymes and testes. *Pasteurella, staphylococci, Corynebacterium* sp and *Actinobacillus* sp have been involved. In the UK, ejaculates rich in leucocytes are not uncommon particularly in young animals with questionable fertility but these infections apparently resolve with no long-term effect on fertility.

Tubular degeneration

Grossly this may sometimes be discerned by a soft feel to the testicle and reduced epididymal distension but more frequently it is recognized by a reduction in sperm density and increased number of morphologically abnormal and immature sperm and constitutes the most common form of reduced fertility. Degenerative changes are found on histopathological examination of testes with, in chronic cases, fibrosis and calcification.

There are many factors and conditions which have been associated with these degenerative changes but quite a number do not cause primary disease in the testes. In the ram the most common cause is elevation of testicular temperature above 41°C (106°F) and the extent of damage is related to the duration of increased scrotal temperature which may arise from high ambient temperature, heat from decomposing litter, excess scrotal wool, scrotal mange or scrotal dermatitis. Systemic disease, foot rot, pain and marked undernutrition have also been implicated in addition to the ageing process itself.

Semen changes may occur as early as three days after an increase in scrotal temperature. Regenerative changes are evident once the cause is removed and semen of near normal density and variable motility, but still with high numbers of abnormal sperm, is produced.

Diagnosis is based on semen evaluation of an animal which probably has reduced testicular and epididymal tone and has previously shown normal fertility. The prognosis depends on the cause of the insult to the testes and spermatozoa and the findings of clinical and semen examination.

Testicular atrophy

This term refers to a marked reduction in size and increased firmness of a previously normal testicle in an animal with an earlier history of satisfactory fertility and represents the end-point of extensive tubular degeneration. It is therefore usually bilateral but can be unilateral when the neighbouring testicle has suffered acute orchitis. Semen samples are of reduced density with increased numbers of abnormal spermatozoa, particularly detached heads and tail deformities.

Conditions of penis and prepuce

Inability to fully erect or extrude the penis, arising from a developmental or congenital defect or adhesions, result in a ram being unable to serve properly. Balanoposthitis (described elsewhere) can be a problem in certain flocks. Preputial prolapse leading to ventral deflection of the erect penis has occasionally been recorded.

Neoplasia

Testicular neoplasms are rare but Sertoli cell tumours and seminomas have been described.

Impotence

Whether due to a psychological or physical condition impotence can be troublesome. Ram lambs often need to learn how to serve properly and it is often useful to run them with an older animal and use them on mature ewes until their performance is normal. However, bullying by older rams should be avoided. Overfatness and painful conditions of back, legs or feet will result in animals being unwilling to seek out and serve receptive ewes.

Treatment depends on identifying the cause and removing or alleviating it. It is sometimes found that rams which are poor workers when young do not improve.

Seasonal effects

Sperm production occurs throughout the year but volume and density reach low levels in the spring. While this seasonal effect is more marked in the hill breeds it should always be taken into account when assessing sperm quality outwith the normal breeding season.

Semen properties

The correlation between semen quality and fertility has been examined by workers in UK and overseas and opinions vary on the value of routine pre-mating evaluation of semen. While collection of semen into an artificial vagina produces the best samples, in the field situation with untrained rams and difficulties in identifying ewes in oestrus, electro-ejaculation using either a transistorized or variable voltage rectal probe is the normal practice. Semen from normal rams is creamy-white and exhibits strong wave motion. Microscopic examination reveals a low percentage of dead sperm (< 10 per cent) and the percentage of abnormal forms is of a similar order or less. Figures in excess of 20 per cent for both dead and abnormal forms are not normal. Logue [3] suggested that two parameters only be used, motility and percentage of abnormal sperm. Any sample with motile sperm and less than 30 per cent abnormals would be considered normal. Ewes served by rams with normal and questionable semen had conception rates of 86 and 79 per cent respectively. Previous workers [6,7] have produced complicated methods of calculating fertility indices which incorporated pH, percentage normal live, percentage abnormal and percentage abnormal necks but these too showed that semen samples had to contain as low as 40 per cent normal live sperm before fertility was adversely affected.

In view of these findings it is important that rams which have no history of infertility and are clinically normal should not be condemned on the result of a single semen sample collected by electro-ejaculation. It is recommended in the pre-mating situation that at least three samples are taken before doubts are raised as to the potential fertility of the animal. The use of a doubtful ram on a small number of ewes ahead of the normal breeding season has much to recommend it as an indication of the likely outcome of more widespread use. If a ram is presented at the start of the breeding season with soft spongy testicles and small, virtually empty epididymes and yet a semen sample of reasonable quality is produced it is probably the case that he will only have functional capacity for a small number of ewes.

There is considerable variation in the number of ewes a particular ram can manage. On average 50–60 ewes per adult ram and up to 25 ewes for a ram lamb are considered normal figures and below these a degree of infertility is present.

When a ram is suspected of being infertile or subfertile then clinical examination and semen evaluation may well help to establish the nature of the problem.

EWE INFERTILITY

The normal conception rate for ewes is 75–80 per cent per service. Information available from the Meat and Livestock Commission indicates that, nationally up to 6 per cent of ewes were barren with the incidence being similar in lowland and upland recorded flocks. There are, however, reports[8] of farms with a much higher number of barren ewes, e.g. 18.5 per cent but figures for individual farms can show a wide variation from year to year.

Developmental defects

These are considered rare and no accurate information is available as to their incidence. Malformations of the genital tract may be found in association with other defects. Ovarian hypoplasia has been recorded. Anastomoses of chorionic vessels are estimated to occur in up to 1 per cent of ovine placentae and therefore a female fetus twin to a male can have freemartin type changes, viz small vagina, enlarged clitoris, blind ending vagina and rudimentary gonads.

Ovarian inactivity

In temperate climates the seasonality of the breeding season is dependent on photoperiodism and genetic differences, e.g. all the year round with Merinos, 6–10 months of the year with Dorset Horns and 4–5 months with mountain breeds.

A degree of suboestrus is common at the start of the breeding season with ovulation occurring without overt oestrus. The sudden introduction of entire or vasectomized rams into a flock where the majority of ewes are in anoestrus will, through the effect of pheromones emitted from skin and wool, stimulate earlier functional and behavioural oestrus over approximately 10 days with peaks first at 18 days and the second 24 days after introduction of rams. The success of 'teasing' varies with breed, strain of ewe and stage of non-breeding season and in some instances when breeding is attempted well ahead of the normal breeding season the teasing effect is lost and animals relapse into anoestrus.[9]

The age at which lambs start to cycle is dependent on age and weight and the number of heats in the first season correlates well with the weight of the animal at first oestrus.

Dietary and environmental effects on ovulation and implantation

The body weight or condition score at breeding time has been shown to have a marked effect on the number of lambs born and the number of overt heats. Ewes should have a condition score of 3–3.5 when rams are first put out and this should be maintained for the first month of pregnancy. Ewes leaner than that should be separated and fed before the breeding season. Ewes losing weight at the time of mating produce significantly fewer lambs than ewes maintaining weight and both produce significantly fewer than ewes gaining weight.[10] This difference can be considered to be composed of differences in ovulation rate and early embryonic loss. Flushing (i.e. increasing the nutritional status) of ewes in store condition for one oestrus cycle prior to mating may increase the ovulation rate by preventing late atresia of large Graafian follicles. This effect is more marked during the early and late breeding season and in ewes compared with gimmers. Flushing has no effect on ovulation rate in ewes in good or poor condition.[11] Loss of weight at mating results in a greater number of instances of total loss of multiple ova. The subsequent breeding of such ewes will result in a repeat of this process.

Up to 30 per cent of fertilized ova can be lost through early embryonic mortality before day 35 of pregnancy. Chromosomal abnormalities and ova with cracked *zona pellucida* may account for 6–8 per cent of this loss.[12] The peak of the embryonic loss occurs from day 15–18 when first placental attachment is occurring. Inadequate nutrition leading to loss of weight at mating results in a higher incidence of total loss of multiple ova. Managemental procedures such as dipping, driving or gathering which result in an elevation in body temperature or other stressful conditions such as adverse weather conditions on their own lead to embryonic loss, or compound the effect of weight loss. It has been found that ewes housed at 35–40 days of pregnancy had higher prolificacy than outwintered ewes.

If total embryo loss occurs before day 12 of the cycle then normal cyclical activity ensues; if it occurs after this, then return to oestrus becomes variable and extended and in the commercial situation, when rams are removed after a period, such ewes might be assumed to be pregnant but are barren. When ova loss is only partial the relative variation in ovulation rate and ova or embryo wastage is not apparent since pregnancy continues but the lamb crop is found to be reduced.

In New Zealand selenium deficiency has been connected with infertility but in the UK there is no evidence at present that a selenium responsive form of infertility exists. Cobalt deficiency has been quoted as being responsible for an increased number of barren ewes and a prolonged lambing season which can be prevented by administering cobalt before the breeding season. Under experimental conditions copper deficiency prevented implantation or induced embryonic loss and fetal death.[13] Suppressed oestrus and reduced conception rates have occurred in ewes being fed kale during the breeding season.

High levels of natural oestrogens in subterranean clover have caused infertility problems in flocks in Australia, but in UK this has not been recorded as troublesome despite the presence of high oestrogen in some grasses.

Fertilization failure

Fertilization rate is generally high for adult ewes but lower for gimmers and this has been considered to be due to behavioural problems resulting in these animals being either mounted only or served late. Generally if fertile rams are left with ewes for three cycles, then this component of barrenness will be minimized.

Infectious causes of barrenness

These are rare when abortion *per se* is excluded. *Toxoplasma* infection in early pregnancy can cause fetal resorption but this and Border disease are discussed in detail elsewhere.

Miscellaneous pathological conditions

Examination of the reproductive tracts of ewes has shown that gross pathological abnormalities can be present in up to 7 per cent of animals. Adhesions to abdominal viscera, even in nullipara, were the most common finding and in such cases ova could not be collected by retrograde flushing of Fallopian tubes.[14] Fertilized eggs, however, could be recovered from animals with ovarobursal adhesions.

Neoplasia

Neoplastic lesions are rarely the cause of ewe infertility.

REFERENCES

1 Edgar D.G. (1963) The place of ram testing in the sheep industry. *New Zealand Veterinary Journal*, **11**, 113–15.

2 Fraser A.F. & Penman J.H. (1971) A clinical study of ram infertilities in Scotland. *Veterinary Record*, **89**, 154–8.

3 Logue D.N. (1981) Fertility testing in rams. *The Veterinary Annual*. 21st issue pp.134–9. John Wright & Sons, Bristol.

4 Lees J.L. (1978) Functional infertility in sheep. *Veterinary Record*, **102**, 232–6.

5 Jansen B.C. (1980) The aetiology of ram epididymitis. *Onderstepoort Journal of Veterinary Research*, **47**, 101–7.

6 Hulet C.V. & Ercanbrack S.K. (1962) A fertility index for rams. *Journal of Animal Science*, **21**, 489–93.

7 Hulet C.V., Foote W.C. & Blackwell R.L. (1965) Relationship of semen quality and fertility in the ram to fecundity in the ewe. *Journal of Reproduction and Fertility*, **9**, 311–15.

8 Johnston W.S., MacLachlan G.K. & Murray I.S. (1980) A survey of sheep losses and their causes on commercial farms in the north of Scotland. *Veterinary Record*, **106**, 238–40.

9 Oldham C.M. (1980) Stimulation of ovulation in seasonally or lactationally anovular ewes by rams. Proceedings of the Australian Society of Animal Production, **13**, 73–4.

10 Gunn R.G. & Maxwell T.G. (1978) Paper presented at Winter Meeting of British Society of Animal Production.

11 Haresign W. (1981) The influence of nutrition on reproduction in the ewe. 1. Effects on ovulation rate, follicle development and luteinizing hormone release. *Animal Production*, **32**, 197–202.

12 Robinson J.J. (1981) Fetal development in the ewe and related problems. Paper presented to British Veterinary Association Congress, 17–20 September 1981.

13 Hidiroglou M. (1979) Trace element deficiencies and fertility in ruminants: a review. *Journal of Dairy Science*, **62**, 1195–206.

14 Long S.E. (1980) Some pathological conditions of the reproductive tract of the ewe. *Veterinary Record*, **106**, 175–6.

FURTHER READING

Watt D.A. (1972) Testicular abnormalities and spermatogenesis of the ovine and other species. *Veterinary Bulletin*, **42**,181–90.

Laing J.A. (Ed.) (1979) *Fertility and Infertility in Domestic Animals*. 3e. Bailliere Tindall, London.

ULCERATIVE BALANITIS AND ULCERATIVE VULVITIS

31

D.W. DEAS

Synonyms Ulcerative balanoposthitis

An ulcerative condition of the prepuce, vulva, penis and skin has been described for many years in a number of countries. At an early stage in the description of the disease it was referred to in America as 'Lip and Leg' ulceration and no doubt at that time more than one agent was responsible. Eventually an association between this descriptive term, its venereal transmission and the establishment of ulcers on the genitalia was established. The condition was later described as ulcerative dermatosis (UD)[1] from which a transmissible agent was eventually described. Although, in many respects the genital lesions and the transmissible agent of UD resembled the lesions and the virus of contagious pustular dermatitis (CPD) it was thought that the two agents were distinct.[2] Later however this agent was further identified as the virus of CPD and the disease UD is now accepted as being CPD.[3]

In the UK a comparable ulcerative genital disease was described as early as 1903[4, 5] but penile lesions were not described in detail. Much later in the UK[6] and in Australia[7] full descriptions of the disease ulcerative balanitis and vulvitis were described. These two reports were remarkably similar but in certain aspects the description of the genital lesions was dissimilar to the genital lesions of CPD for long recognized in the UK. No transmissible agent was demonstrated in these two reported incidents.

The incidence of the disease ulcerative balano-posthitis and ulcerative vulvitis in the UK has never been accurately determined but references have, not infrequently, been made by diagnostic laboratories to flock outbreaks from a number of regions.

Cause

The cause of ulcerative balanitis and vulvitis has not been established in the UK. Exhaustive bacteriological examinations of material from affected ewes and rams have, apart from isolates of *Corynebacterium* spps and other bacteria, not resulted in a constant significant isolation.

Similarly exhaustive virological examinations have failed to reveal a cytopathic agent. Electron microscope studies have failed to identify the agent of CPD in scabs or ulcerations repeatedly taken from the lesions on the penises and prepuces of affected rams and from the vulvar ulcerations of affected ewes.

Transmission experiments have been carried out using encrustations from the prepuce, swabs and scrapings from penile ulcers and vulvar lesions to lightly scarified areas of the glans penis and vulvas of susceptible sheep.

In each instance a transient inflammatory reaction was produced but in no instance accompanied by the characteristic deep ulcerations of the field disease.

Bacteriological examination of material from the transmitted lesions again revealed a mixed flora.

Clinical signs

Flock outbreaks of this disease are observed first about 18–20 days after the start of mating. In each outbreak the first sign noted by the shepherd or stock owner is blood on or around the vulvas. The amount of blood present varies between outbreaks but not uncommonly mats or clots on the wool surrounding the vulva.

Ewe A vulvitis is present in about 20–30 per cent of ewes mated but this figure varies between flocks depending on the severity of the disease. The lesions take the form of vulvar ulcerations, often painful and often surrounded by a zone of hyperaemia. The ulcers can be deep and on removal of the scabs reveal a raw bleeding surface. The vulva is often swollen, reddened and inflamed to a varying degree and the ulcerations tend to be confined to the vulvar skin and the immediate lips of the vulva (Fig. 31.1). No lesions have been observed in the vestibule of the vulva/vagina or in the vagina itself. Affected ewes can often be identified at a distance by the continual switching of the tail and by the fact that the tail is often held away from the vulva.

Ram In each outbreak investigated the whole work-

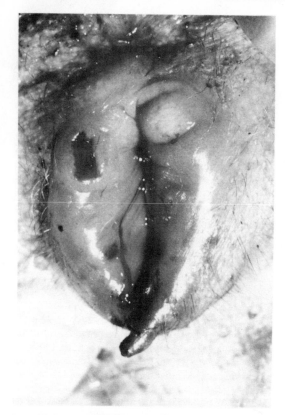

Fig. 31.1 Vulval ulceration and inflammation in a ewe with ulcerative vulvitis.

ing ram stud has been involved. On manual withdrawal of the penis of an affected ram the characteristic lesion can readily be seen involving the soft glans of the penis as a deep punched out ulcer about 1–2 cm in diameter with, in the acute phase of the disease, the crater of the ulcer filled by a large soft blood clot. Not infrequently the ulcerated area can involve the whole of the glans penis with involvement of the urethral process so that the complete top of the penis appears to be involved (Fig. 31.2).

Interference with the ulcer or slight disruption of the blood clot results in copious haemorrhage and it is disturbance of this haemorrhagic blood clot at mating which leads to the presence of blood on and around the vulvas of the ewes.

In the process of healing, the ulcerated area becomes covered with heaped up necrotic material. In uncomplicated cases the ulceration appears to be solely confined to the glans penis and no ulcerations have been observed on the shaft of the penis. This may be due to the fibro-elastic character of the shaft of the penis which makes it less of a predilation site

for the lesion than the soft spongy tissue of the glans. A milky preputial discharge is present. In a small number of cases the preputial orifice is swollen, inflamed and encrusted with scabs which on removal leave a bleeding surface. In badly affected rams the prepuce is so swollen, pendulous and oedematous that is has not been possible to extrude the penis. In these badly affected rams the inflamation is diffuse with seldom any evidence of areas of ulceration affecting the preputial mucosa.

If rams are removed from the ewes and given local or parenteral treatment the penile lesions organize and heal leaving a scar on the glans penis the size of which is dependant on the severity of the original ulcers. If the lesion has been severe, involving the urethral process, the whole tip of the penis is badly scarred.

No testicular abnormalities are noted. If the prepuce is not severely involved the lesion on the penis results only in a slight diminution of the ram's libido.

Rams which have recovered from the penile lesion without too much damage to the urethral process or involvement of the preputial orifice have been used successfully in subsequent matings.

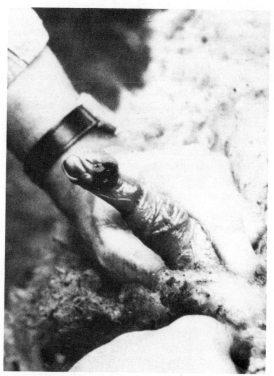

Fig. 31. 2 Ulceration of the glans penis

Epidemiology

The condition of ulcerative balanitis and ulcerative vulvitis does not readily fit the descriptions of diseases of the penis, prepuce, vulva and skin reported from other countries.

In the UK the disease is always associated with the mating period, becomes evident about 2–3 weeks after the rams have been turned out with the ewes and would appear to be spread venereally. In some outbreaks chaser rams known to be free from the disease when turned out with affected ewes readily developed the characteristic penile lesions.

In the outbreaks studied it has not been possible to follow the development of the lesion.

Methods of spread other than by venereal transmission have been investigated. In tailless or in short docked ewes flies can create a problem when mated in the late summer and early autumn and can result in a mild inflamatory reaction on the vulva and around the anus but these lesions are dissimilar to the lesions of ulcerative vulvitis. Similarly preputial encrustations can be produced by flies in ram studs at this time of the year.

Many of the outbreaks of the disease in the UK however occur in October and November in long tailed sheep when flies do not present a problem.

Despite the severity of the lesions in both the ewes and the rams the subsequent lambings are seldom disrupted to any degree. The numbers of barren ewes are within the expected norms for the flocks and there is no increase in the abortion rates.

In a few cases mild scabbing of the face, ears and the bridge of the nose has been present on some rams involved in the outbreaks. Electron microscope studies identified these lesions as due to CPD. It was thought at one time that transmission of this virus might have taken place due to the nuzzling of the vulvar region prior to mounting and mating but this was discounted when no CPD virus was identified in the genital lesions.

The genital disease as described in Australia[7] and referred to as ovine posthitis is as the term indicates a condition affecting the prepuce and preputial cavity and moreover, it is reported that adult wethers are more prone to the disease than rams. The involvement of wethers has never been recorded in the UK.

In outbreaks in the UK there is often a history of purchase of either ewes or rams but these movements have been minimal and are common place in sheep husbandry particularly in the introduction of new rams. Outbreaks of this disease have however been recorded in ram lambs mated to gimmers but in each instance all the stud rams have been run together prior to turn out with the ewes.

REFERENCES

1 Tunnicliff E.A. (1949) Ulcerative dermatosis of sheep. *American Journal of Veterinary Research*, **10**, 240–9.
2 Trueblood M.S. (1966) Relationship of ovine contagious ecthyma and ulcerative dermatosis. *Cornell Veterinarian*, **56**, 521–6.
3 Flook W.H. (1903) An outbreak of venereal disease among sheep. *Journal of Comparative Pathology and Therapeutics*, **16**, 374–5.
4 McFadyean J. (1903) A contagious disease of the generation organs in sheep. *Journal of Comparative Pathology and Therapeutics*, **16**, 375– 6.
5 Roberts R.S. & Bolton J.F. (1945) A venereal disease of sheep. *Veterinary Record*, **57**, 686–7.
6 Webb R.F. & Chick B.F. (1976) Balanitis and vulvovaginitis in sheep. *Australian Veterinary Journal*, **52**, 241–2.
7 Dent C.H.R. (1971) Ulcerative vulvitis and posthitis in Australian sheep and cattle. *Veterinary Bulletin*, **41**, 719–23.

32

PREGNANCY TOXAEMIA

E.J.H. FORD

This disease of sheep in late pregnancy is typified by dullness, anorexia, nervous signs, prostration and eventual death. Groenewald et al [1] referred to an early description of the condition by Steel in his book on 'Disease of Sheep' written in 1890. Sampson [2] listed 24 names which have been given to the same disease and, whilst choice is difficult, 'ovine pregnancy ketosis' is probably the most appropriate name for a disease of pregnant sheep which is accompanied by hyperketonaemia and the clinical signs described below. The condition is widespread and causes a considerable annual loss in most countries.

Clinical disease

The clinical signs have been described by a number of writers.[3, 4, 5] The first sign is that the affected ewe separates herself from the flock and appears depressed. There is loss of appetite and a disinclination to move. The head may be carried in an unnatural position; it may droop, be carried high or to one side, or it may be pushed against a wall. The body temperature is usually unchanged; fine tremors appear over the head and neck and may later become general. Disturbed vision is demonstrated by a poor or absent eye preservation reflex. The ewe becomes unable to rise, is generally dull, and there may be tympanites accompanied by diarrhoea and regurgitation of ingesta through the nostrils. Death soon follows and the mortality approaches 90 per cent.

All writers agree that pregnancy is a constant factor and that the condition is more common in ewes pregnant with twins or triplets than in those carrying single lambs. Roderick et al [6] stated that two differing sets of conditions may bring about the disease, one being an over-fatness and the other an inadequate ration leading to a loss in condition before lambing. A sudden snowstorm at lambing time, or adverse weather are factors which lead to the development of ketosis.

Most workers agree that the onset of parturition during the course of the disease increases the chances of recovery although the lambs are often born dead. Dayus and Weighton [4], describing the disease in New Zealand, stressed the eclamptic nature of some cases,

with champing of the jaws and production of a white froth at the mouth. Twitching of facial muscles and trembling were prominent and the wool could often be pulled out with ease. They regarded the extremely fatty liver as an exaggeration of the fatty infiltration found in all normal pregnant ewes and thought its extreme nature to be the result rather than the cause of the disease. For preventive purposes they stressed the need for adequate carbohydrate intake and the maintenance of an ascending plane of nutrition in pregnancy.

Pathology

At post-mortem examination the uterus will be found to contain two or more developed fetuses, the liver and kidneys are pale and fatty and the adrenals are often enlarged.

Several investigators have examined the degree of fatty infiltration of the liver by histological and by extraction procedures. The extent of the change varies, in some cases the infiltration is so severe that each cell is completely filled by a large globule of fat. Extracted fat ranges up to 35 per cent on a wet weight basis and it seems likely that this may be the maximum amount of fat that the liver cell can contain.

Snook [7] and Dryerre and Robertson [8] examined the livers of normal pregnant ewes and found an increase in histologically detectable fat towards the end of pregnancy but there was only a slight increase in extractable fat. Further evidence on this point was provided by Ford [9] who found that a fall in the plane of nutrition without the production of signs of disease increased the extractable fat but when clinical signs occurred fat content increased even more. Thus the livers of ewes with ovine ketosis differ from the normal only in the severity of the fatty changes.

Many early investigators, e.g. Roderick and Harshfield[10] described fatty infiltration of the tubular epithelium of the kidney. Groenewald et al [1] and Parry and Taylor [11] confirmed the presence of fatty infiltration of the tubular epithelium and also described casts and glomerular degeneration. These authors also noted fat necrosis, fatty changes in the adrenal cortex and atrophy of lymph nodes. It is significant to note that no observer was able to detect microscopical or gross changes in the brain.

As far as the writer is aware there are no other published observations on the adrenal histology of ewes with ketosis nor have lesions in any other organs been described.

Clinical pathology

Sampson et al [12] described hyperketonaemia in pregnancy disease of sheep and postulated that the disease was a result of the pregnant ewe being unable to detoxify the undesirable by-products of the developing fetus because of the fatty state of the liver. The increase in the concentration of aceto-acetic acid and β-hydroxybutyric acid in blood reduced the alkali reserve, lack of appetite followed the acidosis, and emaciation resulted.

Observations on kidney function were made by Parry and Taylor [11] who found raised concentrations of blood urea and plasma creatinine in both experimentally-induced and naturally-occurring ovine ketosis. There was also a reduction in urea and creatinine clearance and in renal plasma and blood flow. All these changes increased in degree as the disease progressed. There was no change in blood pressure and Parry and Taylor were unable to support the theory put forward by Parry [13] that the primary disturbance was a vasomotor change.

Assali et al [14] found increased plasma glucocorticoid concentrations as did Lindner [15] and Reid [16] but not Saba et al .[17]

Experimental disease

In 1937 Roderick et al [18] produced ketosis in pregnant ewes by lowering the plane of nutrition. There was a loss of liver glycogen, increase in liver fat, increased urine acetone and lowered urine pH. They suggested that fat cannot form in the liver when the glycogen content is normal, but that depletion of the glycogen reserve by a lack of carbohydrate in the diet allows fat to accumulate.

Fraser et al [19, 20] extended the experimental approach by studying the effect of reduction in calorie intake on pregnant ewes and showed clearly that ketonaemia could be produced readily in pregnant but not in non-pregnant ewes by reducing intake, and that ketonaemia and clinical signs were more frequent and severe in multiple than in single pregnancies. Fraser et al [19, 20] concluded that ovine pregnancy ketosis, i.e. ketonaemia accompanied by clinical signs, could be produced by a sudden fast in ewes in the last few weeks of pregnancy if the ewes were fat or if they had had a period of an inadequate ration. They also confirmed that supplementary feeding in the last months of pregnancy would prevent ketosis.

Biochemical aspects of ovine ketosis

A number of reviews have been written on the biochemical aspects of ketosis. Some have been of a general nature.[21, 22] Ovine ketosis in particular has been discussed by Reber [23], Kronfeld [24] and Reid.[25] Lindsay [26] in discussing the part played by carbohydrates in ruminant metabolism, considered many findings that are pertinent to pregnancy toxaemia.

The non-pregnant sheep has a daily requirement of 100 g of glucose believed to be derived mainly from propionate and aminoacids by gluconeogenesis. This requirement is increased by one-third for a single fetus near term and obviously by more in the case of twins.[24] Because of the high carbohydrate requirement of pregnant sheep and the absence of a marked increase in propionate absorption from the rumen, there is a drain on oxaloacetate and its precursors in an effort to maintain blood glucose concentrations.

At the same time, there is an increased requirement for tricarboxylic acids to facilitate the catabolism of fatty acids as an energy source. For this reason more acetyl coenzyme A (CoA) needs to be oxidized but condensation to acetoacetyl CoA occurs because of the decreased amounts of available oxaloacetate. This accumulation of acetoacetyl CoA is followed by removal of coenzyme A to give free acetoacetate and beta-hydroxybutyrate, the main ketone bodies found in the bloodstream.

If this were the complete explanation of pregnancy toxaemia the maintenance of blood glucose concentration by intravenous injection of glucose would be a satisfactory method of alleviating the condition. Clinical experience with ketotic sheep, however, shows that this is not so and therefore the fetal-hexose-drain theory of ovine ketosis has been extended in an attempt to explain the poor response to therapy. Reid [25] suggests that the major precipitating cause of the condition is adrenal response to environmental stress. In many circumstances, such as those in which sheep graze on poor quality hill pasture and even in lowland sheep receiving supplementary feeding there is probably an insufficient increase in calorie intake to satisfy the demand of twin fetuses. The resulting hypoglycaemia accentuates the adrenal response and an adrenal steroid diabetes ensues. The interference with carbohydrate utilization would be expected to produce hyperglycaemia, but this is prevented by the continuing hexose requirements of the fetus.[27] The fact that the signs are rarely alleviated when hyperglycaemia occurs, either as a result of repeated therapeutic injections or fetal death, is interpreted by Reid[28] as further evidence for a diabetic state.

Indirect evidence for adrenal hyperactivity is given by the 'break' in the fleece which is noted in ovine ketosis and which is similar to that produced by hydrocortisone or adrenocorticotrophic hormone administration.[29] Lindner[15] cautiously stated that clearance studies are necessary to show whether the increased steroid concentration[15, 16] is due to an increase in adrenal production or to poor hydrogenation and conjugation by the damaged liver.

McClymont and Setchell [5, 30, 31, 32] considered the basis for the clinical signs and concluded that the condition is a hypoglycaemic encephalopathy, there being a relation between the depth of the hypoglycaemia and the severity of the nervous derangement. The early stages of the experimentally-induced process are reversible, particularly by the oral administration of glycerol which reduces the hypoglycaemia, but as the duration of the disease increases so the signs become irreversible.

Prevention and treatment

There is general agreement that a feeding programme for pregnant ewes which provides a gradual increase in energy intake to cover the requirements of the developing fetuses without a reduction in body reserves, will prevent serious outbreaks of pregnancy toxaemia other than those which arise when adverse weather interferes with the flock's access to the food that is provided. Nevertheless, cases can arise in individual animals because factors such as lameness or dental problems interfere with proper food intake.

When treatment is attempted one or more of a number of substances which are likely to improve the supply of glucose to the body tissues are used. These remedies include intravenous 40 per cent glucose solution, oral glycerol or propylene glycol (2–3 oz daily), molasses by mouth or in the food and injections of vitamins A, B and choline. Adrenocorticotrophic hormone, cortisone and synthetic glucocorticoids have been tried. Caesarian section and the use of abortifacients are suggested as a means of reducing the animal's glucose requirements.

There are few reports of controlled trials of therapeutic agents [33] but most of those listed are used because they should be able, on theoretical grounds, to increase the supply of glucose to the tissues. Clinical reports on their use are disappointing. This is particularly true of glucocorticoids[34] in contrast to their value in the treatment of acetonaemia of cattle. Reilly and Ford[35] showed that an injection of betamethasone would produce hyperglycaemia and increase the production of glucose from aminoacids in normal sheep but there are no reports of its clinical efficacy. Similarly Ranaweera et al.[36] investigated the effect of triamcinolone acetonide and found no clinical improvement except in one ewe which lambed soon after treatment. Heitzman et al. [37] obtained encouraging results in the treatment of spontaneous pregnancy toxaemia with the anabolic steroid trienbolone but Ford et al. [38] found that the steroid produced neither increase in the concentration of glucose in plasma, nor clinical improvement but it reduced glucose entry rates. The oral administration of glycerol was shown by Ranaweera et al. [39] to produce a marked hyperglycaemia which was due to gluconeogenesis from glycerol but clinical improvement did not accompany the effects of the glycerol.

The lack of clinical response, in spite of the stimulation of gluconeogenesis by hormones or glycerol, suggests that the abnormalities of nerve function, (hypoglycaemic encephalopathy), may become irreversible at an early state of the clinical disease.

Clearly there is need for further detailed investigation on the treatment of pregnancy toxaemia because, although clinical disease can be prevented by feeding a suitable diet, there will always be some animals needing treatment in spite of the use of preventive measures.

REFERENCES

1 Groenewald J.W., Graf H. & Clarke R. (1941) Domsiekte or pregnancy disease in sheep. I. A review of the literature. *Onderstepoort Journal of Veterinary Science*, **17**, 225–44.
2 Sampson J. (1947) Ketosis in domestic animals. Clinical and experimental observations. *Bulletin of the Illinois Agricultural Experimental Station*, **524**, 407–70.
3 Greig J.R. (1929) The nature of lambing sickness. *Veterinary Record*, **9**, 509–11.

4 Dayus C.V. & Weighton C. (1931) Clinical observations on pregnancy toxaemia in ewes. *Veterinary Record*, **11**, 255–7.

5 McClymont G.L. & Setchell B.P. (1955) Ovine pregnancy toxaemia. I. Tentative identification as a hypoglycaemic encephalopathy. *Australian Veterinary Journal*, **31**, 53–68.

6 Roderick L.M., Harshfield G.S. & Merchant W.R. (1933) Further observations on the functional pathology of pregnancy disease of ewes. *Cornell Veterinarian*, **23**, 348–53.

7 Snook L.C. (1939) Fatty infiltration of the liver in pregnant ewes. *Journal of Physiology*, **97**, 238–49.

8 Dryerre H. & Robertson A. (1941) The effect of pregnancy on tissue lipids in the ewe. *Journal of Physiology*, **99**, 443–53.

9 Ford E.J.H. (1962) The effect of dietary restriction on some liver constituents of sheep during late pregnancy and early lactation. *Journal of Agricultural Science*, **59**, 67–75.

10 Roderick L.M. & Harshfield G.S. (1932) Pregnancy disease of sheep. *Bulletin of the North Dakota Agricultural Experiment Station*, No.261.

11 Parry H.B. & Taylor W.H. (1956) Renal functions in sheep during normal and toxemic pregnancies. *Journal of Physiology*, **131**, 383–92.

12 Sampson J., Gonzaga A.C. & Hayden C.E. (1933) The ketones of the blood and urine of the cow and ewe in health and disease. *Cornell Veterinarian*, **23**, 184–207.

13 Parry H.B. (1950) Toxaemias of pregnancy in the domestic animals, with particular reference to the sheep. In *Toxaemias of Pregnancy*. (Eds.Hammond J., Browne S.J. & Wolstenhome G.E.W.) (Ciba Foundation Symposium) pp.85–93. Churchill, London.

14 Assali N.S., Holm L. & Hutchinson D.L. (1958) Renal hemodynamics, electrolyte excretion and water metabolism in pregnant sheep before and after the induction of toxemia of pregnancy. *Circulation Research*, **6**, 468– 75.

15 Lindner H.R. (1959) Blood cortisol in sheep: normal concentration and changes in ketosis of pregnancy. *Nature, Lond.*, **184** suppl. No.21. 1645–6.

16 Reid R.L. (1960) Studies on the carbohydrate metabolism of sheep. XI. The role of the adrenals in ovine pregnancy toxaemia. *Australian Journal of Agricultural Research*, **11**, 364–82.

17 Saba N., Burns K.N., Cunningham N.F., Hebert C.N. & Patterson D.S.P. (1966) Some biochemical and hormonal aspects of experimental ovine pregnancy toxaemia. *Journal of Agricultural Science*, **67**, 129–38.

18 Roderick L.M., Harshfield G.S. & Hawn M.C. (1937) The pathogenesis of ketosis: pregnancy disease of sheep. *Journal of the American Veterinary Medical Association*, **90**, 41–50.

19 Fraser A.H.H., Godden W., Snook L.C. & Thompson W. (1938) The influence of diet upon ketonemia in pregnant sheep. *Journal of Physiology*, **94**, 346–57.

20 Fraser A.H.H., Godden W., Snook L.C. & Thompson W. (1939) Ketonemia in pregnant ewes and its possible relation to pregnancy disease. *Journal of Physiology*, **97**, 120–27.

21 Campbell J. & Best C.H. (1956) Physiologic aspects of ketosis. *Metabolism*, **5**, 95–113.

22 Krebs H.A. (1960) Biochemical aspects of ketosis. Proceedings of the Royal Society of Medicine, **53**, 71–80.

23 Reber E.F. (1957) Pregnancy disease in ewes: observations relating to treatment and the metabolic pathways involved. *North American Veterinarian*, **38**, 353–5 and 359.

24 Kronfeld D.S. (1958) The fetal drain of hexose in ovine pregnancy toxemia. *Cornell Veterinarian*, **48**, 394–404.

25 Reid R.L. (1968) The physiopathology of undernourishment in pregnant sheep with particular reference to pregnancy toxaemia. *Advances in Veterinary Science*, **12**, 163–238.

26 Lindsay D.B. (1959) The significance of carbohydrate in ruminant metabolism. *Veterinary Reviews and Annotations*, **5**, 103–28.

27 Reid R.L. & Hogan J.P. (1959) Studies on the carbohydrate metabolism of sheep. VIII. Hypoglycaemia and hyperketonaemia in undernourished and fasted pregnant ewes. *Australian Journal of Agricultural Research*, **10**, 81–96.

28 Reid R.L. (1960) Studies on the carbohydrate metabolism of sheep. IX. Metabolic effects of glucose and glycerol in undernourished pregnant ewes and in ewes with pregnancy toxaemia. *Australian Journal of Agricultural Research*, **11**, 42–57.

29 Lindner H.R. & Ferguson K.A. (1956) Influence of the adrenal cortex on wool growth and its relation to 'break' and 'tenderness' of the fleece. *Nature, Lond.*, **177**, 188–9.

30 McClymont G.L. & Setchell B.P. (1955) Ovine pregnancy toxaemia. II. Experimental therapy with glycerol and glucose. *Australian Veterinary Journal*, **31**, 170–4.

31 McClymont G.L. & Setchell B.P. (1956a) Ovine pregnancy toxaemia. III. Further evidence of the correlation of depth of hypoglycaemia and induction of symptoms. *Australian Veterinary Journal*, **32**, 22–5.

32 McClymont G.L. & Setchell B.P. (1956b) Ovine pregnancy toxaemia. IV. Insulin induced hypoglycemic encephalopathy in the sheep and its implications as regards pathogenesis of the disease. *Australian Veterinary Journal*, **32**, 97–109.

33 Thomson G.G. (1956) Observations on some treatments of ovine pregnancy toxaemia. *New Zealand Veterinary Journal*, **4**, 136–44.

34 Holm L.W. (1958) Studies on the treatment of ovine pregnancy toxemia with corticosteroids and ACTH. *Cornell Veterinarian*, **48**, 348–57.

35 Reilly P.E.B. & Ford E.J.H. (1974) Effect of betamethasone on glucose production and on gluconeogenesis from amino acids in sheep. *Journal of Endocrinology*, **60**, 455–61

36 Ranaweera A., Ford E.J.H. & Samad A.R. (1979) The effect of triamcinolone acetonide on plasma glucose and ketone concentration and on the total entry rate of

glucose in twin pregnant hypoglycaemic ketotic sheep. *Research in Veterinary Science*, **26**, 12–16.

37 Heitzman R.H., Herriman I.D. & Austin A.R. (1977) The response of sheep with pregnancy toxaemia to trenbolone acetate. *The Veterinary Record*, **100**, 317–18.

38 Ford E.J.H., Samad A.R. & Pursell H. (1979) The effect of trenbolone acetate on glucose metabolism in normal and ketotic sheep pregnant with twins. *Journal of Agricultural Science*, **92**, 323–7.

39 Ranaweera A., Ford E.J.H. & Evans J. (1981) Gluconeogenesis from glycerol by ketotic sheep pregnant with twins. *Research in Veterinary Science*, **30**, 303–8.

33   VENTRAL HERNIA AND VAGINAL PROLAPSE

B. MITCHELL

Ventral hernia and vaginal prolapse are invariably associated with pregnancy in the ewe. During pregnancy there is naturally an increase in bulk of the intra-abdominal contents. Adverse consequences of this are herniation of intra-abdominal structures through any points of weakness in the integrity of the muscular wall of the abdomen and pelvis.

Cause

The relationship between normal pregnancy and herniation of the abdominal contents is complex. Undoubtedly increased intra-abdominal tension is one factor but this is common to most ewes. More specific causes are factors which encourage weakening of the muscles of the abdominal wall and the ligamentous support within the pelvis. The latter may have been damaged by previous difficult parturitions or continuous 'pressing' by ewes attempting to dislodge severely traumatized or inflamed intra-pelvic structures.

The sites of herniation through the muscles of the abdominal wall are either in the lower flank, involving the fibrous aponeurosis of the external and internal oblique muscles, or in the posterior midline due to rupture of the prepubic tendon at the insertion of the *rectus abdominis* muscle on to the anterior border of the pelvis. Weakening of the flank muscles can be associated with undernutrition or direct trauma as, for example, when ewes are jostled against gateposts. Ischaemia of the abdominal wall from local obstruction of blood vessels results in degeneration and rupture of the supporting structures (T. Hulland—personal communication).

Clinical signs

Ewes with ventral hernia show two types of clinical pattern. First, rupture of the prepubic tendon allows the abdomen to sag almost to the ground. Depending on the weight of the abdominal contents the animal may be unable to rise or, if up, it can walk only with difficulty. The catastrophy is usually sudden and absolute. The only support left on the floor of the abdomen is skin and associated fibrous tissue. Local evidence of damage and tissue fluids may be obscured by the overlying udder. The second picture, lower flank hernia or 'fallen side', is much more variable. The onset is often progressive and the extent magnifies as pregnancy proceeds. The ewe is not distressed unless the disruption of the flank is very severe in which case the ewe will be unbalanced when rising and walking. Palpation of the flank reveals a general area of weakness rather than a discrete hernia ring unless the cause has been a precise injury. Diagnosis of the condition is self-evident from the clinical signs.

Prolapse of the vaginal wall occurs in the last 1–3 weeks of pregnancy. At first it is observed only when the ewe is lying and causes no particular distress returning to normal when the ewe stands. As the extent of the prolapse progresses intermittently over a few days the bladder is included in the herniated structures, the urethra becomes obstructed and the bladder distends. The ewe's straining exacerbates the condition. Even when the prolapse is replaced, because the structures are cold and congested, the ewe continues to strain and a very intractable situation is established. Sudden rupture of the vaginal wall, although unusual, will result in eventration of the intestines and death of the ewe.

Treatment

There is no satisfactory treatment for ventral hernia. Ewes with ruptured prepubic tendons usually require to be killed for humanitarian reasons unless management is directed at salvaging the lambs. Ewes with flank hernias should have restricted exercise till they lamb after which the reduction of intra-abdominal tension greatly alleviates the condition. These ewes, if otherwise fit, will be able to rear their lambs after which they should be culled.

Prolapse of the vaginal wall invarably requires treatment. This is directed at cleaning the everted surface with a warm mild antiseptic and reducing the prolapse by replacing the bladder and cervix into the abdomen. Any difficulty in achieving this is usually because of the distended bladder. It is best emptied by gentle compression and manipulation to free any flexion of the urethra. Once replaced, the vaginal wall should be smoothed out and retained in place till the congestion and temperature of the tissues have returned to almost normal. Further prolapse is very likely unless steps are taken to prevent it.

Closure of the perineal area by suturing the lips of the vulva is not satisfactory because the sutures tear out leaving bad lacerations. An alternative is to tie 4–6 strings across the vulva fixing them by non-slip knots to locks of wool adjacent to the vulva. A plastic support ('Savewe': Alfred Cox [Surgical] Ltd., Surrey, UK) can be fitted in the vagina to prevent the cervix everting. Shaped as a T, having a body (16 cm long) with two sidearms (13 cm long), it is fixed in position by string from the sidearms, passed laterally round the giggots and anchored to a rope circling the abdomen just anterior to the back legs. With all of these forms of support the ewe must be closely watched for the first signs of impending parturition because all restraint must then be removed. Elective Caesarean section after 142 days gestation is an alternative to constant scrutiny.

If straining continues after parturition it may result in prolapse of the uterus, which should be replaced after washing its surface. Retention sutures inserted as deeply as is practicable alongside the vulva and across the vagina are successful now that the intra-abdominal tension is reduced. The author's preference is to use umbilical tape or braided nylon to fashion two horizontal mattress sutures each about 4 cm wide. Ten cm lengths of rubber tubing (obsolete endotracheal tube) are placed on either side of the vulva to prevent the suture material cutting the skin. The rubber tubes are prepared by cutting three 2 mm holes; at the centre and approximately 1 cm from each end. Each mattress stitch is fashioned as a rectangle, the suture material being passed on two sides through the lumina of the tubes, emerging from the holes and being inserted through the skin and vaginal walls to complete the stitch. Too much tension is avoided to obviate ischaemic necrosis of the vaginal wall. When these sutures are removed ten days later, the uterus will be sufficiently involuted not to prolapse again. Ewes which have had prolapses should be culled.

Prevention

Measures to prevent ventral hernia and prolapse of the vagina are directed at minimizing intra-abdominal tension during pregnancy. Since the growth of lambs must not be compromised the intake of bulky diets such as turnips, sugar beet and hay should be controlled by feeding a high concentrate: low roughage diet during the last month of pregnancy. Movement of sheep should be unhurried and jostling avoided.

34

MASTITIS

A.J. MADEL

In 1887 the famous French veterinarian, Edmund Nocard, who worked with Pasteur, outlined the clinical course of mastitis in sheep and described how the experimental reproduction of the infection with *Staphylococcus aureas* could kill a ewe within 48 h. In the introduction to a paper on the disease he stated that mastitis in sheep was *not* well-known to veterinarians despite its importance as 'the plague of cheese-making flocks'.[1] Nearly 100 years later many veterinary surgeons working with sheep in Britain are infrequently presented with cases of mastitis. The lack of interest in the condition is reflected by the paucity of reports in the English literature of investigations into the disease.

In dairy cattle the herdsman is likely to see, in the course of a year, a wide spectrum of clinical forms of mastitis; occasionally he will see cows with an acute severe mastitis in which gangrene and sloughing of the affected quarter occur, whilst less severe acute cases and chronic cases are likely to be encountered fairly frequently. By contrast, the shepherd will see only a small part of this spectrum. Many types of mastitis may well exist and probably depress the level of milk production but the shepherd is likely to recognize only two clinical forms of the disease, an acute and a chronic form. Early in lactation an acute severe mastitis, in which gangrene and sloughing of the affected udder half are common sequelae is easily recognized; this condition is often called 'black garget'. This form of the disease is extensively described in the literature particularly by French workers. In the autumn, most shepherds inspect each ewe before the breeding season and during this examination the udder will be carefully checked. Sheep which have udders containing lumps, suggestive of abscesses, are likely to be culled. The literature on this form of chronic mastitis is particularly sparse.

Acute severe mastitis and chronic mastitis have entirely different clinical signs and occur at widely different times of the sheep's reproductive cycle. They will therefore be described separately in this chapter.

ACUTE SEVERE MASTITIS

Cause

A great variety of organisms have been reported in the literature as causing mastitis but a review of the disease emphasized that *Staph aureus* is the commonest cause of the condition.[2] There have been several reports which state that the gangrenous process is brought about by a mixed infection of *Staph aureus* and *Clostridium perfringens (welchii)* but in an extensive review of the subject Pegreffi concluded that *Staph aureus* on its own was quite capable of producing a gangrenous mastitis in sheep.[3] Pasteurellae have also been reported as causing mastitis.[4] In a recent investigation of 23 acute clinical cases the importance of *Staph aureus* and pasteurellae was confirmed (Table 34.1).[5] Cases due to *Staph aureus* and *Pasteurella haemolytica* are indistinguishable on clinical and pathological grounds.

Table 34.1 The results of bacteriological examinations of 23 cases of acute severe mastitis*

Staphylococcus aureus	14
Staphylococcus aureus and *Pasteurella haemolytica*	4
Pasteurella haemolytica	2
Streptococcus zooepidemicus	1
Corynebacterium pyogenes	1
No organisms isolated	1
Total	23

*Madel A.J. unpublished observations

The predisposing factors which result in a small proportion of the flock (usually 1–5 per cent) succumbing to acute severe mastitis are poorly understood. Vigorous sucking by lambs, causing lesions on the teats, is often cited as an important factor [6] but there is no experimental evidence for this hypothesis of cause and effect; in the early phase of the infection when milk yield is rapidly diminishing, lambs may well damage teats in their attempts to obtain milk. However, any teat lesion is likely to be a rich source of pathogens which can predispose to acute severe mastitis and this consequence is commonly seen when orf lesions affect the teats. Flockmasters often state that cases occur following cold winds and in his book 'The Sheep', published in 1837, William Youatt advised not exposing the udder to the effects of chilling by excessive crutching (removal of fleece from the perineum).[7] This must still be good advice despite the experimental findings that chilling of the cow's udder has equivocal effects on existing mammary infections.[8]

Clinical signs

There are few clinical descriptions of the disease in the literature. The best is undoubtedly that by Nocard [1] and the account which follows is based on his account and on personal observations.

Acute severe mastitis can develop in ewes of any age and usually occurs within the first few weeks of lactation. The earliest sign is often a hind leg lameness or a stiff gait in a ewe which is separated from the rest of the flock. Alternatively, suspicion of mastitis is aroused because the lamb(s) appears hungry and weak.

On clinical examination the ewe is usually depressed and has a temperature of about 40.5°C (105°F). One half of the udder is grossly distended to at least one and a half times the size of the unaffected half. During the first two days of the infection the enlarged gland is hot and red and then gradually becomes cold and dark purple; at this stage the body temperature falls to about 36°C (97°F). The milk is initially thin and contains small clots but within the course of 2–3 days becomes thinner and yellow; there is usually very little secretion and its expression causes pain. Oedema of the tissues of the adjacent inner thigh and ventral abdominal wall are common sequelae.

Approximately half of the affected ewes die at this stage. If the animal survives, the affected gland gradually sloughs leaving foul-smelling suppurating areas which can take many weeks to heal.

Pathology

At necropsy of a ewe with acute mastitis the affected udder half is likely to be considerably enlarged. In Fig. 34.1 the udder from a field case of mastitis has been cut open; the affected left side was greatly enlarged and weighed 1670 g compared with the right which weighed only 280 g. Fig. 34.2 shows the sectioned gland from another field case in which several abscesses have begun to develop in the ventral part of the parenchyma. The rapidity with which lesions develop is evident following experimental infection of the mammary gland with *P. haemolytica* (type A); in one ewe which was killed 29 hours after intramammary inoculation severe haemorrhage and necrosis were observed in the cistern area although no clinical signs of mastitis had been observed in the animal.[5]

The principal histological lesions in acute mastitis are those of haemorrhage and necrosis of the gland.

Diagnosis

The history and clinical signs enable a correct diagnosis to be made easily. Bacteriological examination of the secretion or of the gland post-mortem, although unnecessary for the purpose of making a clinical diagnosis may help in building an epidemiological picture of the condition on a particular farm and in determining the antibiotic sensitivity of the bacteria isolated.

Epidemiology

The annual incidence of the disease rarely exceeds 5 per cent. The possibility of vigorous sucking by the lambs, orf infection or excessive chilling of the udder by cold winds have already been discussed as possible predisposing factors. Although the epidemiology and transmission of the disease are unclear, the level of hygiene in the lambing areas probably influences the likelihood of infection.

Treatment and prevention

The course of the disease is rapid and the effects of the infection on the udder are so severe that by the time the ewe is presented for treatment the changes in the udder are irreversible. Large doses of a broad spectrum antibiotic given intravenously or intramuscularly may save the ewe's life. Intramammary infusion of antibiotics is useless. A high standard of nursing must be given for several weeks until the gangrenous process has fully resolved. Necrotic areas of the gland should be cut off when it is obvious that

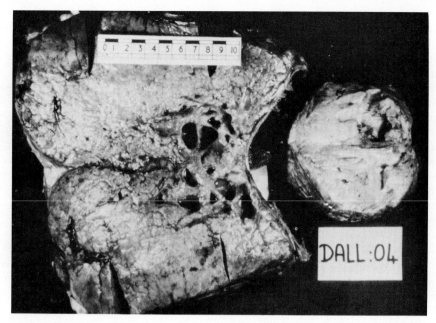

Fig. 34.1 The sectioned mammary glands from a field case of mastitis. The left gland weighed 1670 g, the right 280 g.

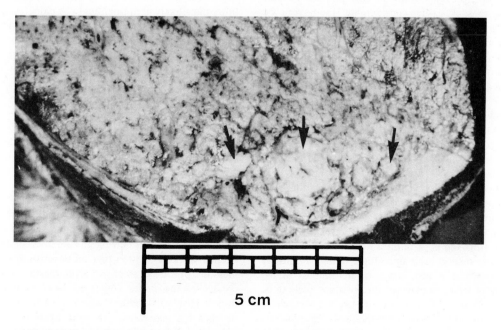

Fig. 34.2 The sectioned mammary gland from a field case of mastitis. The arrows indicate abcesses which have begun to develop in the ventral part of the parenchyma.

sloughing is inevitable.

In several countries a variety of vaccines are available but none can be purchased in Britain. Although most of these vaccines, which are administered parenterally, improve considerably the ewe's chances of survival, their protective effects on the udder are often disappointing. Trials with intramammary vaccines have been encouraging. A review of mammary vaccination[2] indicates the complexity of the subject and although European workers continue to devote a great deal of effort to producing a satisfactory anti-staphylococcal vaccine the chances of any pharmaceutical company marketing a product in Britain seem very slim. Accordingly, the subject is not expanded in this chapter.

Control

It is worth stressing again that the course of this type of mastitis is rapid; affected ewes can die within 24 h of the onset of clinical signs. A high standard of shepherding is therefore essential as a ewe noticed sick in the early stages of the disease can, with appropriate antibiotic therapy and nursing, be saved. Cleanliness of the lambing areas is important. Many shepherds routinely draw milk from newly lambed ewes to ensure that the plug often present in the teat canal is removed. Although this practice enables lambs to suck easily, considerable trauma can be sustained by the teat-end by over-zealous attempts to draw milk.

CHRONIC MILD MASTITIS

Introduction, clinical signs and diagnosis

The annual replacement rate of breeding ewes in 517 recorded flocks, each having between 400 and 450 females, in 1979 was 21.7 per cent (Meat and Livestock Commission, private communication). The replacement rates in the previous five years were very similar and it is estimated that the figure for the national ewe flock is about 20 per cent. Ewes are culled from the flock in the autumn when shepherds routinely inspect each animal; the udders are briefly palpated and if any lumps or deformities in the glands or teats are detected the ewes are likely to be culled. Criteria of abnormality vary between shepherds but the common objective is to cull any ewe which is judged unlikely to provide sufficient milk at the next lactation. The diagnosis of abnormality is largely based on palpation; the majority of abnormalities are abscesses [9] and these are felt as hard lumps either in the body of the gland or in the cistern at the base on the teat (Fig. 34.3). Absence of a teat,

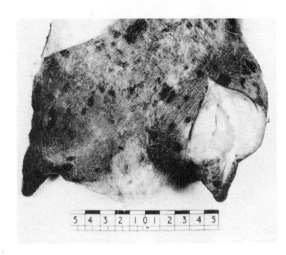

Fig. 34.3 The udder of a culled ewe. The skin around the base of the teat has been removed to expose the abcess in the cistern.

extensive warts on the teats or a generalized calcification of the gland, felt as an intense hardness, are the other most likely abnormalities to be palpated. Sheep exhibiting any of these lesions rarely show signs of illness. However, their detection is easy; any doubtful cases should be turned and examined more thoroughly.

Cause

Most of the abnormalities detected in the udders are the result of bacterial infection.[9] However, the time at which infection occurs, the prinicipal organisms involved and the factors which render the udder susceptible to infection are items which have not been thoroughly investigated. It seems likely that the udder is particularly susceptible to infection in the days immediately following the removal of lambs. French workers showed that the severity of experimentally-produced staphylococcal mastitis was increased if the udder contained large volumes of milk.[10] Distended udders which are dripping milk through the teat canals are more likely to become infected by bacteria than udders where the teat canals remain closed.

Incidence

Estimations of the incidence of udder abnormalities can be made in two ways; either a figure can be obtained from examination of reliable flock records or an assessment can be made from examining culled ewes at a slaughter house. In one report, it was stated that in a flock of 549 ewes, 25 (4.5%) were culled because of 'post-weaning' mastitis [11] and a further investigation showed that of 469 ewes from three flocks 21 ewes (4.5%) were similarly culled.[12] In a similar study, it was reported that about 60 of a flock of 700 ewes (8.6%) were culled each year because of damaged udders.[13]

In an abattoir study, the udders of 635 ewes were examined.[14] Following palpation and inspection of each udder, 50 per cent were classified as abnormal on the basis that they contained either a hard lump, over 2.5 cm in diameter, or were hot and oedematous. In a further abattoir study [9] the udders of 1650 culled ewes were examined and using broadly similar criteria 12.8 per cent were recorded as having clinically obvious lesions; 77 per cent of these lesions were abscesses.

Control

Immediately a ewe's lambs have been weaned she should be kept on a very restricted diet with little water for 1–3 days; a straw yard is the best place to keep sheep during this drying-off period. Milk should only be drawn from the udder if distension is causing obvious distress to the animal and then it must be done gently with clean hands.

There are reports of attempts to reduce the level of post-weaning mastitis by the infusion of long acting antibiotics into the udders at weaning. A reduction of the incidence of mastitis from 4.5 per cent in one year to 0.18 per cent the following year was recorded, when 'dry-ewe' therapy was initiated using a long-acting cerate containing 1 g procaine penicillin and 0.5 g dihydrostreptomycin sulphate divided between the udder halves.[11] A similar reduction from 8.6 per cent to 0.14 per cent was reported using 500 mg of benzathine cloxacillin per half udder.[13] In a clinical trial there was an incidence of 1.5 per cent in 462 sheep treated with a preparation containing 1 g procaine penicillin and 0.5 g dihydrostreptomycin sulphate divided between the udder halves and 4.5 per cent in untreated controls.[12]

The technique of infusing antibiotics into a ewe's udder is straight-forward but does require care. One person turns the sheep and a second cleans the teat-end with a cotton wool swab soaked in surgical spirit before infusing the antibiotic. The majority of intramammary preparations used for treating cows have nozzles which can be easily inserted into a ewe's teat-end. If the infusions are performed by only one person working in dirty conditions the chances of introducing infection are high.

This system of prophylaxis is only practicable on farms where whole groups of lambs are weaned simultaneously. If lambs are randomly removed from a flock it can be difficult to identify the relevent ewes and therefore treatment cannot be given. The procedure is fairly costly in labour as it takes two people a day to infuse between 150 and 200 ewes. However, if the work reduces the numbers of culled ewes by only two or three it can be justified economically. The extra handling of the flock can easily be combined with another routine task.

CONTAGIOUS AGALACTIA

Contagious agalactia is a syndrome of sheep and goats in which the mammary glands, conjunctiva and joints are infected by mycoplasma species. The disease is economically important in Mediterranean, Adriatic and African countries; it has not been reported in Britain.

Any one of three mycoplasma species may cause the disease, *M. agalactia*, *M. mycoides* subsp *mycoides* (large colony form) and *M. capricolum*. The disease occurs in ewes soon after lambing and causes mastitis, arthritis and keratoconjunctivitis; in rams the predominant sign is arthritis. The route of infection is probably oral or by the teat canal and is followed by a generalized infection. Following experimental infection with *M. agalactia* the organism localizes in the eyes, joints and udder in 1–6 days.[15]

The mastitis is acute and severe resulting in a progressive decline in lactation until there is a total and permanent agalactia of the affected gland. The arthritis produces a prolonged and painful lameness; keratoconjunctivitis is a frequent though not invariable clinical sign. The case fatality rate is 10–30 per cent. Sheep which recover show a solid resistance to further clinical attacks but continue to excrete the organism for long periods.[15]

The disease has not been reported in Britain but this may reflect a failure to look for evidence of infection. Until sufficient animals in Britain have been screened and more is known about their mycoplasmal flora it is clearly advisable that potential imports of small ruminants be serologically tested for evidence of infection with mycoplasma.

REFERENCES

1 Nocard M.E. (1887) Note sur la mammite gangreneuse des brebis laitières. *Annales de L'Institut Pasteur,* **1,** 417–27.

2 Landau M. & Tamarin R.H. (1974) Ovine Mastitis Research 1963–73, A Review. *Refuah Veterinarith,* **31,** 134–47.

3 Pegreffi G. (1963) Les mammites infectieuses de la brebis et de la chèvre. *Bulletin de l'Office International des Epizooties,* **60,** 1009–40.

4 Tunnicliff E.A. (1949) Pasteurella mastitis in ewes. *Veterinary Medicine,* **44,** 498–502 and 506.

5 Madel A.J. Unpublished information.

6 Blood D.C., Henderson J.A. & Radostits O.M. (1979) *Veterinary Medicine: a textbook of the diseases of cattle, sheep, pigs and horses.* 5e,p.379. Bailliere Tindall, London.

7 Youatt W. (1837) Breeding—the generative and urinary systems. *The Sheep.* p.498. Baldwin & Cradock, London.

8 Brown R.W., Thomas J.L., Cook H.M., Riley J.L. & Booth G.D. (1977) Effect of environmental temperature stress on intramammary infections of dairy cows and monitoring of body and intramammary temperatures by radiotelemetry. *American Journal of Veterinary Research,* **38,** 181–7.

9 Madel A.J. (1981) Observations on the mammary glands of culled ewes at the time of slaughter. *Veterinary Record,* **109,** 362–3.

10 Le Gall A. & Plommet M. (1965) Observations sur la croissance du staphylocoques et la réaction leucocytaire au cours des premières heures de la mammite experimentale de la brebis. *Annales de Biologie animale Biochimie Biophysique,* **5,** 113–30.

11 Gibson I.R. & Hendy P.G. (1976) Mastitis in dry ewes. *Veterinary Record,* **98,** 511–12.

12 Hendy P.G., Pugh K.E., Harris A.M. & Davies A.M. (1981) Prevention of postweaning mastitis in ewes. *Veterinary Record,* **109,** 56–7.

13 Buswell J.F. & Yeoman G.H. (1976) Mastitis in dry ewes (prevention with cloxacillin). *Veterinary Record,* **99,** 221–2.

14 Herrtage M.E., Saunders R.W. & Terlecki S. (1974) Physical examination of cull ewes at point of slaughter. *Veterinary Record,* **9,** 257–60.

15 Cottew G.S. & Leach R.H. (1969) In *The Mycoplasmatles and the L-phase of Bacteria.* (Ed. Hayfluk A) pp.527–70. Appleton, New York.

SECTION 5

DISORDERS OF MINERAL
METABOLISM

35 DEFICIENCIES OF MACRO-ELEMENTS IN MINERAL METABOLISM

A.J.F. RUSSEL

The essential major or macro-elements are calcium, phosphorus, magnesium, potassium, sodium, chlorine and sulphur. Those most likely to be of importance in sheep production, particularly in pastorally-based systems of management, are calcium, phosphorus and magnesium. Deficiencies of the other essential macro-elements occur only rarely and are generally regarded as being of lesser practical importance. The biological significance, major dietary sources, signs of deficiency and the principal means of prophylaxis are outlined briefly for each element.

The requirements of sheep for the major elements vary according to their age and weight as well as being affected to a large extent by factors such as stage of pregnancy or lactation and fetal number or level of milk production. The presentation of the dietary requirements of these various classifications of sheep for each essential macro-element is outwith the scope of this publication, and for this information the reader is referred to the most recent authoritative source.[1]

POTASSIUM

As with virtually all essential elements, potassium fulfils a number of important roles in the animal body. Its main function, together with sodium and chlorine, is in the osmotic regulation of body fluids and in the maintenance of acid-base balance. Whereas sodium is the principal cation of extracellular fluids, potassium is found mainly within the body cells.

The concentration of potassium in grass and other forages is very high, and the intake of potassium by grazing animals is generally greater than that of any other element. Except in very intensively-managed systems where sheep are continuously housed and fed exclusively on concentrate diets, dietary potassium intakes are invariably greater than requirements and thus potassium deficiency is not encountered under normal farming conditions.

The use of potassic fertilizers has been considered to be a predisposing factor in hypomagnesaemia by causing reductions in both the magnesium content of herbage and the absorption of magnesium from the alimentary tract. Recent evidence[2] suggests that the latter effect may be due more to an imbalance of the sodium:potassium ratio rather than to a high intake of potassium *per se*.

SODIUM AND CHLORINE

As indicated in relation to potassium, the principal function of sodium in the animal body is in the extracellular fluids where it operates in osmotic regulation and the maintenance of acid-base balance. Chlorine is also involved in these functions but has a further important role in gastric secretions as a constituent of hydrochloric acid and various chlorides.

In contrast to potassium, the sodium and chloride concentrations in most herbage plants are low and deficiencies of both elements can occur in sheep, although the incidence is comparatively rare. The clinical signs of sodium deficiency are not pathognomonic. A prolonged deficiency results in pica, anorexia and a progressive loss of condition, generally seen throughout the flock, and which could also be indicative of other conditions such as cobalt deficiency and chronic nematode parasitism or fascioliasis. Plasma sodium concentrations decline only when the affected animals are 'in extremis', although urinary sodium excretion decreases at an earlier stage. The rapid response to the addition of common salt (sodium chloride) to the diet is probably the only retrospective indication of a deficiency. Responses to the provision of salt, either as a salt lick, as a constituent of a general mineral supplement, or as an additive to concentrates, are more likely to be due to the alleviation of a sodium deficiency than to the correction of a lack of chlorine.

The provision of salt licks or mineral supplements containing salt, to which sheep have free access, does not entail any risk of sodium or chlorine toxicity, but salt poisoning can occur when sodium chloride is incorporated in supplements, such as feed-blocks, as a means of regulating dietary intake. When foodstuffs of this type are used, adequate water must be freely available at all times.

SULPHUR

The principal function of sulphur is as a constituent of the aminoacids cystine, cysteine and methionine but it also occurs in insulin and in certain vitamins. Sulphur is generally considered to be particularly important in sheep nutrition as the sulphur content of wool is about 0.04 (4%), incorporated in the cystine molecules.

The main sources of sulphur in sheep diets are the above three sulphur-containing aminoacids which are constituents of the animal's normal dietary protein. Rumen micro-organisms can also use inorganic sulphur, usually provided in the form of sodium sulphate.

Sulphur deficiency causes depressed wool production and poor fleece characteristics, including a lack of crimp caused by delayed keratin formation. These signs are also a feature of prolonged and severe copper deficiency. Despite the importance of sulphur and the responses which have been demonstrated to the provision of additional sulphur to sheep, sulphur deficiency as such is not normally given consideration as it would invariably arise as a secondary effect of a more serious protein deficiency. A lack of sulphur can, however, limit production when sheep are fed diets containing urea or other forms of non-protein nitrogen, and in such situations it is recommended that sodium sulphate be incorporated in the diet at a ratio of 1:9 by weight with urea.

Sulphur can combine with molybdenum in the rumen to form tetrathiomolybdates which reduce the absorption of copper and thus lead to an induced copper deficiency. This occurs, however, only in the presence of relatively high concentrations of dietary molybdenum and there is little risk of inducing copper deficiency by the addition of sulphur to diets low in protein and supplemented with non-protein nitrogen.

CALCIUM AND PHOSPHORUS AND THEIR RELATIONSHIP TO VITAMIN D

Vitamin D   It is not possible to discuss deficiency of calcium or phosphorus without reference to vitamin D which plays an essential interactive role in calcium and phosphorus metabolism.

The animal obtains its vitamin D from the diet as vitamin D_2 and from insolation as vitamin D_3, both forms having equal biological potency in mammals. The concentration of vitamin D_2 in sun-dried herbage is higher than in fresh material but, unless deliberately supplemented, the diet probably contributes only a small proportion of the total requirement. Vitamin D_3 is synthesized from precursors in the skin and its production is subject to seasonal variation. In winter in the UK production probably ceases and vitamin D status may become borderline, especially at parturition.[3]

Spectacular progress towards the full understanding of the mechanisms of the action of vitamin D has been made in the last decade. It has been shown that vitamin D is, in fact, a hormone precursor which is converted to the main circulating form, 25-hydroxyvitamin D (25-OHD) in the liver. Subsequent metabolism of 25-OHD in the kidney under precise physiological control produces the active hormonal compound, 1,25-dihydroxyvitamin D (1,25-diOHD) which, by stimulating bone resorption and intestinal absorption of calcium and phosphorus, increases the concentrations of these elements in plasma. The role of vitamin D in the regulation of calcium and phosphorus metabolism has been reviewed.[4]

A deficiency of vitamin D in growing lambs causes rickets, characterized by bone deformities, enlarged joints and lameness. In adult sheep the deficiency can cause osteomalacia with its associated signs of weak and easily fractured bones. In practice a low phosphorus intake also appears to be a prerequisite for these conditions to be manifested. Their occurrence can be prevented by proper attention to the mineral and vitamin constituents of the diet.

Calcium   Calcium forms a larger proportion of the mineral content of the animal body than any other element. Almost all of it is contained in the skeleton and the teeth, but it is also an important constituent of cells and body fluids, and plays an essential role in certain enzyme systems. The skeleton constitutes a large and important reserve of calcium and of phosphorus and ewes can tolerate moderate deficiencies of calcium for considerable periods without untoward effects on production or health. In the long-term, however, there must be opportunities for these reserves to be replenished if production and health are to be maintained.

Animal by-products, such as fish meal, and green

crops, including grass and particularly legumes, have high concentrations of calcium, but cereal grains and root crops are poor sources of this element. Milk is an excellent source of calcium for unweaned lambs.

Recent estimates of the calcium requirements of sheep[1] are considerably lower than, and in many cases less than half, previously published figures. This revision of calcium requirements is due, first, to a recognition that faecal endogenous losses are considerably lower than was formerly believed; secondly, to the demonstration of the fact that sheep absorb calcium according to their needs, and thirdly, because it has now been shown that in most diets a very high proportion of the calcium present is in a readily available form. Much of the new knowledge in these areas has been contributed by work conducted at the Moredun Institute.[5, 6, 7, 8]

The economically most important condition caused by a calcium deficiency in sheep is parturient paresis, more commonly referred to as hypocalcaemia or 'lambing sickness'. Unlike bovine parturient paresis, which invariably occurs at parturition, the condition can be found in ewes some weeks or even longer either before or after lambing. Its occurrence is often associated with stressful conditions, e.g. the gathering of ewes for pre-lambing vaccination. Ewes in the early stages of hypocalcaemia show incoordination and muscular tremors, but without the tetanic spasms associated with hypomagnesaemia. They generally fall with hind limbs extended and thereafter remain recumbent, with mild tympany, and pass into a comatose state. If untreated the condition is usually fatal, although death is not as rapid as in hypomagnesaemia, generally occurring from a few hours to two days after the development of the first signs.

It is important to distinguish between the clinical signs of hypocalcaemia, as outlined above and those of hypomagnesaemia (see below) and pregnancy toxaemia. As with some forms of hypocalcaemia, pregnancy toxaemia occurs in the final weeks before lambing and many of the symptoms are similar, although ewes suffering from pregnancy toxaemia generally live longer. Serum calcium and plasma 3-hydroxybutyrate determinations on blood samples from affected individuals and representative numbers of the flock are of value in confirming diagnoses. At post-mortem examination pregnancy toxaemia cases have a characteristic white liver indicative of fatty degeneration, as opposed to the absence of any macroscopic lesions in cases of hypocalcaemia. The rapid response of the hypocalcaemic ewe to calcium therapy contrasts with the poorer prognosis and frequent lack of response to treatment of the ewe with pregnancy toxaemia.

Plasma calcium concentration is controlled primarily by parathyroid hormone (PTH), 1,25-diOHD and calcitonin, although it is also affected to a lesser extent by a number of other hormones. Hypocalcaemia induces an increased secretion of PTH which in turn stimulates synthesis of 1,25-diOHD in the kidney. The latter, acting in concert with PTH, increases bone resorption, while also increasing absorption of calcium and phosphorus from the small intestine. The hyperphosphataemia which might result from these two responses is prevented by a further selective action of PTH on the kidney which increases tubular reabsorption of calcium and decreases that of phosphorus. In hypercalcaemia, on the other hand, secretion of PTH is reduced and that of calcitonin increased. The principal action of calcitonin is to reduce bone resorption, thus reducing plasma concentrations of both calcium and phosphorus. Any consequent hypophosphataemia is prevented by the kidney, independent of vitamin D and PTH, which under these conditions reduces urinary excretion of phosphorus.

The hypocalcaemia of parturient paresis is caused by a breakdown of the homeostatic effects of these controlling mechanisms rather than by an inadequacy of calcium reserves. Elevated concentrations of PTH and 1,25-diOHD are invariably observed in cases of parturient paresis, but there is conflicting evidence for the increased concentration of calcitonin.[9, 10, 11] The significance of the latter is still subject to debate but it has been suggested that it may be important in explaining the inability of the animal to respond to the calcium mobilizing hormones.[9]

In cattle, and particularly dairy cows, it is considered that the primary cause of parturient paresis is the sudden and markedly increased requirement for calcium imposed by the onset of lactation, but as the condition can occur in ewes some weeks prior to parturition, when the demand for calcium is not excessively high and is increasing in a progressive manner, it is possible that there may be other, as yet unrecognized, predisposing factors. It may not be coincidental that the vitamin D status of ewes is at a minimum at parturition.[3]

Ewes suffering from parturient paresis respond very rapidly to the administration of calcium, generally given as a 200 g/l (20% w/v) solution of calcium borogluconate. (This demonstrates that it is in fact the hypocalcaemia and not the hypophosphataemia which is the principal cause of the symptoms of parturient paresis.) Treated ewes generally resume normal behaviour and eating within an hour or less, but should be inspected frequently as the condition can recur, necessitating further treatment.

Despite the effectiveness of the treatment of even comatose ewes, prevention is, as always, preferable to cure. The principle of prevention of parturient paresis is to ensure the ready availability of sufficient calcium to meet the very rapidly increasing requirements. This can be achieved either by supplying supplementary calcium, e.g. by parenteral administration or by strategies designed to increase the absorption of dietary calcium and the mobilization of skeletal calcium reserves. The administration of parenteral calcium on a flock basis is impracticable. Likewise injection of 25-OHD or 1 α-hydroxyvitamin D (a synthetic analogue of 1,25-diOHD) which after metabolism to 1,25-diOHD leads to a stimulation of intestinal absorption of calcium, is not feasible in commercial sheep production although it has been used with some success in cattle. Such treatment is expensive and, because of the relatively short biological half-life of these compounds, the lambing date must be predictable with some precision. Injections of vitamin D itself incur the risk of toxicity unless careful control of dose and seasonal timing is maintained, and there are not sufficient experimental data available at present to assess the value of such treatment.

At one time it was considered that hypocalcaemia in ewes was best prevented by feeding high levels of calcium and phosphorus during late pregnancy. It is now realized, however, that this approach is likely to be counter-productive. Excessive dietary intakes of calcium are likely to suppress the mechanisms which facilitate the mobilization of calcium from skeletal reserves, and can thus induce rather than prevent hypocalcaemia when the demand for calcium from reserves is increased. The more modern approach is to feed diets relatively low in calcium and normal in phosphorus until shortly before lambing. The objectives of this are to reduce calcitonin secretion, and, by stimulating parathyroid hormone production, to increase calcium absorption from the intestine and induce a moderate mobilization from skeletal reserves. Dietary calcium and phosphorus intakes should then be increased at parturition. This strategy allows the increased requirements for calcium and phosphorus at the onset of lactation to be met from the increased intake of these elements, supplemented if necessary by some further withdrawal from skeletal reserves.

There is, however, some debate as to the actual levels of calcium intake required for the most effective operation of this strategy. The most recent review[1] quotes dietary calcium requirements for 75 kg ewes as increasing from 1.9–3.9 g/day during the final 12 weeks of pregnancy, rising thereafter to 6.5 or 8.8 g/day in ewes producing 2 or 3 kg milk/day respectively. The advisory services are as yet understandably cautious about recommending such comparatively low levels in practice and one source [12] advocates daily intakes of some 13 g in the final month of pregnancy for ewes of a similar weight, with a commensurate increase following lambing.

Parturient paresis in the bovine[10, 11] and the relevance of calcium and phosphorus homeostatic mechanisms to this condition in sheep and cattle [9] have been reviewed recently.

Phosphorus   Most of the phosphorus in the body is found in association with calcium in the skeleton and teeth, and the two elements are traditionally considered in relation to each other. Phosphorus, however, is also important in its own right as a constituent of saliva and a number of major metabolites including phospholipids, phosphoproteins, nucleic acids, hexosephosphates and adenosine di- and triphosphate. Phosphate deficiencies have been identified in many areas of the world and are considered by some to be more common and economically important in ruminants than deficiencies of any other element.

Feeding-stuffs with high phosphorus contents include cereals, milk, fish meal and meat by-products containing bone. Hay, straws and crop by-products such as sugar-beet pulp are poor sources of phosphorus. A high proportion of the phosphorus in cereals is present in the form of phytates which are not readily available to simple-stomached animals. Phytates can, however, be hydrolized by rumen micro-organisms and it has been shown that phytate-phosphorus is as readily utilized by sheep as any other source of the element.

As with calcium, the results of recent studies, to which work at the Moredun Institute has made a major contribution[13, 14], have shown that endogenous faecal losses of phosphorus are much lower than earlier work had indicated, and present estimates of those are about one-third of the formerly used values. Accordingly, the most recent estimates of phosphorus requirements of sheep[1], and particularly those of pregnant and lactating ewes, are considerably lower than previously quoted.

The absorbability of milk-phosphorus by suckling lambs is very high and probably in excess of the 0.95 value used in most calculations. The coefficient of absorption decreases with age, and the commonly used values are 0.73 for weaned lambs up to 1 year of age and 0.60 for older sheep.

Because phosphorus is required with calcium for skeletal development when the supply of vitamin D is limited, phosphorus deficiency produces rickets in growing lambs and osteomalacia in adult sheep. In chronic cases of phosphorus deficiency affected animals have stiff joints, frequently show muscular weakness, and have a general poor performance and ill-thrift. Phosphorus deficiency is also one cause of pica, but this is seen more commonly in cattle than in sheep. The deficiency is considered to cause infertility in cattle, and may also do so in sheep, although the evidence for this is less clear. Sheep generally suffer less than cattle from phosphorus deficiency and this has been attributed to their more selective grazing habits by which they select those plants and parts of plants containing higher concentrations of the element.

Adult sheep can tolerate moderate phosphorus deficiencies for considerable periods without showing any signs and without apparent adverse effects on performance, despite the fact that dietary phosphorus deficiencies are reflected relatively quickly in decreased concentrations of serum or plasma inorganic phosphate. Hypophosphataemia also occurs in ewes with parturient paresis, but this is an effect rather than a cause of the condition which, as noted above, responds rapidly to the correction of the associated hypocalcaemia. Phosphorus deficiency, as diagnosed by low circulating concentrations of inorganic phosphate, responds readily to the provision of additional dietary phosphorus, given usually as a mineral additive to concentrates or as a 'straight' mineral mix.

Phosphorus deficiency has also been implicated in the condition known as 'cappie' or 'double scalp', in which the frontal bones of the skull are extremely fragile and can, by pressure, be easily broken into the frontal sinus. At one time this was not uncommon in northern England and in parts of Scotland. The cause was never positively identified, but it was considered to be due primarily to a deficiency of phosphorus, frequently associated with severe helminthiasis. The condition is now comparatively rare, being found only occasionally in the northernmost counties of Scotland. It is more likely to be caused by a general low level of juvenile nutrition, perhaps including deficiencies of calcium, phosphorus and protein, but cannot be prevented or cured solely by the provision of additional dietary phosphorus or by anthelmintic treatment.

Traditionally the roles of calcium and phosphorus in the body have been considered together, and estimates of dietary mineral requirements have generally included recommendations on calcium: phosphorus ratios. This view is now less strongly held, although it is possible to formulate diets with more than the minimum recommended quantities of both elements, but in abnormal ratios, which will impair growth rate. It is also known that the proportion of calcium can have important effects in low phosphorus diets. The most recent review[1] concludes that it is not possible to state with certainty the optimal calcium: phosphorus ratio for animal performance, or indeed whether such a ratio exists. However, in all recommended allowances the calcium: phosphorus ratio falls within the range of 1:1 to 2:1 and this is generally accepted as both safe and satisfactory.

MAGNESIUM

The quantity of magnesium in the animal body is less than that of any other macro-element, but, as with calcium and phosphorus, the greater part of it is contained in the skeleton. The remainder is found in the soft tissues and fluids where it fulfils an important function as an activator of some 300 enzyme systems including those utilizing adenosine triphosphate. Magnesium is stored in the skeleton and in growing lambs on magnesium-deficient diets relatively large amounts of the element can be mobilized from this source. In mature sheep, however, skeletal magnesium is not readily available. Because of this a dietary deficiency of magnesium can very rapidly lead to hypomagnesaemia and tetany, particularly in grazing ewes in early lactation, and it is important that attention is paid to the magnesium content of the diet at that time.

Vegetable proteins, especially cottonseed and linseed, are good sources of magnesium, and legumes are generally richer in magnesium than is grass. It is, however, imprudent to rely on magnesium-rich foods as the only source of the element and as a general rule lactating ewes grazing spring herbage should receive supplementary magnesium.

Magnesium deficiency can occur in sheep in a chronic form in animals subjected to poor nutrition during winter, but the acute form of hypomagnesaemia, also known as 'grass tetany', 'grass staggers', 'lactation or magnesium tetany', is the more common and by far the more important. It generally occurs in ewes within the first month or six weeks after lambing. Affected ewes exhibit hyperaesthesia and trembling, especially of the facial muscles. Some ewes appear unable to move, and other walk in an uncoordinated manner with a characteristic stiff-limbed gait. They collapse on one side and show repeated tetanic spasms, with all four limbs rigidly extended, and with a characteristic extension of the head and neck. Death can be very rapid without the coma typical of hypocalcaemia. The condition develops extremely quickly, and very often the first indication of hypomagnesaemia is the finding of a dead ewe which had appeared normal some hours previously.

Differential diagnosis from hypocalcaemia is based on clinical signs, the fact that hypomagnesaemia in its tetanic form generally occurs only in the post-lambing period and from blood analysis. Other causes of central nervous disturbance which have to be distinguished from hypomagnesaemia include listeriosis, cerebrocortical necrosis and louping-ill in tick-infested areas.

The condition has been studied extensively and it is now considered that acute hypomagnesaemia is due principally to a reduced absorption of magnesium which may be associated with, and exacerbated by, a redistribution of the element within the body.[2] It is now thought that the previously postulated increase in magnesium excretion is not implicated, nor is the condition a result of an increased magnesium requirement.

The absorption of dietary magnesium is low, average values from a number of studies being of the order of 0.24–0.29 (24–29%). The variances associated with these mean figures are, however, considerable due to large between-animal and between-diet differences, and in the calculation of dietary allowances a value of 0.17 (17%) has been recommended [1] to provide a margin of safety. In sheep the major site of magnesium absorption is the rumen and, although absorption has been shown to occur in the large intestine, this is considered to be of little importance and cannot compensate for reduced absorption in the proximal part of the digestive tract. Magnesium is actively transported across the rumen by a process in which sodium and particularly the sodium: potassium ratio, is important. Spring grass can be low in sodium, especially if treated with potassic fertilizers, and this can lead to an imbalance of the rumen sodium: potassium ratio which will adversely affect magnesium absorption. Spring grass is also high in protein and there is evidence that elevated rumen ammonia concentrations from the degradation of this protein can likewise reduce magnesium absorption. Further evidence indicates that the high protein content of such herbage also results in a dietary energy: nitrogen imbalance, leading to a reduction in voluntary herbage intake which exacerbates the energy deficit and in turn reduces both magnesium intake and absorption.

Hypomagnesaemia can impair calcium and phosphorus homeostasis and is not infrequently accompanied by hypocalcaemia and hypophosphataemia. In such cases the evidence concerning parathyroid response is conflicting. In some cases an inappropriately low circulating PTH concentration has been reported; other workers have shown expected elevated concentrations without a response from the target tissues. The underlying mechanisms are not yet fully understood.

There is now a considerable body of evidence to indicate that magnesium can be removed from the circulation to other body compartments and that this can cause clinical hypomagnesaemia. This is frequently associated with increased lipolysis, and factors which elevate plasma non-esterified fatty acid concentrations can predispose to tetany. Such factors include dietary energy deficiency and stressful conditions such as unaccustomed handling, transportation and low temperatures. The turning-out to pasture of housed ewes shortly after lambing constitutes a stress and as such increases the risk of hypomagnesaemia. If weather conditions are adverse and the pasture has been fertilized with nitrogen and potassium, the risk must be greatly increased. Plasma or serum magnesium concentrations of less than 0.7 mmol/l (17 mg/l) indicate that animals are at risk and, although clinical signs of tetany may not appear until concentrations have fallen to around 0.2–0.3 mmol/l (5–7 mg/l), the decrease to such levels can occur very rapidly.

There are a number of measures which can be taken to lessen the risk of hypomagnesaemia. It has been advocated[2] that adequate sodium be provided to stabilize the rumen sodium: potassium ratio, and that the diet should have a high fibre content to encourage rumination and salivation and a high energy content to prevent excessive rumen ammonia concentrations. This also promotes volatile fatty acid and carbon dioxide production which facilitates magnesium absorption. A high energy diet also serves to minimize lipolysis, but at the same time efforts must be made to avoid the imposition of physiological or environmental stress.

In addition to these husbandry measures, supplementary magnesium should also be given when there is a risk of hypomagnesaemia. This is generally provided in the form of magnesium oxide (available commercially as calcined magnesite) supplied at 7 g/ewe/day. To encourage ewes to consume this it is usually incorporated in a concentrate, and such is the potential loss from hypomagnesaemia that it is normally considered economic to feed concentrates solely as a carrier of the daily magnesium allowance, even when supplementary energy and protein are not required. Alternative, but less satisfactory, forms of supplying magnesium at times of risk include the foliar dusting of herbage with calcined magnesite and the provision of a high-magnesium mineral mixture either as a powder or liquid to which ewes have free access. The addition of magnesium to drinking water is a less effective means of administration when ewes are lactating.

While every effort should be made to supply ewes at risk with magnesium in sufficient quantity and suitable form to prevent hypomagnesaemia, care must be exercised to ensure that the correct amount is not exceeded. Excessive magnesium can cause scouring and unnecessarily prolonged feeding can produce other undesirable effects. It is also important that diets fortified with magnesium should not be fed to male sheep as this can result in the formation of magnesium ammonium sulphate calculi in the urethra, causing urolithiasis and the possibility of death following rupture of the bladder.

Ewes affected with hypomagnesaemia, even in the comatose stage, can be treated successfully with a combined injection of magnesium and calcium. This is given intravenously as a 200 g/l (20% w/v) solution of calcium borogluconate with 39 g/l (3.9% w/v) magnesium chloride. An additional dose is given subcutaneously, and on occasion the use of a muscle relaxant may be indicated. Recovery is generally rapid, but as with hypocalcaemia, relapses are not uncommon and the treatment may have to be repeated, and here the prognosis is poor.

The field of magnesium metabolism and hypomagnesaemia has been recently reviewed.[2, 15]

General

In theory deficiencies of the essential macro-elements can be avoided by careful attention to dietary intake in the light of the considerable knowledge which now exists on mineral requirements. The reality, however, is much less straightforward. The vast majority of sheep in this and other countries are kept at pasture for at least the greater part of the year, where they selectively graze plants and parts of plants. This makes it impossible in commercial practice to esti-mate with any confidence their intake of minerals or any other nutrient.

Sheep can, however, tolerate moderate deficiencies of most elements for a considerable period without serious effects on their health or productivity, always provided that there is opportunity at some stage to replenish their depleted mineral reserves. The most important exception to this generalization is, as indicated above, with respect to magnesium where provision of supplementary magnesium should be made at certain critical times. Special attention should also be paid to the provision of calcium, and in particular to the avoidance of providing high levels of dietary calcium in anticipation of a later increased demand. In many management systems sheep receive some form of supplementary feeding before and during lambing and this provides opportunities for some degree of regulation of dietary intakes of magnesium and calcium.

As far as the other major elements are concerned, the possibility of a mineral deficiency should always be considered when investigating any signs of ill-thrift, or where a poor level of production exists.

Acknowledgements

The author gratefully acknowledges the information on vitamin D provided by Dr. B.S.W. Smith, Moredun Research Institute.

REFERENCES

1 Agricultural Research Council (1980) The nutrient requirements of ruminant livestock. Royal Commonwealth Agricultural Bureaux, Farnham.

2 Martens H. & Rayseiquier Y. (1980) Magnesium metabolism and hypo-magnesaemia. In *Digestive Physiology and Metabolism in Ruminants*. (Eds. Ruckebusch Y. & Thivent P.) pp.447–66. MTP Press, Lancaster.

3 Smith B.S.W. & Wright H. (1981) Seasonal variation in serum 25-hydroxyvitamin D concentrations in sheep. *Veterinary Record*, **109**, 139–41.

4 De Luca H.F. (1979) The vitamin D system in the regulation of calcium and phosphorus metabolism. *Nutrition Reviews*, **37**, 161–93.

5 Field A.C. & Suttle N.F. (1969) Some observations on endogenous loss of calcium in the sheep. *Journal of Agricultural Science*, **73**, 507–9.

6 Sykes A.R. & Field A.C. (1972) Effects of dietary deficiencies of energy, protein and calcium on the pregnant ewe. 1. Body composition and mineral content of the ewes. *Journal of Agricultural Science*, **78**, 109–17.

7 Sykes A.R. & Dingwall R.A. (1975) Calcium absorption during lactation in sheep with demineralized skeletons. *Journal of Agricultural Science,* **84**, 245–8.

8 Field A.C., Suttle N.F. & Nisbet D.I. (1975) Effects of diets low in calcium and phosphorus on the development of growing lambs. *Journal of Agricultural Science,* **85**, 435–42.

9 Care A.D., Barlet J.P. & Abdel-Hafaez H.M. (1980) Calcium and phosphate homeostasis in ruminants and its relationship to the aetiology and prevention of parturient paresis. In *Digestive Physiology and Metabolism in Ruminants.* (Eds. Ruckebusch Y. & Thivend P.) pp.429–46. MTP Press, Lancaster.

10 Allen W.M. & Davies D.C. (1981) Milk fever, hypomagnesaemia and the Downer Cow Syndrome. *British Veterinary Journal,* **137**, 435–41.

11 Littledike E.T., Young J.W. & Beitz D.C. (1981) Common metabolic diseases of cattle: ketosis, milk fever, grass tetany and downer cow complex. *Journal of Dairy Science,* **64**, 1465–82.

12 Meat and Livestock Commission (1981) *Feeding the Ewe.* 2e, MLC, Milton Keynes.

13 Field A.C., Sykes A.R. & Gunn R.G. (1974) Effects of age and state of incisor dentition on faecal output of dry matter and on faecal and urinary output of nitrogen and minerals, of sheep grazing hill pastures. *Journal of Agricultural Science,* **83**, 151–60.

14 Sykes A.R. & Dingwall R.A. (1976) The phosphorus requirement of pregnant sheep. *Journal of Agricultural Science,* **86**, 587–94.

15 Littledike, E.T. and Cox, P.S. (1979). Clinical mineral and endocrine relationships in hypomagnesaemic tetany. In. Rendig, V.V. and Grunes, D.L. (eds). Grass tetany: proceedings of a symposium sponsored by Divisions C-l, C-2 and C-6 of the Crop Science Society of America. (American Society of Agronomy, Special Publication, No. 35). Madison, Wisconsin, American Society of Agronomy.

DISORDERS RELATED TO TRACE ELEMENT DEFICIENCIES   36

N.F. SUTTLE AND K.A. LINKLATER

There is considerable disagreement about the extent to which trace element deficiencies limit the health and productivity of sheep. Confusion arises because both the clinical signs of disorder and the biochemical criteria used to support diagnoses are often non-specific. Affected animals may simply appear unthrifty, i.e. 'pining', and healthy animals may be found with subnormal blood biochemistry. The surest diagnosis is often a response to a specific trace element supplement. This chapter outlines the commonest manifestations of the most widespread deficiencies but every veterinarian should be on the look out for hitherto unrecognized pathological and physiological consequences of deficiency. The lesions produced by a particular deficiency may vary according to the stage of development at which it occurs, the presence of other deficiencies and the stresses imposed by environmental conditions.

COBALT (Co) DEFICIENCY

Clinical signs

The clinical signs of Co deficiency are non-specific. Lambs are particularly susceptible, exhibiting growth retardation, debility, emaciation and a watery discharge from the eyes. In severe cases there is loss of appetite and anaemia: the latter is normochromic and normocytic and reflected by pale visible mucous membranes. A dramatic drop in serum B_{12} levels can occur after weaning and in the UK, lambs are most likely to become deficient in autumn. As far as the adult ewe is concerned an animal debilitated by Co deficiency may show secondary effects in the form of infertility and poor mothering ability. At present there is no evidence that Co deficiency in the dam directly affects the viability of her offspring. Associa-

tions between Co deficiency and cerebro-cortical necrosis (CCN) have been suggested and neuropathological lesions have recently been described in animals depleted under experimental conditions.[1]
No correlation between the incidence of CCN and liver Co or serum B_{12} levels was found in a survey in south east Scotland, however (K. Linklater, unpublished data).

Pathology

In severely-affected animals the carcass is extremely emaciated with no body fat depots. There is bone marrow hypoplasia and haemosiderin deposits are present in spleen and sometimes in liver. In New Zealand an ovine white liver syndrome in which livers are grossly enlarged, fatty and friable, has been associated with vitamin B_{12} deficiency. Histological examination has shown fatty changes in hepatocytes and bile ductules and mesenchymal proliferation in portal areas.[2] Whether this condition is a direct consequence of vitamin B_{12} deficiency or a toxic liver disease against which Co or vitamin B_{12} is protective has still be be resolved.

Cause

The only known role for Co is as a constituent of vitamin B_{12} which has two coenzyme functions in the body. Co deficiency is therefore synonymous with vitamin B_{12} deficiency. Relationships between dietary Co and animal vitamin B_{12} status are influenced by:
1. the concentration of Co in the diet
2. ingestion of extraneous Co, particularly as soil which generally contains far more Co than the herbage it supports
3. the efficiency with which Co is incorporated into true B_{12} by the rumen microflora
4. the efficiency of B_{12} absorption.
The latter may be impaired by competition from physiologically useless analogues which are also formed in the rumen and also by factors such as parasitic infestation altering the synthesis of intrinsic factor and damaging the absorptive capacity of the mucosa. Nevertheless there is an approximate relationship between incidence of disease and Co intake; dietary concentrations below 1.87 μmol (0.11 mg) Co/kg DM being regarded as nutritionally inadequate.[3] Dietary concentrations are in turn broadly related to the available Co level in the soil and levels

below 5.1 μmol (0.3 mg) /kg soil, extractable with 0.34 mol/l acetic acid are regarded as deficient.[4]

Inadequate B_{12} absorption may lead to the reduced activity of two vitamin B_{12} coenzymes. *Methylcobalamin* assists methyl transferase which recycles methyl groups via homocysteine, thus promoting methionine synthesis. Methionine supply ultimately influences DNA synthesis and B_{12} deficiency may thus impair red cell maturation.[5] Anaemia is, however, a late sign of deficiency in all species, due largely to the slow turnover of erythrocytes and it is possible that the development of other cell types is impaired before anaemia is detectable. *Deoxyadenosylcobalamin* assists methyl malonyl CoA mutase and thus performs a key role in the energy metabolism of ruminants, facilitating the metabolism of propionate via succinate and the tricarboxylic acid cycle. The prominence of anorexia as an early symptom of deficiency in sheep and cattle has been related to impaired clearance of propionate from the bloodstream. Methylmalonic acid (MMA) also accumulates and becomes incorporated into branched-chain fatty acids which may impair lipid metabolism in both the liver and CNS.

Diagnosis

There are many conditions which can be confused with Co pine especially undernutrition, parasitic burdens and deficiencies of other trace elements. Diagnosis therefore requires supportive biochemical data. Concentrations of vitamin B_{12} in the bloodstream are of limited value since the vitamin may be stored in the blood and depleted to very low levels before a functional deficiency occurs in the tissues.[6] While levels of 0.295 pmol (400 pg)/ml in serum were once regarded as diagnostic of deficiency, lower thresholds (0.150 pmol or 200 pg/ml) are now adopted. A further complication surrounds the analytical procedures used. It is not known if analogues of B_{12}, extensively synthesized in the rumen, enter the circulation. If they do then non-specific microbiological and protein-binding assay methods give misleading indications of B_{12} status.

A functional deficiency of the two vitamin B_{12} coenzymes can be detected by the appearance of the abnormal metabolites formiminoglutamic acid (FIGLU) and MMA in urine and almost certainly the same metabolites will be detectable in the bloodstream. MMA levels may however decline as the anorexia of Co deficiency develops. Vitamin B_{12} concentrations in the liver correlate well with both response to Co therapy and urine MMA excretion in lambs and levels below 0.09 μmol (0.12 mg)/kg wet weight can be regarded as deficient.

Due to the inadequacies of many of the biochemical parameters, response to treatment remains the most conclusive method of confirming a diagnosis of Co deficiency. This may be conveniently done by treating a small group of sheep, say 20, and monitoring the clinical and weight gain responses.

Prevention and treatment

The capacity to store vitamin B_{12} in the body is limited and continuous rather than discontinuous methods of treatment are preferred. Several methods are available and the one selected will depend on the economic, dietary and general husbandry factors pertaining to each circumstance.[7]

Oral dosing Vitamin B_{12} synthesis by the rumen microflora increases dramatically when Co is given orally and doses of 34–119 μmol (2–7 mg) per week are highly effective. However, this frequency of treatment is rarely practicable. While single large doses of 4.25 mmol (250 mg) per sheep at monthly intervals prevent the development of severe deficiency problems they are not adequate for optimum performance. By far the best oral method is the administration of a Co-containing bullet of high specific gravity which slowly releases Co in the rumen. These should not be given to lambs before weaning as an insoluble coating of calcium phosphate can form on the bullet preventing the release of Co. The risk of regurgitation is minimal provided the bullets are carefully administered and the animals are kept from grazing for a few hours after dosing.

Injection Intramuscular or subcutaneous injections of the hydroxycobalamin analogue of vitamin B_{12} give a rapid response but they have to be repeated frequently and dose rates of 0.18–0174 μmol (250–1000 μg) per month have been suggested. If animals are to be kept for any length of time therapy should be continued by another method. Toxicity due to overdosing with vitamin B_{12} injections in sheep has not been reported.

Feed supplementation Where sheep are being fed concentrates, mineral mixes containing as little as 680 μmol (40 mg) Co/kg will provide an adequate Co intake when they are included at the rate of 25 kg per tonne provided the concentrate forms at least 10 per cent of the total food intake. The risks of poisoning by overdosing with Co are minimal since lambs have been known to tolerate levels of 17 mmol (1000 mg)/kg DM in their diets.

Pasture application of cobalt sulphate The Co content of herbage can be increased to safe levels by a single application of 2–3 kg/ha of hydrated cobalt sulphate in a low volume spray or incorporated in a granular fertilizer at the time of manufacture. Where set stocking is practised, it has been found preferable to apply 6 kg cobalt sulphate/ha to one third of the grazing area. The effect of Co fertilization often lasts for 4 years or more but is governed by soil conditions, and is short-lived on soils of high pH and high manganese content. Co should not be applied too soon after liming and it may be prudent to have the manganese status of the soil checked. The price of Co salts has fluctuated widely in recent years and this has tended to moderate their use as fertilizers.

COPPER (CU) DEFICIENCY

Clinical signs

The earliest sign of Cu deficiency is the loss of wool crimp (steely wool). It can be found in the absence of other abnormalities and transient deficiencies show up as a band of uncrimped wool of low tensile strength. Another early manifestation of Cu deficiency in sheep is swayback (enzootic ataxia) in lambs. Swayback takes two forms, a congenital one which is apparent at birth and a delayed one which does not become apparent until 6–8 weeks of age. Affected lambs are incoordinated and have a tendency to sway on their hind legs. The congenital form is often more severe and some lambs may be so badly affected that they cannot get up. At the other extreme, the delayed form may be inapparent unless the sheep are driven (see Chapter 16).

Cu-deficiency has recently been reported in Scottish Blackface lambs on improved hill ground. The deficiency was imposed after birth and live weight-gain was depressed, anaemia was evident, bones were susceptible to fracture and fleeces were grey in colour, stringy and sparse in appearance. Susceptibility to infection may also have been increased.[8]

Pathology

In lambs severely affected with congenital swayback, there is cavitation of the cerebral white matter and marked internal hydrocephalus. In less severe cases, brains and spinal cords may be grossly normal and characteristic lesions only detected histologically. These consist of chromatolysis in the brain stem and bilaterally symmetrical demyelination in the spinal cord. The critical site moves from the cerebrum in neonatal ataxia to the spinal cord in delayed swayback, coinciding with the peak periods of myelin formation in those parts of the CNS and suggesting that impaired myelin synthesis is an important factor. The contrasts between the CNS lesions in congenital and delayed swayback, and their absence in growth-retarded lambs [8], illustrates the importance of the stage of development at which Cu deficiency is imposed.[6]

In the cases of ill-thrift reported above [8], post-mortem lesions were non-specific and similar to any other 'pine' condition viz emaciated carcasses, almost total lack of body fat, and haemosiderin deposits in spleen and liver. There were fractures of the long bones and osteoporosis was detected histologically. Hypochromic anaemia is a late consequence of deficiency and it may be microcytic in the lamb and macrocytic in the ewe.

Cause

The unique susceptibility of ruminants to clinical Cu deficiency on natural diets is related to events in the rumen which commonly leave less than 10 per cent of the ingested Cu in an absorbable form. Scottish Blackface ewes on permanent pasture in summer absorb only 2 per cent of the Cu in the herbage and the value falls to 1 per cent in autumn. Ingestion of soil as a leaf contaminant may be partly responsible for the drop in absorption in the autumn and winter and the high incidence of swayback after mild winters. Other important factors are the small increases in herbage molybdenum (Mo) (up to 41.6 μmol or 4 mg/kg DM) and sulphur (S) (up to 125 mmol or 4 g/kg DM) which often follow liming and pasture improvements; these can cause significant reductions in Cu absorption by sheep. At these low levels of absorption, dietary concentrations of 157 μmol (10 mg) Cu/kg DM or more are required and few pastures exceed this level. Grass conserved as hay yields twice as much of its Cu to the sheep as the fresh material and dry summer grazings may resemble hay in this respect. With these and other sources of highly absorbable Cu (including cereals and brassicas) dietary concentration of 47–78 μmol (3–5 mg)/kg DM may be adequate. Dietary Cu concentrations are therefore no guide to the cause of deficiency problems.

The functional disorder which results from a lack of absorbable Cu in the diet is hard to define. Cu forms an essential part of at least 10 metalloenzymes which perform quite distinctive functions[6] and it is not known, for certain, which becomes rate-limiting in the diverse pathological manifestations of Cu deficiency. The development of swayback for example may be related to a deficiency of cytochrome oxidase causing anoxia and chromatolysis in the neurones or impaired synthesis of phospholipid and hence of myelin or to a deficiency of dopamine-B-hydroxylase and the subsequent accumulation of catecholamines in CNS. Similarly, abnormal bone matrix formation may be related to a defect in cytochrome oxidase activity in osteoblasts or lysyl oxidase in chondroblasts.

Granulocytes from hyocupraemic ewes and lambs have reduced contents of the Cu: Zn enzyme super-oxide dismutase (SOD) coupled with impaired microbial killing activity in vitro. A cause and effect relationship is possible in that generation of the superoxide radical is involved in microbicidal activity.[6] Evidence of increased susceptibility of the Cu-deficient animal to infection has been reported in small animals.

Diagnosis

Swayback may be confused with several other conditions which produce ataxia. Most common among these is the occurrence of abscesses in the spine following pyaemic spread from the navel, or castration and docking wound infections. These can be differentiated on clinical examination and confirmed at autopsy. Border disease can also cause cavitations of the brain but swayback can be distinguished partly by clinical examination (see 'Diseases of the Nervous System and of Locomotion') together with histopathological examination of the brain and spinal cord and Cu analyses (< 78.5 μmol [5 mg] Cu/kg DM) of affected tissues. Cu concentrations at other sites such as the liver and bloodstream, reflect current Cu status and may be particularly unhelpful in confirming delayed swayback where the lesions may develop several months before clinical disease is detected. How then should such data be interpreted?

Liver concentrations below 157 μmol (10 mg)

Cu/kg DM indicate that Cu depletion has proceeded to the point where Cu reserves are exhausted and this point can be reached rapidly in the newborn in which the size of the liver and its Cu store are small. Plasma Cu concentrations below 9.42 μmol (0.6 mg)/l generally indicate a state of deficiency, but this may not have persisted for sufficient time or been present at the critical times of development necessary for pathological changes or growth retardation to develop. It should be noted that Cu concentrations may be 1.57–3.14 μmol (0.1–0.2 mg)/l higher in whole blood and lower in serum than in plasma from the same animal. Because they decline at a slower rate than plasma Cu, low SOD concentrations in the erythrocyte indicate a prolonged deficiency and low values are more likely to be associated with impaired function and growth responses. Normal SOD levels are 0.4–0.5 iu/mg Hb for lambs and 0.3–0.5 iu/mg Hb for ewes.

Prevention and treatment

Since damage to the CNS and wool fibre in Cu deficiency is irreversible and there may be no compensatory growth in treated lambs, the emphasis must be on prevention. A number of methods are available.

Fertilizers It is impossible to generalize about the effectiveness of this method of prevention. On sandy Aeolian soils in Australia, a single application of 8.25 kg $CuSO_4$/ha in 1963 had doubled the Cu content of herbage sampled 13 years later and responses in liver Cu were found after 17 years.[9] By contrast, on peaty soils in New Zealand, annual applications of Cu are sometimes needed. Annual foliar applications of small amounts of Cu (0.2 kg $CuSO_4$. $5H_2O$/ha) increase herbage Cu concentrations but the extent and persistence of the effect is dependent on subsequent rainfall and in drought conditions Cu toxicosis is a hazard. Application of Cu-rich pig slurry also increases herbage Cu levels.

Supplementation of the diet Oral doses of 4 mmol (1g) $CuSO_4$ $5H_2O$ in solution given 8 and 4 weeks before lambing have been used for many years to prevent swayback. Dosing newborn lambs to prevent delayed swayback may be effective but since they absorb Cu up to 50 times more efficiently than the dam, dose/unit bodyweight is much reduced and 25 μmol (6.25 mg) $CuSO_4$. $5H_2O$/kg liveweight should suffice.

Provision of Cu orally in a single dose, slow release form is now possible using particles of cupric oxide in an appropriate vehicle. Doses of 50 mmol (4 g) CuO/ewe and 12.5 mmol (1 g) CuO/lamb are recommended and in the absence of antagonists such as Mo, the effect on the ewe may last for several years.

Continuous Cu supplementation via compound feeds intended for sheep is forbidden in Europe although risks of poisoning are only significant for in-wintered stock. Supplementation of home-grown supplementary feeds to a level of 314 μmol (20 mg) Cu/kg DM is an attractive alternative for sheep kept outside.

Injection The subcutaneous or intramuscular injection of chelated Cu is a widely used and generally effective preventative measure but not without hazard. Methionates and glycinates are slowly translocated from injection sites, produce large local reactions which can reduce carcass value and afford variable protection, albeit with virtually no risk of acute toxicity. At the other extreme diethylamine oxyquinoline sulphonate is rapidly translocated and gives no local reactions but its acute toxicity at 15.7–31.4 μmol (1–2 mg) Cu/kg liveweight limits doses to levels which are sometimes ineffective. The CuCaEDTA chelate is intermediate in all respects and acute toxicity is extremely rare at doses of 1 mg/kg bodyweight. For swayback prevention the decision to inject should be based on the presence of hypocupraemia (plasma Cu < 9.42 μmol [0.6 mg]/l) in mid or late pregnancy; one hypocupraemic ewe in a sample of 10 per cent of the flock may merit the injection of the whole flock.

With all methods of prevention, maintaining the Cu status of the ewe does not necessarily protect her lamb from growth retardation on improved pasture. This is because Cu transfer via the placenta and milk in the ovine is limited. Weaned lambs can be given Cu orally or by injection at rates/unit liveweight that are appropriate for adults but the latter is not recommended in animals soon to be slaughtered. Because of the ever-present risk of toxicity only one method of Cu supplementation should be applied at any time.

Breeding Breeds and individuals within breeds differ in their susceptibility to Cu deficiency due partly to genetic differences in the efficiency with which they absorb Cu from the diet. Thus the Scottish Blackface ewe absorbs Cu 50 per cent less efficiently than the Welsh Mountain and is more susceptible to swayback. Some control over Cu deficiency may therefore be obtained by the appropriate selection of breeds and the culling of susceptible individuals.

SELENIUM (SE) DEFICIENCY

Clinical signs

The most commonly recognized clinical syndrome of Se deficiency in sheep is nutritional myopathy which is also known as white muscle disease (WMD) and stiff lamb disease. Lambs are commonly affected at any time from birth to 6 months of age. Those born to severely deficient ewes may be stillborn or die within a few days of birth, usually because of sudden heart failure. Young lambs may merely be stiff and the symptoms are often confused with those of joint-ill. Older lambs affected with WMD may be recumbent, reluctant to move or collapse when driven. Symptoms can be precipitated by periods of severe weather in animals fed predominantly on roots.

In Australia and New Zealand an 'ill-thrift' syndrome has been attributed to Se deficiency. Affected stock fail to grow, respond favourably to Se therapy but not to vitamin E. In the UK one of the major questions yet to be answered is whether Se deficiency causes reduced animal performance without other signs of disease. In New Zealand infertility occurs in ewes in association with WMD and ill-thrift in the lamb. Up to 30 per cent of the ewes may be infertile due to high embryonic mortality, 3–4 weeks after conception. This syndrome also responds to Se but not vitamin E.[10]

Pathology

It is often difficult to detect lesions in the muscles in congenital forms of the disease but heart lesions are commonly seen as white plaques in the myocardium. In older lambs, affected muscles are pale and may show white striations due to the deposition of calcium. All muscles can be affected but those of the neck, leg and trunk are often most severely affected and the bilateral distribution of affected musculature is striking. The basic lesion is hyaline degeneration followed by coagulative necrosis and calcification. Individual muscle fibres may be oedematous and swollen with loss of cross striations and sarcolemmal proliferation.

Cause

The incidence of Se deficiency is related to the Se status of the diet which in turn reflects the Se status of the soils on which pasture and crops grow. For example vast areas of the mid-west States in the USA[10] and the peat soils in the South Island of New Zealand are deficient in Se. Newborn animals are put at risk when Se in milk declines as a result of the low Se intake by the mother. The nutritional requirement for Se is put at 1.02 μmol (0.08 mg)/kg DM[3], but disease does not inevitably occur at lower concentrations.

The only known function of Se is the destruction of peroxides through the activity of the selenoenzyme, glutathione peroxidase (GSHPx). If peroxides accumulate in the tissues, they cause oxidative damage to cell membranes. Vitamin E is an anti-oxidant and can also protect membranes from oxidative damage. The two nutrients may thus have mutually sparing effects, the two deficiencies can produce similar symptoms and Se deficiency is more likely to occur when vitamin E intakes are low. Deficiencies may only develop to the clinical stage in the presence of factors which stress the cell membrane: these include muscular exercise and the ingestion of polyunsaturated fatty acids (PUFA's) which generate peroxides. The inter-relationships between these factors is summarized in Fig.36.1 and amply illustrated by the

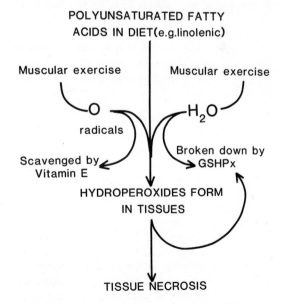

Fig. 36.1 Selenium, as the enzyme glutathione peroxide (GSHPx), and vitamin E both protect the tissues from exogenous and endogenous sources of oxidative stress.

acute myopathy which can affect cattle turned out to graze in the spring.[6] It would be surprising if sheep were not subjected to these same factors although the interactions are not manifested in the form of acute myopathy.

Se deficiency can affect tissues and cells other than muscle. Granulocytes from Se-depleted calves have been found to have an impaired ability to kill engulfed micro-organisms. The reason may be related to the role of GSHPx in removing products from the respiratory burst which accompanies phagocytosis.[6] The membrane of a tissue or cell which is in a particularly active metabolic state may be vulnerable to Se or vitamin E deficiency and this may explain why there are several clinical manifestations of deficiency.

Diagnosis

Several conditions may be confused with WMD, notably joint-ill, swayback, spinal abscesses and, where the intercostal muscles are involved, pneumonia. Animals affected by WMD do not show a febrile response but haematological examination may reveal a neutrophilia. Confirmation in the live animal depends on the measurement of certain enzymes in the bloodstream and the most important of these is the muscle enzyme, creatine phosphokinase (CPK), which leaks into the plasma from damaged muscles. In affected animals plasma CPK activity may reach 5000 iu/l or more compared with normal levels of 100 iu/l or less in plasma promptly separated from erythrocytes. Other factors such as sudden muscular exercise may cause transient rises in this enzyme but these are not as marked as those seen in myopathy. As the half-life of CPK in plasma is relatively short, high levels indicate acute or continuing extensive muscle degeneration. Other enzymes such as the aminotransferases (transaminases), whilst not being specific for muscular damage, remain elevated much longer and are therefore useful as an aid to diagnosis in chronic cases: for example glutamic oxalacetate transaminase (GOT: also known as aspartate aminotransferase) rises from normal levels of less than 50 iu/l to 500 iu/l or greater.

Diagnosis is supported by the detection of low levels of Se or GSHPx in whole blood. Since Se is present predominantly as the enzyme in the erythrocyte, the two parameters are highly correlated and values for one can be predicted from the other. Because factors other than low Se status can precipitate disease, and because the erythrocyte is not the target cell, relationships between blood Se or GSHPx

and clinical disease are arguable. Thus some regard values < 0.635 μmol (0.05 mg) Se/l whole blood (equivalent to 600 units GSHPx/l) as deficient while others use much lower thresholds (< 0.254 μmol (0.02 mg) Se/l in New Zealand). The former should be used to indicate an increased *risk* of deficiency which might be enhanced if plasma vitamin E concentrations were also low (< 0.5 iu [500 μg]/l). Since GSHPx is incorporated into cells of long half-life at erythropoiesis, concentrations in blood respond slowly to changes in Se intake. Plasma GSHPx is present in lower concentrations but responds more rapidly to changes in intake and may be more useful in detecting acute deficiencies, but its diagnostic value has yet to be explored.

Treatment

Clinically-affected animals are most effectively treated by single subcutaneous or intramuscular injections of Se at 2.54 μmol (0.2) mg Se/kg liveweight. A 1.4 per cent solution of hydrated sodium selenite gives 38 μmol (3 mg) Se/ml of solution and is a convenient means of therapy but 2 months should be allowed between Se treatment and slaughter for human consumption. Vitamin E may be given at the same time by injection or as daily oral doses of 50–100 iu of alpha-tocopherol.

Prevention

The incidence of Se-responsive disorders can be reduced by raising intakes of Se and vitamin E and minimizing the oxidative stresses placed upon cells. The use of certain crops, such as turnips, which have particularly low Se contents, should be controlled. Vitamin E levels can be maintained by avoiding factors which destroy the vitamin such as the excessive weathering of hay and the storage of moist grain in the presence of propionic acid or its treatment with alkali. Constituents rich in PUFA's, e.g. oil seeds should be kept to a minimum and management changes, e.g. onset of spring grazing should be gradual. Where such measures are insufficient, supplementation may be practised in a number of ways.[7]

Oral drenches are widely used where short-term supplementation is required. They have the advantage that an accurate dose may be administered but may have to be repeated every 2–3 months. Pellets containing elemental Se in an iron matrix have shown considerable promise in cattle but results in sheep in the UK have been variable due possibly to the use of an inappropriately small Se grain size ($<$ 45 μm). In some areas Se pellets have been shown to provide a continuous supply of Se for more than a year and have the advantage that animal handling is reduced. Disadvantages are that some pellets become dislodged from the rumen and lost from the animal whilst others may become coated with calcium phosphate and stop releasing Se.

Injections are widely used for prophylaxis at the rates given earlier and have the advantage that controlled doses of Se can be given. They can be effective for up to 3 months and can be given at strategic times to prevent the onset of clinical signs. Nutritional myopathy in lambs can be prevented by treating them at birth or their dams in the later stages of pregnancy and WMD in older lambs can be prevented by injecting them before they are folded on turnips in winter.

Se supplementation of the diet can be effected by the addition of mineral mixes to concentrate rations. It is desirable to achieve a final dietary Se concentration of 1.27 μmol (0.1 mg)/kg dry matter and therefore, on Se-deficient farms, the concentrate part of the ration may require much more than this. A mineral mix containing 15 mg/kg Se added to concentrates at the usual rate of 25 kg/tonne produces a concentrate with 4.76 μmol (0.375 mg) Se/kg which is suitable for feeding to ewes at rates of up to 1 kg/head/day. Excessive Se in the diet may be toxic and it is undesirable to exceed 2.54 μmol (2 mg) Se/kg of the total diet for any length of time. It is inadvisable to give Se in licks or free choice minerals since intakes vary widely.

The Se content of herbage can be raised by applying Se to the soil in fertilizers but the amounts required vary from soil to soil. There is also the possibility of some plants accumulating toxic quantities of the element. Therefore until more work is done, soil application is not generally recommended.

IODINE (I) DEFICIENCY

Clinical signs

Although I deficiency is less common than Co, Cu and Se deficiences, sheep are more susceptible to I deficiency than cattle and it is the newborn lamb which is particularly susceptible. The common manifestation is late abortion or neonatal death and severely-affected lambs have visibly enlarged thyroid glands which are up to 50 times heavier than normal. The ewe is rarely affected either clinically or subclinically. Infertility, loss of libido in the male, poor wool growth, milk yield and weight gain may however occur in older animals affected by I deficiency.[11] In view of the retarded development of the CNS recently observed in severely-depleted newborn lambs, behavioural abnormalities may ensue. Enlarged thyroids are often found in lambs grazing brassica crops but this is not thought to be associated with poor performance.

Pathology

Changes in thyroid structure, of the newborn lamb which is deficient in I, occur in the following sequence: cell enlargement, cell proliferation and colloid depletion, invagination and finally collapse and infiltration of follicles.[12]

Cause

Iodine deficiency may arise from a simple lack in the diet which in turn may reflect a low I content in the soil.[11] Alternatively, goitrogenic factors in the diet may induce a deficiency by impairing the metabolism of I. Iodine is an essential constitutent of triiodothyronine which controls oxidative phosphorylation and thus basal metabolic rate (BMR) and protein synthesis. Most body functions are therefore affected in I deficiency and factors which increase BMR, e.g. low winter temperatures, increase I requirements. Goitrogens impair I metabolism in one of two ways, by impairing the uptake of I by the thyroid gland or the iodination of tyrosine residues within the gland. The thiocyanate or nitrate type, found in brassica and legume crops impair uptake while the thiouracil type found in brassica seeds impair iodination.[11]

Diagnosis

Congenital goitre has been diagnosed from the presence of thyroids >2.8 g but histological abnormalities have been found in smaller thyroids. Low I status is reflected by low levels of protein (globulin)-bound iodine and its constituents, triiodothyronine

(T3) and tetraiodothyronine (T4) in the plasma of the ewe and fetus. Since only the free form of T3 is physiologically active and it provides by far the smallest fraction of the total serum I, low levels of T3 plus T4 are poor indicators of functional I deficiency: increased concentrations of thyroid stimulating hormone (TSH) are a better indicator. Milk is an important route of I secretion, reflecting I intake, and concentrations < 0.63 μmol (80 μg)/l are indicative of dietary deficiency.

Prevention

The diet should contain 1.19 μmol (0.15 mg) I/kg DM in summer and 2.38 μmol (0.30 mg)/kg DM in winter.[3] Cereals are a particularly poor source of I containing 0.32–0.71 μmol (0.04–0.09 mg) I/kg and when they, or other goitrogenic crops, e.g. leafy and root brassicas or subterranean clover, form an appreciable part of the ration in late pregnancy I supplementation is particularly needed. I is routinely incorporated into mineral supplements as iodide but less volatile forms such as periodates and organo-iodine compounds are preferable in warm, dry climates. I supplementation is only effective against goitrogens of the thiocyanate or nitrate type and in the absence of quantitative information on their antagonistic effect, a four-fold increase in I requirements should be allowed. Use of feeds containing thiouracil-type goitrogens should be avoided in late pregnancy.

Treatment

All clinical and subclinical effects of I deficiency, including fetal brain retardation, respond to I administration and oral supplementation is adequate.

OTHER TRACE ELEMENTS

It is rare for deficiencies of elements other than Co, Cu, I and Se to cause clinical problems under natural conditions. For these reasons other trace elements are covered only briefly.

Iron (Fe)

Anaemia has been induced by feeding diets of very low Fe content(< 250 μmol [14 mg]/kg DM) to lambs before and after weaning.[3] Although ewes' milk is not particularly rich in Fe, requirements for Fe diminish with age and initial hepatic reserves coupled with adventitious sources of Fe from the environment, e.g. soil and faeces are almost invariably sufficient to prevent deficiency.

Manganese (Mn)

The swollen joints and stiff gait which affect lambs on experimental diets of very low Mn content 14.6 μmol (0.8 mg/kg DM) may result from impaired mucopolysaccharide synthesis due to low glycosyl-transferase activity.[3] Mn is an essential constituent of this and other metalloenzymes. Mn supplementation has improved fertility in ewes on pastures containing 364–546 μmol (20–30 mg Mn)/kg DM in 2 out of 3 seasons[7] but there is no other evidence that Mn deficiency is an important condition in sheep.

Zinc (Zn)

Under experimental conditions Zn deficiency causes growth retardation, parakeratosis, wool loss, excessive salivation and impaired spermatogenesis.[3] With the exception of male fertility, these abnormalities have only occurred on diets of very low Zn content (107–275 μmol [7–18 mg]/kg DM). Pastures rarely contain < 306 μmol (20 mg)/kg DM and sheep are able to absorb Zn very efficiently at low intakes. This may explain why only 3 reports of Zn deficiency affecting sheep under natural conditions have been noted.[3] Although the young rapidly growing lamb has a particularly high requirement for Zn, ewes' milk is rich in Zn with 107 μmol (7 mg)/l, and protection may thus be afforded at the most vulnerable period.

REFERENCES

1 Fell B.F. (1981) Pathological consequences of copper deficiency and cobalt deficiency. *Philosophical Transactions of the Royal Society,* Series B, **294**, 153–69.

2 Sutherland R.J., Cordes D.O. & Carthew G.C. (1979) Ovine white liver disease—an hepatic dysfunction associated with vitamin B$_{12}$ deficiency. *New Zealand Veterinary Journal,* **27**, 227–32.

3 Agricultural Research Council (1980) *The Nutrient Requirements of Livestock.* No.2. Ruminants. 2e, Farnham Royal, Commonwealth Agricultural Bureaux.

4 Council of the Scottish Agricultural Colleges (1982) Trace element deficiencies in ruminants. The report of a study group commissioned by COSAC and the Scottish Agricultural Research Institutes, Edinburgh. E. of Scotland College of Agriculture.

5 Chanarin I., Deacon R., Perry J. & Lumb M. (1981) How vitamin B_{12} acts. *British Journal of Haematology,* **47**, 487–91.

6 Suttle N.F. (1982) Micronutrients as regulators of metabolism. In *The Nutritional Physiology of Farm Animals.* (Eds. Rook J.A.F. & Thomas P.) Longman, London.

7 Reuter D.J. (1975) The recognition and correction of trace element deficiencies. In *Trace Elements in Soil-Plant-Animal Systems.* (Eds. Nicholas D.J.D. & Egan A.R.) pp.291–324. Academic Press, London, New York.

8 Whitelaw A., Armstrong R.H., Evans C.C. & Fawcett A.R. (1979) A study of the effects of copper deficiency in Scottish blackface lambs on improved hill pasture. *Veterinary Record,* **104**, 455–60.

9 Gartrell J.W. (1981) Distribution and correction of copper deficiency in crops and pastures. In *Copper in Soils and Plants.* (Eds. Lonergan J.F., Robson A.D. & Graham R.D.) pp.313–49. Academic Press, Australia.

10 Rosenfield I. & Beath O.A. (1964) *Selenium: Geobotany, Biochemistry, Toxicity and Nutrition.* pp.233–77. Academic Press, London, New York.

11 Underwood E.J. (1977) *Trace Elements in Human and Animal Nutrition.* 4e, pp.271–301. Academic Press, London.

12 Statham M. & Bray A.C. (1975) Congenital goitre in sheep in Southern Tasmania. *Australian Journal of Agricultural Research,* **26**, 751–68.

SECTION 6

DISORDERS OF SKIN AND WOOL

37   PSOROPTIC MANGE

D.W. TARRY

Synonym 'Sheep scab'

Psoroptic mange has been known to farmers for generations, under the name of 'sheep scab' or 'sheep scabies'. It is a highly contagious and extremely unpleasant disease. As with all types of mange it is the result of the activity of a parasitic mite, scarcely visible to the naked eye. Cases are now infrequent in the UK but vigilance must still be maintained. The disease can be easily overlooked in its earliest stages and recent experience has once more shown how rapidly it can be spread at that time. This risk is especially serious when affected animals are sent to markets or are moved into close proximity with neighbouring flocks. The disease generally reaches the critical stage 8–12 weeks after infection and the sheep, which by then has lost most of its fleece, may well die through nervous debilitation and lack of food. This dramatic progression of the disease is a result of the rapid increase in mite numbers, shown in the laboratory to multiply from 25 to over a million in twelve weeks (Kirkwood, A.C., personal communication).

Cause

The parasitic mite responsible for this 'notifiable' disease is *Psoroptes communis* var.*ovis*. Highly host-specific forms of the mite occur on a number of other farm animals, from cattle (var. *bovis*) to rabbits (var. *cuniculi*). There is no evidence that natural transfer between host species has ever taken place. *Psoroptes* is a surface feeder unlike the other 'notifiable' mange mite of sheep, the burrowing species *Sarcoptes scabiei*. *Sarcoptes* has not, however been reported in recent years from sheep in the UK though it is common on other livestock. A further 'sheep mange' species, *Chorioptes ovis*, is however not uncommon in the UK and is often associated with foot mange. It is not notifiable. The identification of these important mites is dealt with later in the chapter.

Clinical signs and epidemiology

As a surface parasite, *Psoroptes* mites in the wool locate the epidermis and pierce this to feed on lymph; a number of red blood cells may also be ingested.[1] Following the punctures, yellowish pustules develop which can be readily detected if the fleece is parted. As these expand they rupture and the exudations form the typical yellowish crusts. The fleece at this stage has become characteristically moist and matted above the scabs, unlike any other form of mange. As further lesions appear around the older ones, the crusts darken and lift away from the skin, with any wool which has not already been rubbed off as a response to the intense irritation (Plate 5). The sheep will now be hypersensitive and will 'mouth' characteristically when touched but some may react in a more exaggerated manner by rolling on the ground, or even with an involuntary paroxysm. At the peak of the disease, about 12 weeks from infestation, a number of animals will die. The rest, in a debilitated condition, eventually recover.

The mites, in an active infestation, occur in enormous numbers. Each female deposits up to 100 eggs, and the emerging larvae take only 11 days to develop and become fully reproductive individuals. Contrary to earlier suggestions, it has been found that on a full-fleeced sheep this rate of increase can occur throughout the summer. The mites however, are very vulnerable to desiccation and the summer populations may be greatly reduced because of the general reduction in fleece humidity in open pasture during summer weather, and especially from the effects of exposure to sunshine and evaporation following the annual shearing in late spring. Although mites may be equally active anywhere in the fleece they are thus more readily found during the summer months in skin folds and crevices such as the perineum, the inguinal, interdigital and infraorbital fossae, and the scrotal area. Serious spread of disease is consequently not ruled out during the summer; in the months July–August of 1977 and 1978 there was a total of 11 outbreaks in Britain. Experimentally also, the disease is transmitted readily during the summer months.

Because *Psoroptes* mites are closely adapted to the environment of the host fleece, most will die within a week if they are removed from the host or left as a residual infestation in sheep pens. It has been shown, however [2] that ovigerous females are more resistant and under ambient winter conditions may survive for

20 days. Similarly eggs under cold conditions may still hatch after 3 weeks but it is generally assumed that affected premises are free of infection after 20 days, and the British Veterinary Association [3] suggested that new stock may, in practice, be safely moved in after only 8 days on the basis that field outbreaks have never been known to recur after that period. Others have been more cautious and periods of disuse of 17 days [4] and of 30 days have been recommended. Recently, workers in USA have advised on the basis of extensive experimental work, that infected premises be vacated for at least 2 weeks, and preferably one month.[5]

Economic effects

Sheep scab in the UK is a notifiable disease, already eradicated in most of the country, but the potential loss to the national flock if it were not controlled is enormous. It was recently shown that the loss in value as a result of reduced weight gain from a flock of only 100 gimmer ewes, infested in October, was in the region of £1000 in 14 weeks.[6] The mean difference in weight gain compared to that of healthy animals was 30 per cent. Such a loss does not include wool value which would have become negligible, nor the potential mortality as a number of sheep would certainly have died, if the trial had continued. A similar effect on weight gain has also been noted in infested calves.[7]

Diagnosis: recognition of *Psoroptes ovis*

Although the symptoms of sheep scab infection are readily recognized when an outbreak occurs viz. loss of wool, matted fleece, hypersensitivity and a tendency for nibbled wool to be left on the teeth, the final diagnosis must still depend on the identification of the mite itself. Skin scrapings for examination of the mites must be taken from the moister, marginal parts of the lesion, where the mites are still active, after clipping away any remaining wool. Scrapings are carefully digested in 10 per cent caustic potash solution, centrifuged and mounted for scanning under the microscope. An examination of the scrapings before treatment, under a good light, may however save time by revealing live mites which are just visible to the naked eye. These can then be mounted individually for examination using a fine needle to lift and manoeuvre them. Fortunately

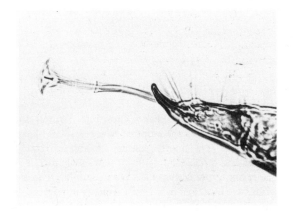

Fig. 37.1 (a) The sheep scab mite, *Psoroptes*, pedicel and sucker (x 632)

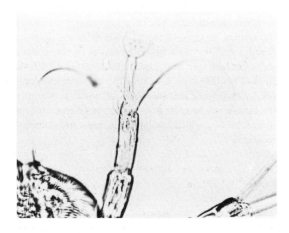

Fig. 37.1 (b) The mange mite, Chorioptes, pedicel and sucker (x 610) (Crown copyright)

Psoroptes has certain characteristic features (Fig. 37.1). The possession of a long three-segmented stalk (pedicel) to the suckers of at least two pairs of legs is sufficient to confirm the identification. In *Chorioptes* the pedicel is short and unsegmented. Less clearly seen is the shape of the sucker which is funnel-shaped in *Psoroptes* but clearly bell-shaped in *Chorioptes*, the other mange mite which can be relatively common on sheep in some parts of Britain. *Chorioptes* is largely confined to the pasterns and feet, although it occasionally affects the scrotum and inguinal area. *Sarcoptes*, perhaps only likely to be found on imported animals, has a conspicuously rounded body with sturdy legs which do not extend beyond the body outline. The 'pedicels' are long and unjointed. A further distinct feature of *Sarcoptes* is the prescence of tooth-like 'dorsal spines' which are not found on any other parasitic mite. Severe skin irritation and wool-loss may be the result of other ectoparasite infestations [8] and the causative agent must always be identified.

The history of sheep scab control

According to Anglo-Saxon records sheep scab was certainly recognized in the 10th century, but legislation for its control was not introduced until the end of the 19th century and it was not until 1903 that compulsory powers for sheep dipping were introduced in Britain. Although these early treatments were not persistent enough to kill larvae emerging some days later from mite eggs, this marked the beginning of a long campaign to overcome the disease. Before this decision, the number of UK outbreaks annually was often between 2–3000; in 1905 it fell to 947 and never again rose above the 1000 mark.[9] To overcome the problem of the hatching of mite eggs, unaffected by most pesticides, in the fleece after treatment, a second dipping within 7–14 days of the first was introduced in 1914. This technique was called 'double dipping' and remained in use until the introduction of the persistent organo-chlorine insecticides in 1948 rendered it more or less obsolete. Lindane, (the active constituent now being known as HCH, or hexachlorocyclohexane) was shown to give protection against reinfection for at least 8 weeks, and 'single dipping' could thus be effectively introduced for sheep scab eradication. Soon afterwards, by following a carefully supervised dipping programme under the Sheep Scab Orders, the disease was eradicated from Scotland and England by 1951 and from Wales in 1952.

Sheep scab was not seen subsequently in the UK until January 1973 when a new generation of veterinarians had to deal with the resurgence of the problem, and 27 outbreaks were soon confirmed on the Yorkshire-Lancashire border. The disease was thought to have been reintroduced with sheep imported from Ireland, and the mites could well have been very inconspicuous if it was in the almost symptomless 'latent phase' the previous summer. Since the spring of 1973 outbreaks have appeared in many parts of Britain, being especially persistent in parts of the west country, mainly as a result of market contacts. The disease reached a peak in 1975 when the number of outbreaks rose to 103. The introduction of compulsory dipping for the entire country in 1976 subsequently resulted in a fall in the number of outbreaks, and the total for 1980 was only 34. Scotland has experienced a late rash of outbreaks in 1981, after the treatment restrictions had been relaxed there, but there is every reason now to believe that eradication of this disease is once again within sight.

Control

Sheep scab in the UK is a notifiable disease under the Sheep Scab Order (1977) and anyone suspecting the existence of the disease must report this at once to the police or to the nearest veterinary officer of the Ministry of Agriculture. Treatment of a flock in which the disease has been confirmed by the Ministry of Agriculture is mandatory using a sheep dip of an approved type.

For 30 years gamma-HCH was the only pesticide approved as a single dip for sheep scab, at an effective concentration of 50 mg/l. It is still very widely used but has certain disadvantages, notably the fear of an organochlorine residue build-up in outbreak areas after continued use. With the active participation of the Ministry of Agriculture's Weybridge laboratory the Veterinary Products Committee has now been able to license the use of diazinon, the first organophosphorus pesticide available for this purpose. Diazinon does not carry the same risks of cumulative residues in meat products and the environment, and it has now also been proved to be effective as a single dip, if correctly used. At Weybridge Laboratory it was shown [10] that a minimum maintenance concentration of 100 mg/l in the dip wash was essential to kill the mites completely. This is not necessarily the same as the original dilution, because the various available formulations have different, and considerably higher, initial concentrations. The active ingredient is progressively 'stripped out' by treatment of a number of sheep until the minimum level is reached and further pesticide is added. Sheep withdraw up to 5 gallons of dip as they climb out, and the wash draining back contains very little active pesticide. If there were no depletion there could be no protection. The majority of field failures can be attributed to a failure to maintain the correct minimum concentration of available pesticide, by 'replenishment' at the correct stage. As with gamma-HCH, it was found that a minimum immersion period of one minute was required with diazinon if eradication of the disease was to be ensured. Full penetration of the fleece was not possible in a shorter time. Spraying and showering treatments also were not suitable, as active mites in the axillae and inguinal folds for example did not always receive an effective dose of insecticide.[11]

The resistance problem

The use of gamma HCH in some countries has been accompanied by the development of resistant strains of *Psoroptes* mites.[12, 13, 14, 15, 16] This presumably arises through the constant reinfestation of dipped sheep with fleeces eventually containing low-level sublethal residues of pesticide, as a result of contact with infested sheep. These may be either animals missed in the round-up for dipping, or those not effectively treated, through failure on the part of farmers to appreciate the importance of maintaining a correct concentration. Several organophosphorus dips have been in use against sheep scab for some time in these countries, the difficulties of ensuring an adequate treatment often being overcome by mandatory repeat treatments.[17] The risk of the continued reappearance of resistance under the conditions found in countries such as Argentina and Brazil will, however, be difficult to avoid. Recently in Brazil mites were found to be resistant to gamma HCH at 85.4 per cent and to diazinon at 20.5 per cent; amitraz was found to be a possible substitute but may only provide relief for a few years.

There has been no established case of acaricide resistance in *Psoroptes* in the UK, and in view of the relatively brief persistence of diazinon and the attention given to eradicating local outbreaks effectively so that reinfestation is avoided, resistance is now most unlikely to occur.

The future

Few of the new pesticides constantly being synthesized have useful acaricidal properties. Of these, only a small number are suitable and persistent enough to be considered for sheep scab treatment. It has, for example, been suggested that the avermectins may be effective against mange mites including *Psoroptes*.[18,19] If their use were practicable, both efficacy and facility of treatment could be improved by this new systemic method which could overcome not only the problem of access to mites protected from dipping fluids by scab crusts, but also that of the adverse effects of dipping large numbers of sheep under often extreme winter conditions.

As more is understood about the biology of the sheep scab mite and about effective dipping, the more it becomes possible to ensure eradication and to prevent re-infestation of UK flocks. Transfer of *Psoroptes* mites from other hosts such as rabbits and calves has been achieved occasionally in the laboratory [20, 21] but probably is of little real significance and has never been demonstrated or even seriously suspected in the field. The most serious risk for some years to come in the UK will be of re-infestation, carried with imported sheep from countries where eradication is not yet complete. The most important preventative measure for the next few years will be careful surveillance and treatment of these animals, and on our farms alertness on the part of the whole farming community. Early and coordinated control action is an essential factor in the prevention or limitation of sheep scab outbreaks.

REFERENCES

1 Wright F.C. & DeLoach J.R. (1981) Feeding of *Psoroptes ovis* (Acari: psoroptidae) on cattle. *Journal of Medical Entomology*, **18**, 349– 50.

2 Lapage G. (1968) In *Veterinary Parasitology*. 2e. Oliver & Boyd, London.

3 British Veterinary Association (1949) *Diseases of Farm Livestock*. II 2e. (BVA publication no. 6) BVA, London.

4 du Toit P.J. (1923) Sweating sickness in calves. Ninth and tenth reports of the Director of Veterinary Education and Research, Department of Agriculture, pp.233–50. Union of South Africa.

5 Wilson G.I., Blachut K. & Roberts I.H. (1977) The infectivity of scabies (mange) mites *Psoroptes ovis* (Acarina: Psoreptidae), to sheep in naturally contaminated enclosures. *Research in Veterinary Science*, **22**, 292– 7.

6 Kirkwood A.C. (1980) Effect of *Psoroptes ovis* on the weight of sheep. *Veterinary Record*, **107**, 469–70.

7 Fisher W.F. & Wright F.C. (1981) Effects of the sheep scab mite on cumulative weight gains in cattle. *Journal of Economic Entomology*, **74**, 234–37.

8 Tarry D.W. (1974) Sheep scab: its diagnosis and biology. *Veterinary Record*, **95**, 530–2.

9 Watson J. (1977) Review of the history of sheep scab and its control in Great Britain. Proceedings of the British Crop Protection Conference Pests and Diseases, **3**, 955–9.

10 Kirkwood A.C. & Quick M.P. (1981) Diazinon for the control of sheep scab. *Veterinary Record*, **108**, 279–80.

11 Kirkwood A.C., Quick M.P. & Page K.W. (1978) The efficacy of showers for control of ectoparasites of sheep. *Veterinary Record*, **102**, 50–4.

12 Rosa W.A.J., Niec R., Lukovich R. & Nunez J.L. (1969) Present resistance of *Psoroptes ovis* to gamma benzene hexachloride. (tr) *Gaceta Veterinaria*, **31**, 373–8.

13 Ault C.N., Romano A. & Miramon R.E. (1962) Resistance of *Psoroptes ovis* to benzene hexachloride. (tr) *Revista de Medicina Veterinaria, Buenos Aires*, **43**, 357–60.

14 Rosa W.A.J. & Lukovich R. (1970) Response of different *Psoroptes ovis* strains to dips containing lindane or diazinon. (tr) *Revista de Medicina Veterinaria. Buenos Aires*, **51**, 127–9.

15 Lastra H.C. (1972) Brief review of acaricides for sheep in use in Argentina. (tr) *Gaceta Veterinaria*, **34**, 531–9.

16 Laranja R.J. (1978) Organochlorine and organophosphorus resistance in a strain of *Psoroptes ovis*. (tr) *Boletim do Instituto de Pesquisas Veterinarias 'Desiderio Finamor'*, **5**, 5–11.

17 Suvurov P.S. (1970) 1. Treatment of psoroptic mange in cattle with chlorofos (trichlorphon) dust. II. Acaricidal action of Ciodrin (crotoxyphos) ftalaphos (phosmet) and chlorophos on *Psoroptes bovis*. *Trudy Vsesoyvznogo nauchno-issledovatel'skogo instituta veterinarnoi sanitari (i ektoparazitologii)*, **35**, 373–8 & 379–82.

18 Wilkins C.A., Conroy J.A., Ho P., O'Shanny W., Malatesta P.F. & Egerton J.R. (1980) Treatment of psoroptic mange with avermectins. *American Journal of Veterinary Research*, **41**, (12). 2112–13.

19 Lee R.P., Dooge D.J.D. & Preston J.M. (1980) Efficacy of invermectin against *Sarcoptes scabiei* in pigs. *Veterinary Record*, **107**, 503–5.

20 Meleney W.P. (1967) Experimentally induced bovine psoroptic acariasis in a rabbit. *American Journal of Veterinary Research*, **28**, 892–4.

21 Zeilasko B. (1979) Studies on the epizootiology of psoroptic mange *(Psoroptes Ovis)* in sheep and cattle. Thesis, Hanover Veterinary University, German Federal Republic. 44pp.

38 CONTAGIOUS PUSTULAR DERMATITIS

J.A.A. WATT

Synonyms Orf, contagious ecthyma, scabby mouth

Contagious pustular dermatitis (CPD) is a virus disease which primarily affects the lips and coronet of younger animals though all ages can be affected. Lesions may, less commonly, appear on other hair-covered areas and CPD may take the form of a venereal infection in breeding animals. CPD chiefly affects sheep, goats and related species, e.g. musk ox. Though the disease can affect other species, including man, these are usually isolated cases and the lesions do not tend to spread from the site of primary infection. Distribution is world-wide.

Cause

The disease has been recognized as an entity since the last century and was shown to be caused by a specific virus in 1923.[1] The virus is a member of the Poxviridae of the genus *Parapoxvirus*. The virus survives in artifically-dried scab in the refrigerator for many years and has been reported to survive for at least 6 months in scab under field conditions.[2] The virus can be grown on tissue cultures.[3, 4]

Clinical signs

In Britain the disease tends to show a seasonal incidence, the first peak occurring in spring or shortly after lambing and the second in summer and early autumn. In the former the disease affects principally the lips of the lambs and the udders of the ewes, while in the latter, the lesions on the mouths of the lambs are more localized and are common on the middle of the lips though in individual cases they may be more extensive. Additionally, in older lambs, the lesions are common on the coronet where they are frequently proliferative and persistent. Such lesions are sometimes referred to as 'strawberry foot rot' which some consider is caused partly or entirely by infection with *Dermatophilus congolensis*. This site appears to be the primary one in this type, the lips becoming infected from rubbing or licking the foot or leg lesion.

The first signs seen in the post-natal lamb are usually small scabs at the commissures of the lips but these quickly spread and the scabs increase in thickness and hardness preventing the lamb from forming an efficient seal for sucking. The severity of the lesions varies from one outbreak to another and from lamb to lamb. The disease is usually characterized by very rapid spread. This results from ewes which have developed udder lesions from their affected lambs refusing to allow their lambs to suck. While the ewes are at the feeding trough these hungry lambs will attempt to steal milk from other ewes, thus spreading the virus. A possible serious secondary effect of these lesions in the ewe, is the risk of acute staphylococcal mastitis.

The duration of the disease in the individual lamb, varies considerably as does the severity but healing is usual within 2–4 weeks. The course of the disease in the flock is, of course, much longer.

The lesions in lambs usually commence at the junction of skin and mucous membrane along which they spread, though occasionally a stomatitis develops, particularly in lambs being reared artificially.

The disease is readily transmitted by the application of active scab to a light scarification on susceptible lambs. Within 24–48 h, depending on the weight of infection in the inoculum, a marked erythema develops along the line of the scarification with the formation of tiny papules. These develop to pustules, over the next 24 h which then break to form a thick scab over the affected area.

Secondary bacterial infection plays a significant part in the severity of the lesions, the most important invaders being staphylococci, alpha haemolytic streptococci and corynebacteria. In some cases the lesion is complicated by invasion by *Dermatophilus congolensis* which can also complicate diagnosis. When the lesions affect the buccal cavity, *Fusobacterium (Sphaerophorus) necrophorum* may be a secondary invader which has been reported to spread to the viscera. This organism has also been observed in the venereal form of the disease.

Losses suffered in lambs with the buccal form of the disease are largely the result of their inability to feed or the refusal of the ewe to suckle and may, in the absence of facilities for artificial feeding, result in severe losses.

Mastitis in the ewe as a result of teat lesions is usually a very acute staphylococcal mastitis which results in the loss of the affected half udder and not infrequently the death of the ewe.

The late summer outbreaks do not result in such severe disease but cause unsightly sores on the mouth and feet with loss of condition and such lambs cannot, of course, be marketed. The mode of spread of the disease, with the coronet being the most common site of the primary lesions, suggests that the tough seed stalks of grass, stubble etc. cause the abrasions which give access to the virus which is deposited on the grass by infected lambs. The condition runs the same course as described in sucking lambs but the secondary lip lesions may not appear until those on the feet are well advanced, thus prolonging the course of the disease in the individual animal. In the flock the condition may be present clinically for up to six weeks or even longer with all stages of infection present. In some cases the proliferative changes may become very severe resulting in a marked 'strawberry'-like lesion with peripheral swelling and oedema. Such lesions are slow to heal and result in marked loss of condition.

A venereal form of the disease has been recognized since 1903 in England, since when it has been described on the continent of Europe and in America. Tunnicliff and Matiseck (1941)[5] demonstrated a filterable virus which they considered to be distinct from that causing CPD. Many reports comment on infection of the lesion with *Fusobacterium necrophorum* and in some cases in Britain CPD virus has been isolated. The disease manifests itself as ulcers at the vulva of the ewe and the prepuce and sometimes the penis of the ram. Trueblood in 1966[6] reported that examination of viruses from cases diagnosed as 'ulcerative dermatitis' and CPD respectively showed these viruses to be indistinguishable. This leaves unresolved the problem of whether one or two viruses are involved.

The venereal disease usually appears soon after the rams are turned out and spreads rapidly within the flock. In the ewe it begins as small pustules at the vulva on the skin/vaginal mucosa junction and in the ram at the preputial orifice. At both sites the lesion shows a tendency to spread and form shallow ulcers which may become infected by *Fusobacterium necrophorum* giving a 'floor' of necrotic tissue to the ulcer. Less commonly the lesion is proliferative. The rams rapidly become reluctant to mate and the sequel can be disruption of breeding, with a consequent long drawn out lambing.

Pathology

The initial lip or teat changes commence within 48 hours of infection.[7, 8] The superficial nucleated cells of the epidermis, i.e. the granular and outer Malpighian layers, undergo swelling, with cytoplasmic vacuolation and shrinkage of the nuclei. There is no vesicle formation, however. Concurrently, the inner Malpighian layer displays increased mitotic activity and becomes thickened, with elongation of the rete pegs. The underlying dermis is hyperaemic and contains infiltrates of mononuclear cells.

Over the next 3–4 days, the vacuolated cells gradually become necrotic. During this stage granular intracytoplasmic inclusions may be seen in the degenerating cells; these persist until the dead cells are shed.[8] The dermis is now heavily infiltrated with neutrophils. These spread into the dying epidermal layer to form multilocular pustules which soon coalesce above a highly proliferative prickle-cell layer. For a short time, an attenuated stratum corneum encloses the pustules, but this soon ruptures and by about 8 days after infection a thick scab consisting of exudate, cellular debris and contaminant bacteria, has formed. This scab is firmly attached, and if forcibly removed, bleeding results from the raw spongy base. Intense proliferative activity beneath the scab causes parakeratosis with incomplete keratinization, contributing to the thickness of the scab. The dermal inflammatory reaction soon takes on a granulomatous character; the scab is eventually shed and healing results, usually by about a month after infection.

Lesions inside the mouth or buccal cavity do not form scabs but appear as raised reddened or greyish areas surrounded by narrow, intensely hyperaemic zones.

Diagnosis

Diagnosis is generally easy on clinical grounds but, on occasion, further tests are necessary. The most satisfactory is the examination of moist active scab material by electron microscopy for the presence of virus particles. The oldest, which is still useful where access to an electron microscope is difficult, is direct transmission of scab from the suspect lesion to a scarified area on a susceptible lamb. The site usually chosen is the skin inside the thigh. However, if a small area is plucked on the flank and one or two superficial linear scarifications made, application of a suspension of the suspect scab to this area, will if orf is present, cause the typical papule pustule reaction in 4–6 days, while, should the lesion be due to *Dermatophilus congolensis*, the hair follicles will be infected evenly over the area, reacting in 2–3 days.

The presence of the typical CPD virus particles are then readily demonstrable by electron microscopy which is not always the case in older scab material.[9] The CFT for antigen is also useful for confirmation and is reported to be a sensitive test. Immunodiffusion is also useful. Other tests for the confirmation of the disease are under active investigation.

Conditions which could be confused with CPD are staphylococcal (periorbital) dermatitis and face scab. The former is often very seasonal being seen usually a month or so before lambing when box feeding is employed. The latter is characterized by a serous exudate which rapidly hardens on lips, face and ears, the skin being intact beneath the scabs. This condition appears suddenly and all the cases in such incidents appear over 1–2 days with no subsequent spread. No infective agent has been demonstrated.

Epidemiology and transmission

The spread of the disease within the flock is limited by the need for direct physical contact with an infected animal or active scabs from such an animal. This is of importance in housed lambs in particular and provision of solid divisions between groups of lambs is recommended as a prophylactic measure. In the field, spread is by direct contact with virus and experimental evidence suggests that some slight scarification of the skin is required. Primary lesions, as already discussed, are largely confined to the lips in sucking lambs, coronet and lips in older lambs and vulva and prepuce in adults. The lesions tend to remain localized at the predilection sites of lips and coronet but occasionally lesions occur elsewhere on the face and legs.

Once an outbreak has occurred in a flock recurrence is rarely seen till a new crop of susceptible lambs is born and it has been generally postulated that infection arises from shed scabs containing the highly resistant virus. However, sequential studies, which have shown that the virus content of scabs decreases rapidly as healing proceeds, appear to conflict with this view [9], though scabs from active lesions also contaminate the environment. The venereal type does not tend to recur in subsequent years in traditionally-managed flocks.

Control

Various treatments have been tried but results have generally been disappointing. Emollient preparations containing broad-spectrum antibiotics may be useful in certain cases, e.g. on scabs on the lips of lambs. Antibiotics in aerosol form have been claimed to shorten the course of the disease but in the larger flocks it is unlikely that the results achieved would justify the labour involved.

Control and prevention is by use of a live vaccine applied to a site unsuitable for the production of the more chronic lesion, e.g. inside the thigh, behind the elbow or on the caudal fold. It must be emphasized that the virus is live and virulent and should not be used in closed flocks in which the disease is not known to occur because of the risk of introducing the infection to the flock. The site of choice is the inside of the thigh except before lambing as here persistent scab could infect the udder. In this case the recommended site is behind the elbow. The caudal fold is not recommended because of the possible enhanced risk of secondary infection. It is difficult to assess the degree of duration of the immunity produced as a number of variables are involved, the most important being the weight of the challenge. Johnston, W.S. and Watt, J.A.A. (unpublished) showed that vaccinal immunity withstood massive challenge at 5 weeks though over the next 3 months, the resistance waned according to the weight of the challenge. The challenge was with decreasing dilutions of the homologous strain of virus.

The time chosen for vaccination should be governed by the expected timing of the clinical disease. The disease affecting suckling lambs does not commonly recur annually in a flock being more sporadic in occurrence. Where the disease does recur annually, ewes should be vaccinated behind the elbow 6 weeks before lambing. The late summer type is, however, much more regular in appearance and here lambs should be vaccinated some 6 weeks before the expected time of the appearance of the disease.

The venereal form of disease does not commonly become endemic in a flock and a routine examination of ewes and rams should be undertaken before they are mixed to ensure that no lesions are present on vulva or prepuce. The appearance of the disease in the flock should, with careful shepherding, be seen early in the outbreak and an attempt made to limit its spread as it may well be confined initially to one group of ewes and 1–2 rams. All rams which have had possible contact should be isolated and the group of ewes carefully examined. Any showing lesions should be removed. The remainder should be iso-

lated for 10 days then re-examined and any showing lesions added to the infected group. Clean rams may then be introduced to the clean ewe group. Where groups are large or mixing has occurred, this procedure is unlikely to be successful. During an outbreak of this form of the disease, vaccination is not recommended.

In the sucking lambs isolation of the affected lambs and their dams reduces the weight of infection but this is rarely worthwhile as the rapid spread renders it impractical. On occasion vaccination may be considered advisable and has been reported to arrest the spread. If vaccination during an outbreak is considered necessary and commercial vaccine is not available, an autogenous vaccine can be prepared and used in the flock. In this case, the limitations in Britain imposed by the Medicines Act must be borne in mind.

REFERENCES

1 Aynaud M. (1923) La stomatite pustuleuse contagieuse des ovins (Chancre du mouton). *Annales de l'Institut Pasteur*, **37**, 498–527.

2 Boughton I.B. & Hardy W.T. (1934) Contagious ecthyma (sore mouth) of sheep and goats. *Journal of the American Veterinary Medical Association*, **85**, 150–78.

3 MacDonald A. & Bell T.M. (1961) Growth of contagious pustular dermatitis virus in human tissue cultures. *Nature, Lond.*, **192**, 91–2.

4 Plowright W., Witcomb M.A. & Ferris R.D. (1959) Studies with a strain of contagious pustular dermatitis virus in tissue culture. *Archiv fur die Gesamte Virusforschung*, **9**, 214–31.

5 Tunnicliff E.A. & Matiseck P.H.A. (1941) A filterable virus demonstrated to be the infective agent in ovine balano-posthitis. *Science*, **94**, 283–4.

6 Trueblood M.S. (1966) Relationship of ovine contagious ecthyma and ulcerative dermatosis. *Cornell Veterinarian*, **56**, 521–6.

7 Abdussalam M. (1957) Contagious pustular dermatitis II. Pathological histology. *Journal of Comparative Pathology and Therapeutics*, **67**, 217–22.

8 Jubb K.V.F. & Kennedy P.C. (1970) *Pathology of Domestic Animals*. 2e. Vol.2,p.595, Academic Press, New York and London.

9 Romero-Mercado C.H., McPherson E.A., Laing A.H., Lawson J.B. & Scott G.R. (1973) Virus particles and antigens in experimental orf scabs. *Archiv fur die Gesamte Virusforschung*, **40**, 152–8.

39

HEADFLY AND BLOWFLY MYIASIS

W. T. APPLEYARD

HEADFLY

Headfly disease of sheep in Britain is associated mainly with the muscid fly *Hydrotaea irritans.* *Hydrotaea* is about the same size and shape as the domestic housefly (*Musca domestica*), but it has an olive green abdomen with yellow-orange bases to the wings. Although this species occurs widely throughout the UK, headfly-related disease is recorded only in the north of England and the Scottish Borders, however the problem appears to be an increasing one.

Cause

Headflies, attracted by nasal, lacrymal and other secretions approach sheep in swarms to feed. The presence of large numbers of flies around the head of a sheep causes great annoyance and irritation forcing the sheep to rub its head against the ground, undergrowth, walls etc., or to scratch itself with its hind feet. Such trauma may produce a break in the skin with exudation of blood or serum which is a further attraction to the flies. Headflies do not possess piercing or biting mouthparts but are equipped with rasping labellae. These, acting at the periphery of a minor abrasion, prevent healing and lead to a rapid expansion of the lesion with the possible loss of large areas of skin over the head (Fig. 39.1).

Minor breaks in the skin which arise apparently spontaneously at the skin/horn junction of young sheep with rapidly growing horns are quickly attacked by headfly and this probably accounts for the greater susceptibility of yearling sheep. Older animals in which horn growth has virtually ceased are more resistant. In rams fighting often results in head wounds which attract large numbers of flies. Horned breeds are more susceptible to attack and those sheep which have their faces covered with hair rather than

Fig. 39.1 Blackface ewe with extensive loss of skin and hair around the horn base, the result of headfly (*Hydrotaea irritans*) activity. Two headflies are feeding on the wound.

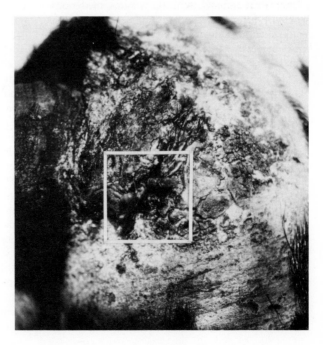

wool, e.g. Scottish Blackface, are particularly vulnerable.

Clinical signs

Affected animals when under attack from flies (usually on warm, still days) shake and rub their heads and may be seen to be trying to avoid the large swarms of flies around them. Closer examination reveals open wounds particularly around the bases of the horns. In some cases the lesions are minor but in others the damage is extensive perhaps involving most of the head. In severely affected flocks up to 50 per cent of the ewes and up to 90 per cent of the lambs show lesions. While the flies remain active, healing of the wounds does not take place. Secondary infection with bacteria is common and on occasion blowfly strike also occurs. Severe systemic reaction may be associated with such complications leading to pyrexia, inappetence and loss of condition. The wounds usually heal following cessation of fly activity but scarring and disfigurement may remain, reducing the value of the animal.

Diagnosis

This is based on the recognition of the typical lesions at the base of the horns affecting a substantial proportion of a flock, grazing in mid to late summer. In individual animals, particularly rams, the lesions should be distinguished from trauma associated with fighting and from blowfly strike although it should be remembered that blowfly strike may follow headfly damage.

Epidemiology

The life cycle of the fly involves four stages: egg, larva, pupa and adult. Eggs which are laid from late July onwards hatch within 7 days and the larvae feed and grow until late Autumn when development ceases. The following spring, feeding and development resume and pupation takes place in May. A minimum of 4 weeks later the adult hatches and following a meal of vertebrate protein egg-laying commences in July, continuing until early September.

Feeding activity of the flies occurs mostly between late June and September and is closely dependent on weather conditions. Flies will not take to the wing on windy days (wind speed in excess of 6 m/s) and are most active at temperatures of 18–22°C with a relative humidity of 70–80 per cent. Flies are active only during the day and will feed on a wide range of host species other than sheep.

The larvae and adults are very prone to desiccation and tree-cover is the favoured habitat providing protection from the sun, with the flies venturing out for short distances from the trees to feed. Eggs are laid in soil with a dense cover of vegetation (usually on the edge of woodland) where the conditions are damp and warm and the apparent increase in the prevalence of the headfly problem may be due in part to the increase in afforestation of upland areas. More intensive stocking of such areas may also be responsible for part of the increase in the problem since a positive correlation exists between flock size and severity of headfly disease.

Control and treatment

Various control methods have been suggested but all have proved less than totally effective or of limited practical application. Housing the sheep on warm, still days prevents attack but there are few situations in which this would be feasible. Other possibilities are to avoid grazing pastures with extensive tree cover close by during the summer months or to fence off the pasture at least 15 m out from the cover. Protective cloth headcaps have been used and shown to be highly effective but these devices are costly, time-consuming to apply and may be lost.

A number of commercially available repellents were compared in a controlled trial.[1] Results indicated that 0.05 per cent crotoxyphos and pine tar oil applied as a cream to the base of the horns and the head gave a substantial reduction in the incidence of damage. A compound code-named GH74 (1,1 bis [p-ethoxyphenyl]-2-nitroprone) used as a 0.125 per cent dip also gave a significant reduction in the severity and extent of the lesions as did Marshall's 'Anticap', a commerical preparation of various animal oils. Other repellents marketed for the purpose gave little or no reduction in the incidence of lesions.

In a later trial, permethrin, a synthetic pyrethroid with far greater photostability than the naturally-occurring compound, was evaluated.[2] The permethrin was sprayed on the head and back of the sheep as a 0.1 per cent emulsion at two weekly intervals throughout the fly season. This gave a significant reduction in the incidence and severity of lesions but again failed to eliminate the problem totally. Use of effective insecticides applied to the sheep may over a period of several years be of value in reducing the population of flies by killing the females before egg-laying occurs. At present there is no totally effective, practical chemical preventative treatment and agents which have shown a degree of success have required regular, repeated application throughout the fly season to achieve their success.

Treatment of sheep with damaged heads presents similar difficulties. Topical or, in severe cases, parenteral treatment with broad spectrum antibiotic controls secondary infection but healing does not take place whilst the flies continue to feed at the lesions. Steps must therefore be taken to prevent the flies from reaching the lesions either by housing the sheep, using protective headcaps or effective repellents until healing is complete or until fly activity has ceased. In severe cases housing of badly affected animals may be the only effective treatment.

At present there is no means of controlling the fly population away from the sheep. Breeds of sheep do vary greatly in susceptibility and in the long term keeping less susceptible breeds, such as those which are naturally polled and with wool covered heads, may offer the best solution in areas where headflies are a problem.

BLOWFLY MYIASIS

Synonyms Cutaneous myiasis, strike

Myiasis is the infection of living tissues with the larvae of dipteran flies. Blowfly myiasis is frequently classified in terms of the anatomical area which has been attacked, e.g. body strike, poll strike, breech strike, etc. Under British conditions body strike is the most frequently encountered type.

Cause

The larvae of three types of fly commonly parasitize sheep in Britain. These are *Lucilia sericata* (greenbottle), *Phormia terrae-novae* (black blowfly) and *Calliphora erythrocephala* (bluebottle). Elsewhere in the world different species of fly produce disease but the underlying causes and symptoms are similar. Blowflies are not obligatory parasites and multiply successfully without any requirement to feed on living matter. Replication is rapid with the complete life cycle taking as little as 7 days under favourable conditions. Female flies lay over 1000 eggs and thus given suitable environmental conditions a very rapid build up in the fly population may occur.

Clinical signs

Affected animals are restless, reluctant to graze and may bite or kick at the struck area. On close examination a foul-smelling area of moist brown wool is apparent and in the early stages maggots are visible in the wool. The maggots are 10–14 mm long, grey-white, yellow or slightly pinkish in colour with cylindrical segmented bodies which taper anteriorly. In more advanced cases the wool falls out revealing an irregular wound which exudes foul smelling liquid. The larvae, at this stage, are invading the tissues and only their posterior parts are visible.

Pathology

Primary flies (*Lucilia* and *Phormia*) lay their eggs on damaged or soiled areas of fleece. The larvae, after hatching, crawl to the skin which they lacerate using anterior hooks and secrete proteolytic enzymes which digest the tissues. Secondary flies (*Calliphora*) themselves unable to initiate a strike, are attracted to the damaged area and usually secondary bacterial invasion follows. Death may be the result of septicaemia or possibly associated with absorption of toxins from the liquefied proteins.

Diagnosis

The condition is obvious and observation of wounds infested with maggots confirms the diagnosis.

Epidemiology

Two main factors affect the onset of myiasis:
1. the prevalence of flies;
2. the susceptibility of the sheep
The weather conditions of late spring and summer favour rapid development of flies with a massive build up-in the fly population from early summer onwards. The area which is struck must first be soiled or damaged in some way to attract the flies. Soiling of the fleece with faeces is likely to follow an outbreak of diarrhoea and is predisposed to by the anatomical conformation of certain breeds e.g. Merino. In

Britain high humidity leading to mycotic infection of the fleece commonly precedes blowfly strike.

Under British conditions of husbandry strike occurs from June to September in both hill and lowland sheep. Clipped adults and young lambs with short fleeces are not often attacked but as the fleece length increases so does the risk of strike particularly in humid weather. In hill sheep at the beginning of the season there is a first wave of damage affecting the unclipped adults. A second wave of attack occurs in August and September affecting adults and lambs which both have long fleeces by that time. In lowland flocks where the clipping of the adults is earlier, first cases are usually in July, in lambs.

Control

Control is based on management practices to reduce the susceptibility of the sheep and the use of insecticides to kill the flies and larvae. Diarrhoea is commonly due to parasitic gastroenteritis and appropriate steps should be taken to control parasite burdens and prevent fleece soiling from this cause. Shearing the sheep before the commencement of the fly season where practicable will be of value and once flies are known to be active care should be taken to avoid wounds, e.g. castration or docking. A major factor in breech strike in Merino sheep is the wrinkled area of skin around the breech which is prone to soiling with urine and in Australia, Mules' operation is carried out routinely to remove surgically the surplus skin.

Various insecticides have been in use for many years but it was not until the 1940s that really effective compounds were developed. The first were the chlorinated hydrocarbons—dieldrin, aldrin, DDT etc. which gave good control of blowfly (and other ectoparasites) with a long period of activity after application. Certain compounds in this group proved highly toxic in the environment and were withdrawn in the UK in the mid 1960s but in fact by that time strains of blowfly resistant to the chlorinated hydrocarbons had already appeared and they would have soon become much less effective. Of the chlorinated hydrocarbons gamma-HCH remains in widespread use particularly for the control of sheep scab. Organophosphorus (OP) compounds replaced the chlorinated hydrocarbons and are still widely used although OP resistant strains of blowfly have now arisen. They are used singly or mixed with other OP's or chlorinated hydrocarbons. Examples include chlorfenvinphos, diazinon, bromophos ethyl etc.

More recently synthetic pyrethroids have been developed which appear highly promising as safe, effective insecticides for use in sheep dips affording protection lasting 12 or more weeks. Examples of the pyrethroids are permethrin, cypermethrin and cypothrin. A somewhat different approach has been the use of drying agents which consist of mixtures of zinc and aluminium oxides with steroids and fatty acids. When applied to the fleece these reduce the moisture content by up to 30 per cent for a period of 10–12 weeks. This reduces the risk of fleece rot considerably and a reduction of 75 per cent in the incidence of strike was claimed in a field trial carried out in Australia. [3] These agents may prove to be of value in Britain where wool rot is a major predisposing cause of strike and they have the advantage of not creating resistant strains of insects.

Treatment

Affected animals should have the wool removed from the area surrounding the damaged region and as many of the maggots as possible should be physically removed. A larvicidal insecticide (most OP compounds are larvicidal) should be applied together with topical and, if necessary, systemic antibiotic therapy. The animal should also be protected from further attack, either by housing or the use of effective insect repellents, until the lesions have healed.

REFERENCES

1 French N., Wright A.J., Wilson W.R. & Nichols D.B.R. (1977) Control of headfly on sheep. *Veterinary Record,* **100**, 40–3.
2 Appleyard W.T. (1982) Field assessment of permethrin in the control of sheep headfly disease. *Veterinary Record,* **110**, 7–10.
3 Hall C.A., Martin I.C.A. & McDonell P.A. (1980) The effect of a drying agent (B26) on wool moisture and blowfly strike. *Research in Veterinary Science,* **29**, 186–9.

FURTHER READING

Watts J.E., Murray M.O., Graham N.P.H. (1979) The headfly strike problem of sheep in New South Wales. *Australian Veterinary Journal,* **55**, 325–34.

40

PYODERMAS

F.M.M. SCOTT, J. FRASER AND G.R. SCOTT

Staphylococcal bacteria are readily isolated from the skin and orifices of healthy sheep.[1] Certain strains have pathogenic potential and can induce a benign folliculitis in sucking lambs [2], a mammary impetigo in suckling ewes or a severe facial dermatitis.[3] In addition, they rapidly colonize fresh wounds and other dermal lesions such as those resulting from an infection with Parapoxvirus (orf).

STAPHYLOCOCCAL FOLLICULITIS

Synonym Plooks

Staphylococcal folliculitis is a benign pyoderma of young lambs manifested by the development of transient pustules in perilabial and perineal skin.

Cause

The causal bacteria are beta-haemolytic, coagulase positive staphylococci. Several different phage types of staphylococci appear to be involved.

Clinical signs

Clinical signs are minimal. The pyoderma is manifested by the emergence of pustules, either singly or in a crop, in the skin around the lips or under the tail. Ewe-lambs, particularly, are prone to develop lesions in the perineal region (Fig. 40.1). Occasionally both sites are affected simultaneously. The pustules are circular, range in diameter from 2–5 mm and are surrounded by a ring of erythema. They persist for less than three weeks and heal without scar formation. If the affected skin is pigmented, the site of the healed lesion is conspicuous for some time as an irregular, smooth, white patch. Fresh crops of pustules commonly develop within three weeks. There is no systemic disturbance.

Pathology

The gross pathology is limited to changes at the

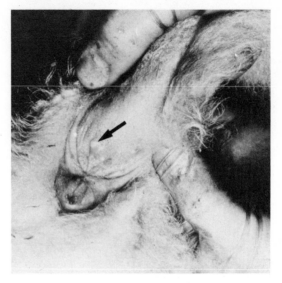

Fig. 40.1 Staphylococcal folliculitis of the perineum.

affected sites. Histopathologically the lesion is a pyogenic folliculitis underlying an inflamed, ulcerated epidermis (Fig. 40.2). Sebaceous glands are not affected.

Diagnosis

The condition is usually observed when sucking lambs are handled for other reasons. Although confirmatory measures are seldom attempted the causal organism is readily isolated by plating the pustule contents on 5 per cent sheep or ox blood agar. After the plates are incubated at 37°C for 24 h they are examined for bacterial colonies and the type of haemolytic activity is noted. Smears from single colonies are stained and examined for Gram-positive cocci. Coagulase production is assessed in a tube-test using fresh rabbit plasma.[4]

Epidemiology and transmission

Lambs are known to acquire staphylococci within a few hours of birth [1] and fully developed pustules have

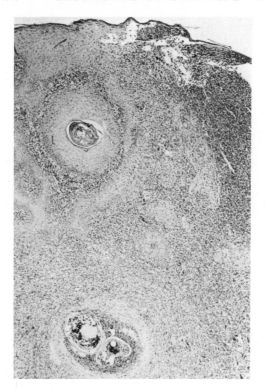

Fig. 40.2 Histopathological section of staphylococcal folliculitis showing an ulcerated epidermis and affected and unaffected follicles (H & E x 80) Photographed by Mr. R.C. James.

been observed in lambs less than 24 h old. The majority of affected lambs however first show clinical signs 3–4 weeks after the onset of lambing in the flock, the earliest cases occurring at the perineal site in ewe-lambs.

The pattern of the annual outbreaks is typical of a 'virgin-soil' epidemic of a contagion with the peak incidence occurring 2–3 weeks after the appearance of the first cases, i.e. 6–7 weeks after the start of lambing. The prevalence is always higher in flocks lambed and housed indoors. Ticks are not responsible for this infection.

Treatment

The condition is trivial and intervention is seldom attempted but occasionally the topical application of antiseptic ointments is worthwhile.

MAMMARY IMPETIGO

Synonym Udder impetigo

Impetigo is a superficial pyoderma of nursing ewes characterized by the formation of fleeting pustules of the skin of the udder.

Cause

The organism most often associated with udder pustules is a coagulase-positive *Staphylococcus*.

Clinical signs

The impetigo follows colonization of superficial abrasions on the surface of the udder. Pustules form rapidly and rupture easily with encrustation. They are discrete and vary from 2–5 mm in diameter. Healing occurs within 10 days but fresh infections lead to new pustules. There is no systemic disturbance.

Pathology

Histopathologically the pustule is revealed as being very superficial and it is found to contain numerous neutrophil polymorphs and cocci in a serofibrinous exudate.

Diagnosis

The condition is usually diagnosed fortuitously. Confirmation, if necessary, follows isolation of the causal *Staphylococcus* by conventional methods. [4]

Epidemiology and transmission

An essential percursor of mammary impetigo is a superficial abrasion, the cause of which is generally attributed to the sucking lamb that may or may not itself have folliculitis. Mammary impetigo however is not the source of lamb folliculitis but it has been linked to the subsequent development of acute pyogenic mastitis which usually occurs in the first 6 weeks after lambing.

Treatment

Topical applications of antiseptic ointments usually suffice.

STAPHYLOCOCCAL DERMATITIS

Synonyms Facial or periorbital eczema, facial dermatitis, eye scab, necrotic ulcerative dermatitis (NUD).

Staphylococcal dermatitis is a severe pyoderma of sheep evident as black scabs over deep facial ulcers filled with necrotic debris.

Cause

The cause of staphylococcal dermatitis of sheep is attributed to a coagulase positive, haemolytic *Staphylococcus aureus* which possesses alpha and beta haemolysins.[3] Most, but not all isolates, also produce delta haemolysins. The bacterium is sensitive to many antibiotics but resistant to trimethoprim.

Clinical signs

Severely affected sheep present an alarming and distressing appearance (Figs. 40.3a and 40.3b). The lesions are usually restricted to the head and particularly affect the skin over the nasal and maxillary bones, around the eyes and ears, and at the base of the horns. Suppurative ulcers form which become capped with black scabs and surrounded by a zone of alopecia. The ulcers are irregularly shaped ranging in diameter from 20–60 mm and contain necrotic debris. The lesions bleed easily and forceable removal of the scab reveals a deep crater with lytic contents. There is little or no systemic disturbance but if lesions occur around the eyes the local damage may be severe enough to interfere temporarily with vision. On rare occasions the eye can become infected causing permanent loss of sight. When lesions occur at the base of the ears necrotic damage to the dermal and subcutaneous tissues may result in drooping of the ears. Lesions heal from the periphery where a ring of new skin slowly advances to enclose the ulcer over a period of 5–6 weeks.

Sheep injected with infected tissue suspensions containing a variety of bacteria may have fever lasting for 48 h after exposure but pyrexia is uncommon in sheep injected with pure cultures of

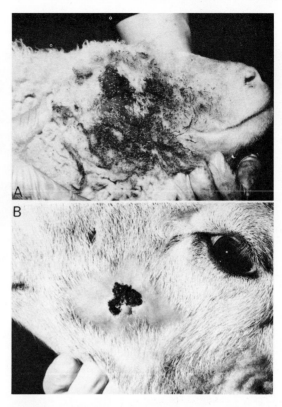

Fig. 40.3 (A) Staphylococcal dermatitis of sheep showing a severe lesions around the eye.
(B) Healing lesion of staphylococcal dermatitis. Note the zone of alopecia around the central scab.

the *Staphylococcus*. An inflammatory reaction starts at the site of infection and develops into a raised circumscribed swelling filled with foul-smelling pus which leaks through a central hole. A scab forms in 1–2 weeks which is surrounded by a hairless areola. Generally complete resolution takes 5–6 six weeks but may be delayed in cases of severe infection.

Pathology

Ulcers involve most layers of the dermis. Histological examination of experimentally-induced lesions[5] shows that abscesses develop in the dermis and subcutis dissecting along loose fascial planes. Cellular infiltration occurs predominantly by neutrophils and macrophages. The tissue may be distended as if by oedema but no fluid exudate has ever been observed. The overlying epidermis becomes necrotic which allows pus to exude and subsequently the

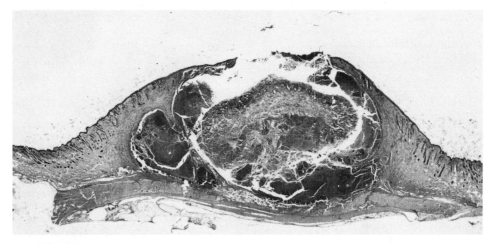

Fig. 40.4 Histopathological section of an experimentally induced lesion of staphylococcal dermatitis. (Photograph by courtesy of Mr. K.W. Angus) (x 5.5)

deeper dermal and subcutaneous tissues become necrotic (Fig. 40.4). The lesion heals by granulation and encapsulation of the primary abscess and any satellite abscesses. The cellular infiltration at this stage is predominantly mononuclear and any necrotic muscle fibres are replaced by collagen.

Diagnosis

Most cases are diagnosed clinically but confirmation may be obtained following isolation of the causative organism by conventional methods.[4] Differential diagnosis with regard to orf can usually be resolved by removing the scab; in staphylococcal dermatitis a deep thick-walled ulcer is revealed whereas in orf there is often an area of exuberant granulation tissue.

Staphylococcal dermatitis differs from the facial eczema that occurs in sheep in New Zealand and the 'yellowses' that affects lambs and hoggs in the United Kingdom. These conditions are characterized by hepatitis and photosensitization with skin lesions restricted to the unpigmented areas and parts not covered with wool. A dermatitis of sheep has been described in America [6] from which staphylococci were isolated. The organism was identified as *Staphylococcus albus* but no details were given regarding its classification.

Epidemiology and transmission

Staphylococcal dermatitis has emerged as a problem in the UK in the last two decades. The ulcers can

develop in sheep of all ages but adult ewes are most often affected just before lambing. The disease is contagious and its spread through a flock has been attributed to infection of abrasions of the head occasioned by feeding at troughs. Ulcers at the base of the horns result from infection of minor wounds caused by fighting.

Control and treatment

Severe cases can be treated by flushing and washing the ulcerated areas and applying a topical antibiotic preparation. Affected sheep may also be injected intramuscularly with a long acting antibiotic. Minor cases will heal without attention.

In-lamb ewes should be allowed sufficient trough space when being hand-fed, to minimize contact between heads and limit aggressive behaviour when feeding.

REFERENCES

1 Watson W.A. (1965) The carriage of pathogenic staphylococci by sheep. *Veterinary Record,* **77,** 477–80.
2 Scott G.R., Lawson J.B., Laing A.H., McPherson E.A. & Ritchie J.S.D. (1981) Staphylococcal folliculitis in sucking lambs. (In preparation.)
3 Scott F.M.M., Fraser J. & Martin W.B. (1980) Staphylococcal dermatitis of sheep. *Veterinary Record,* **107,** 572–4.

4 Cowan S.T. (1974) The staphylococci. In *Cowan and Steel's Manual for the Identification of Medical Bacteria*, 2e, pp.47–50. Cambridge University Press.

5 Fraser J., Scott F.M.M., Angus K.W. & Martin W.B. (1982). Experimental re-infection of the skin of sheep with *Staphylococcus aureus. Veterinary Record*, **111**, 485–6.

6 Hardy W.T. & Price D.A. (1951) Staphylococcic dermatitis of sheep. *Journal of the American Veterinary Medical Association*, **119**, 445–6.

41 PHOTOSENSITIZATION

E.J.H. FORD

Synonyms Yellowses, saut, plochteach, alveld, facial eczema, geeldikkop.

Photosensitivity diseases occur in animals throughout the world. This chapter contains a general account of the condition and the mechanisms by which it may arise. The different types of the disease are illustrated by selected examples and a brief description is given of its occurrence in the British Isles, followed by some suggestions on how its appearance in a particular flock may be investigated.

Cause

Photosensitization arises when a photosensitized animal is exposed to sunlight. A state of photosensitivity is produced when a photodynamic substance is present in the tissues.[1] Such a substance absorbs energy from light and, in the presence of molecular oxygen, the cell membranes of the exposed tissues become severely damaged by an abnormal rate of oxidation. Each photodynamic substance absorbs energy from light of a particular wavelength, usually in the visible range, which is known as the 'action spectrum' for that substance. The condition must be distinguished from sunburn, a normal reaction of unprotected skin to ultraviolet irradiation. Photodynamic agents play no part in the causation of sunburn.

Clare[2, 3] classified photosensitization in to three main types, on the basis of the way in which the photodynamic agent entered the circulation and reached the skin. In his first type, called primary photosensitization, the photodynamic agent is a substance, not normally present in the diet, which enters the bloodstream after being absorbed from the digestive tract. Poisoning by one of several species of *Hypericum*, but especially by St. John's Wort, *Hypericum perforatum*, is an example of this type. The plant has a world-wide distribution and its leaves contain a red fluorescent pigment named hypericin (action spectrum 590–610 nm) believed to be a mixture of polyhydroxy derivatives of helianthrone, which is absorbed from the alimentary canal of animals when grazing on pastures containing *Hypericum*. An account of recent work with the plant has been given by Araya and Ford.[4] Another cause of primary photosensitization is buckwheat (*Polygonum fagopyrum*), which contains fagopyrin with an action spectrum of 580–590 nm.

Clare's second type of photosensitization is that due to aberrant pigment synthesis. This type is exemplified by congenital porphyria (pink tooth) which has been reported in cattle in South Africa[5], in Britain[6] and in pigs and cattle in Denmark[7] but not yet, as far as the writer is aware, in sheep. The condition is inherited as a recessive factor and is due to the production of excessive amounts of uro- and coproporphyrin during the synthesis of haemoglobin. These porphyrins stain the bones and teeth and their presence in the circulation sensitizes the affected animals to sunlight.

The third type of photosensitization is described as hepatogenous because it is invariably associated with a derangement of biliary excretion, due either to structural or functional changes in the liver. These liver changes interfere with the excretion of phylloerythrin. This is a photodynamic agent, activated by light of 300–450 nm wavelength. It is produced by the microbial degradation of chlorophyll in the gastrointestinal tract of herbivorous animals. The key role phylloerythrin in hepatogenous photosensitization was first described by Rimington and Quin.[8]

The hepatogenous type is the most common type of photosensitization in animals, as there are many substances which can interfere with the excretion of phylloerythrin in the bile. Many of these are plant toxins, for example, lantadene from *Lantana camara*[9], icterogenin from *Lippia rehmanni* and an unknown toxin from dubbeltjie, *Tribulus terrestris*. The condition can also be caused by drugs, e.g. dimidium bromide[10], by algae, e.g. *Microcystic flosaquae*, by aphis[11] and by fungal toxins of which the most important is sporidesmin; a substance found in the spores of *Pythomyces chartarum*. This fungus grows on rye grass pastures and the disease produced by the ingestion of spores is a serious problem in sheep in New Zealand, where the condition is known as 'facial eczema'.[12] The fungus also causes losses in cattle in Australia.

An inherited defect in phylloerythrin excretion has been described in Southdown sheep[13] and in Corriedale sheep.[14]

Clinical signs

Although photosensitization has many different causes and can be classified according to the origin of the photodynamic agent[2], the clinical effects are characteristic of the species of animal and arise in those parts of the body in which the skin is least protected from exposure to light by a thick or pigmented coat. In sheep, the ears in particular are affected, as are the face, eyelids, lips, lower part of the limbs and the midline of the back at the parting of the fleece in long-woolled sheep. The ears become swollen and oedematous and develop a characteristic drooping appearance. The affected skin dries and cracks and the dead skin sloughs. In particular, the tips of the ears curl and a part, or all of the ear may be lost from necrosis.

Skin lesions in white pigs and keratitis in calves and pheasants after the oral administration of the anthelminthic phenothiazine, have been reported.[15, 16] Phenothiazine is metabolized in the digestive tract to its sulphoxide which is absorbed and can be found in the aqueous humour of calves but not of sheep. In dosed animals which have absorbed the sulphoxide, radiation of wavelengths up to 390 nm, which penetrate the cornea but not the skin, produces the keratitis. This condition does not arise in sheep because they are able to metabolize phenothiazine sulphoxide to phenothiazone.

Photosensitization of sheep in the British Isles

The initial onset of clinical signs is remarkably rapid. A susceptible animal exposed to sunlight develops, within a few hours, a painful swelling of the unprotected areas of skin as mentioned previously. The animal is severely distressed, loses its appetite and rubs itself vigorously in an attempt to relieve the irritation. After the development of oedema there is often seepage of serous fluid through the skin. This dries to produce a yellowish crust, which is particularly noticeable on the ears and face of white sheep and may be the origin of the popular name of 'yellowes'. This stage is followed by necrosis of the affected skin. Appetite slowly returns and skin sloughing, followed by regeneration, takes place during a period of several weeks. In the author's experience, many of these cases are hepatogenous in origin, because an increase in the concentration of phylloerythrin in plasma is detectable in the initial acute phase, but the agent which causes the temporary interference with the excretion of phylloerythrin in the bile has not been identified.

The disease is of widespread but sporadic occurrence, and may affect individual animals or several sheep in one flock. It occurs mainly in hill grazing areas and has acquired many local names, such as 'saut' in Cumbria, 'plochteach' in Perthshire and 'yellowses' in many areas, particularly when white faced sheep are affected. In many parts of Scotland cases arise each year on certain farms. It is seen in young lambs about the end of May when they are beginning to ingest appreciable amounts of herbage. The incidence increases during June and falls to zero by early July; there may be a 10 per cent morbidity rate on some hirsels. After the acute phase, most lambs gradually regain their appetite, but do not grow as well as unaffected animals. Lesions of the eye and surrounding tissues interfere with vision and some lambs are drowned in bogs and burns.

Investigation of 'plochteach' showed that the type was hepatogenous[17], although structural changes detectable in the liver by light microscopy were slight. Attempts to isolate an infectious agent from the tissues of affected lambs were unsuccessful. A similar disease in Norway called 'alveld' has been attributed by Ender [18] to ingestion of the Bog asphodel (*Narthrecium ossifragum*), a common plant of wet hill pastures. However, it has not been possible to reproduce 'plochteach' by dosing with large quantities of macerated *Narthrecium* collected, at different stages of maturity, from Scottish grazings.

So far no specific cause of the impaired ability of affected lambs to excrete phylloerythrin has been identified. Because many of the affected flocks are 'hefted' to hill grazing and are, therefore, virtually closed units and because of the yearly incidence in young lambs at the commencement of grazing, it is tempting to suggest that there may be a hereditary disposition to the disease as has been described in Southdowns and Corriedales. Further investigation on flocks with a regular annual incidence could be rewarding.

Diagnosis and investigation of photosensitization

As the clinical signs are more characteristic of the species affected than of the cause of the photosensitization, they are more an indication of the nature of the disease than a help in elucidating the cause. In this respect the age of the affected animal may be helpful: if the condition occurs in a young animal on first exposure to sunlight and a similar susceptibility has been seen in related animals, the congenital type should be suspected. On the other hand, previous exposure, without ill effect, would rule out the congenital type. Recent administration of a drug, or access to pastures containing *Hypericum*, would suggest the primary type. The most common type, i.e. hepatogenous, only occurs when animals are grazing and its existence can be confirmed by the detection of a high (>10 $\mu g/100$ ml) concentration of phylloerythrin in serum or plasma. The sample should be collected in the acute phase of the disease, as the inappetence caused by the irritation and pain of the skin lesions reduces the intake of chlorophyll. This is followed by a rapid fall in phylloerythrin production and hence in the concentration in plasma.

In some cases of hepatogenous photosensitization, but not all, there may be a rise in the concentration of bilirubin in plasma. Jaundice may develop, but its onset would be later than that of the initial clinical signs. The release into plasma of liver-specific enzymes, e.g. glutamate or sorbitol dehydrogenases or gamma glutamyl transferase, is an aid in the detection of damage to hepatic cells. The absence of an increase in the activity of such enzymes however, would not rule out the possibility of the hepatogenous type of photosensitization, because some cases may be due solely to a reduced ability to excrete phylloerythrin without any other coincident liver damage.

REFERENCES

1 Blum H.F. (1964) *Photodynamic Action and Diseases Caused by Light.* Hafner Publishing Co., New York.

2 Clare N.T. (1952) Photosensitization in diseases of domestic animals. Review series No. 3, Farnham Royal, Commonwealth Agricultural Bureaux.

3 Clare N.T. (1955) Photosensitization in animals. *Advances in Veterinary Science*, 2, 182–211.

4 Araya O.S. & Ford E.J.H. (1981) An investigation of the type of photosensitization caused by the ingestion of St. John's Wort (*Hypericum perforatum*) by calves. *Journal of Comparative Pathology*, 91, 135–41.

5 Fourie P.J. (1936) The occurrence of congenital prophyrinuria (pink tooth) in cattle in South Africa (Swaziland). *Onderstepoort Journal of Veterinary Science and Animal Industry*, 7, 535–66.

6 Amoroso E.C., Loosmore R.M., Rimington C. & Tooth B.E. (1957) Congenital porphyria in bovines; first living cases in Britain. *Nature, Lond.*, 180, 230–1.

7 Jørgensen S.K. & With T.K. (1955) Congenital porphyria in swine and cattle in Denmark. *Nature, Lond.*, 176, 156–8.

8 Rimington C. & Quin J.I. (1934) Studies on the photosensitization of animals in South Africa. VII. The nature of the photosensitizing agent in Geeldikkop. *Onderstepoort Journal of Veterinary Science and Animal Industry*, 3, 137–57.

9 Gopinath C. & Ford E.J.H. (1969) The effect of *Lantana camara* on the liver of sheep. *Journal of Pathology*, 99, 75–85.

10 Ford E.J.H. & Boyd J.W. (1962) Cellular damage and changes in biliary excretion in a liver lesion of cattle. *Journal of Pathology and Bacteriology*, 83, 39–48.

11 Mohamed F.H.A., Imbabi S.E. & Adam S.E.I. (1977) Heptagenous photosensitization in horses due to *Aphis craccivora* on lucerne. *Bulletin of Animal Health and Production in Africa*, 25, 184–7.

12 Cunningham I.J., Hopkirk C.S.M. & Filmer J.F. (1942) Photosensitivity diseases in New Zealand. 1. Facial eczema: its clinical, pathological and biochemical characterization. *New Zealand Journal of Science and Technology*, 24, 185a–98a.

13 Clare N.T. (1945) Photosensitivity diseases in New Zealand. IV. The photosensitizing agent in Southdown photosensitivity. *New Zealand Journal of Science and Technology*, A27, 23–31.

14 Cornelius C.E., Arias I.M. & Osburn B.I. (1965) Hepatic pigmentation with photosensitivity: syndrome in Corriedale sheep resembling the Dubin–Johnson syndrome in man. *Journal of the American Veterinary Medical Association*, **146**, 709–13.

15 Whitten L.K., Clare N.T. & Filmer D.B. (1946) A photosensitized keratitis in cattle dosed with phenothiazine. *Nature, Lond.*, **157**, 232.

16 Clapham P.A. (1950) Keratitis in pheasants following treatment with phenothiazine. *Journal of Helminthology*, **24**, 61–2.

17 Ford E.J.H. (1964) A preliminary investigation of photosensitization in Scottish sheep. *Journal of Comparative Pathology*, **74**, 37–44.

18 Ender F. (1955) Aetiology of photosensitization in lambs. *Nordisk Veterinaermedicin*, **7**, 329–77.

42   DERMATITIS AND PRURITIC CONDITIONS

A.O. MATHIESON

MYCOTIC DERMATITIS

Synonyms Actinomycotic dermatitis, dermatophilosis, lumpy wool, rain-rot, strawberry foot-rot.

Mycotic dermatitis is a skin disease widespread in many sheep flocks where in a mild form its presence often goes unnoticed. The severity of the disease varies greatly amongst individual sheep; fine-woolled breeds being most susceptible but severe cases do occur in coarse-woolled sheep. The condition is of considerable economic importance because affected fleeces are usually downgraded.[1]

Cause

The causal organism, *Dermatophilus congolensis* is not a fungus as the disease nomenclature suggests but a bacterium of the order *Actinomycetales*. It exhibits fine branching and multiseptate, coarse, hyphal filaments which divide transversely and longitudinally to form cuboidal packets of Gram-positive coccoid cells arranged in 2–8 parallel rows within the branching filaments (Fig. 42.1). The organism is not host specific and causes dermatitis in a wide range of mammals throughout the world of which the most

important is bovine streptothricosis.[2] *Dermatophilus congolensis* infection is also recognized as a zoonosis particularly of workers in animal industries.

Clinical signs

Three forms of mycotic dermatitis are recognized on the basis of the distribution of the lesions. One, which is seen mostly in lambs, affects mainly the ears (Fig. 42.2) but other hairy parts of the head particularly the nose may be affected. Initially the lesion consists of a focus of hyperaemia followed by papule formation and exudation resulting in small heaped-up scabs about 0.5 cm in diameter. In mild cases these cause little discomfort but when secondary bacterial infection is present the ears become swollen and painful.

In the second form, termed 'lumpy wool' *Dermatophilus* infection of the skin of the dorsal part of the body especially over the lumbar region and flanks produces an intense irritation. When the infection is extensive and severe, considerable fleece damage and wool loss occurs as a result of the animal rubbing. The denuded areas are covered by hard, thick scabs formed from dried serum exudate. Accidental removal of these scabs reveals raw bleeding areas attractive to flies. In the less severe form, most commonly seen in Britain, the disease follows a more

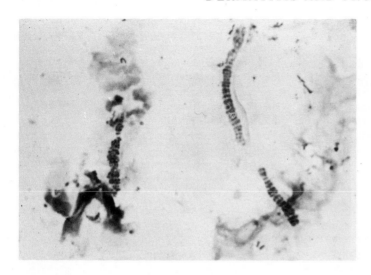

Fig. 42.1 Filaments of *D. congolensis* x 800.
By courtesy of Mr. E.A. McPherson

transient, intermittent course in which wool loss is not a feature and the lesions can only be observed when the fleece along the back of the animal is parted to reveal moist scab formation due to serum exudation at the base of the wool fibres. Healing usually occurs spontaneously about 7–14 days following the onset of the attack, the scab about 0.3 cm thick becomes detached and hardens binding bundles of wool fibres together. These amber-coloured scabs are carried upwards away from the skin surface as the wool fibres grow (Plate 6). Frequently past evidence of the occurrence of clinical mycotic dermatitis in a flock is only revealed during shearing when the hard scabs can be seen or felt in the fleece. 'Strawberry foot-rot' is the term used to signify raised, scab-covered lesions of the lower limbs, mainly in the region of the coronet, in which *Dermatophilus* can be demonstrated.

When lesions are present on the ears, face or limbs it is not uncommon to find contagious pustular dermatitis virus present also. Secondary infection of ear and face lesions with opportunist organisms, such as *Pseudomonas spp.*, may also exacerbate the condition and attract fly invasion.

Pathology

The coccal forms of *Dermatophilus congolensis* found in dry scabs when exposed to moisture develop long flagellae providing active motility. Invasion of the skin by these infective forms, termed zoospores, is probably facilitated by prior damage to the lipid zone. The organism proliferates in the living epidermis by extending branching filaments which penetrate hair and wool follicles inducing an acute inflammatory response which serves as a barrier to invasion of the dermis. Successive layers of epidermis become infected resulting in a scab composed of alternate strata of cornifying epidermis and exudate. There is little evidence of the development of any strong immunity.[3]

Diagnosis

A provisional diagnosis of *Dermatophilus* infection based on clinical appearance is usually readily confirmed by making impression smears from the underside of moist scabs, fixing and staining with

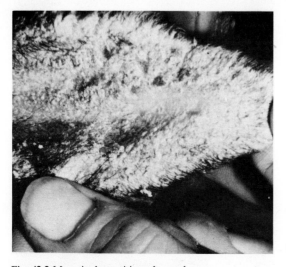

Fig. 42.2 Mycotic dermatitis scabs on the ear.
By courtesy of Mr. E.A. McPherson

Giemsa or Gram to demonstrate characteristic branching mycelium and coccoid forms. The organisms may also be demonstrated in smears using an acridine orange stain and examining for fluorescence in ultraviolet light. Where the scab is tending to become dry, several small pieces from the underside should be placed on a slide and moistened with the least amount of water that will completely soften them. Squash the material to form a thin layer for staining but avoid excessive disruption of intact filaments. Stained smears are examined without a coverglass under oil immersion. Alternatively fresh scab can be embedded in paraffin wax, sectioned and stained.

In older scab material the hyphal forms may be absent and where the viability of the organism is lessened the cocci may stain Gram-negative. For such cases and where secondary bacterial infection is present a fluorescent antibody technique has been described. Soaking macerated scab in physiological saline stimulates the release of zoospores which can be collected in 2–4 hours from the surface of the saline for inoculation of blood agar plates to be incubated at 37°C. The addition of 1 per cent polymyxin to the medium will inhibit other bacterial contaminants. After incubation for 24 h pin-point colonies surrounded by zones of beta-haemolysis appear. Following several days incubation the colonies are waxlike, convex, with a wrinkled surface and are, characteristically, embedded in the medium and vary in colour from greyish-white to orange. In glucose or serum broth, small fluffy spherical colonies form at the bottom of the medium which remains clear or slightly turbid. Stained smears of fluid cultures demonstrate the Gram-positive branching filaments. Specific identification can be confirmed on the basis of biochemical characteristics. [4]

Epidemiology and transmission

The coccoid form of *Dermatophilus congolensis* can survive in dry scabs for a long period whilst the longevity of the infective zoospore stage is relatively short. Chronic infection of the ear surface of individuals within a flock appears to be the main means by which the disease is perpetuated although, due to the lack of host specificity, other affected mammals can provide a reservoir of infection. The appearance of clinical disease can be associated with periods of wet weather when it seems that constant wetting weakens the natural skin, fleece or hair barrier. However, any factor reducing the efficiency of the protective lipid layer, such as excessive dryness after shearing or the action of detergent wetting agents, predisposes to infection. Subsequent wetting by heavy rain stimulates liberation of zoospores from reservoir sites and facilitates their spread over the back area. It is not uncommon for individuals to experience successive attacks of infection each of about one week in duration. Infection in lambs is most likely to occur during sucking when the dam's fleece is wet, the ears and nose being the primary site. From the ears, infection is slowly transmitted over the body. During the first 6 weeks of life the lamb's skin contains little lipid and is therefore highly susceptible. Insect vectors may also transmit infection between individual animals.

Control, prevention and treatment

Dermatophilus infection of sheep is so widespread that elimination of infection from a flock is not a practical proposition. In flocks where the disease is a problem it is advisable to use both for summer and for autumn/winter protection a dip containing either 0.5 per cent zinc sulphate or 1 per cent potassium aluminium sulphate (alum) to control infection. Occasionally the use of a pre-winter dip without either of these ingredients precipitates an outbreak of mycotic dermatitis 10–12 days post-dipping. These compounds prevent the establishment of infection by coagulating the flagellae of the zoospores. Dusting of the fleece with finely powdered alum has been shown to be an effective method of controlling infection but is too laborious a procedure in all but very small flocks.

Treatment of severe back infections by topical application of powdered alum, as has been recommended, is largely ineffective due to adherence of the dense scabs to the skin in which unchecked filamentatous invasion is occurring. Effective cures of such cases have been obtained by parenteral treatment with 70 mgm streptomycin and 70 000 iu penicillin per kg bodyweight. Similar therapy is recommended for the treatment of severe infection of the ears and face. When the lesions have resolved, dusting of the previously affected areas with powdered alum prevents early reinfection.

FLEECE ROT

Synonym Canary stain

Discolouration of wool is a significant cause of down-grading of fleeces. The cause of yellowing of wool or canary stain, as it is frequently termed by wool merchants in Britain, has not been clearly established but is thought to result from the development of a moist eczema in which non-specific, pigment-producing bacteria proliferate. Most commonly the affected band of wool fibres has a bright yellow appearance but greenish-blue or brown discolouration can occur. As with mycotic dermatitis, appearance of the condition can be correlated with periods of heavy rainfall. Both conditions may be closely linked as *D. congolensis* has been demonstrated from typical wool samples affected by canary stain. These pigment-producing organisms may be no more than opportunist bacteria superimposed on a mild mycotic dermatitis infection, in which case control measures advocated for mycotic dermatitis are likely to reduce the incidence of fleece discolouration.

RINGWORM

Ringworm infection in sheep is a relatively rare occurrence and infection usually arises from contact with inanimate objects contaminated by other livestock.

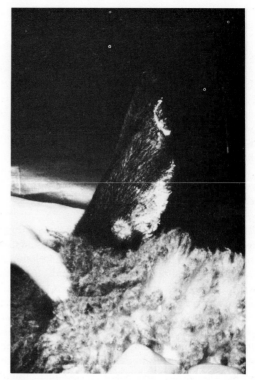

Fig. 42.3 Ringworm lesions on ear of sheep.
By courtesy of Mr. P.K.C. Austwick; Crown copyright.

Cause

The pathogen mainly reported from cases occurring in Britain has been *Trichophyton verrucosum* but other dermatophytes may be encountered.[5]

Clinical signs

Crusty wart-like lesions on the face and ears (Fig. 42.3) accompanied by intense pruritis have been described most frequently but lesions on the back have also been noted. Clinically the disease resembles mange and careful examination and diagnosis is necessary to differentiate ringworm from sheep scab.

Diagnosis

Microscopic examination of skin scrapings and hair plucked from the affected area reveals the characteristic spores but specific identification requires inoculation of suitable media.

Treatment

Affected sheep have been observed to recover spontaneously within 4–6 weeks which by comparison with the longer duration of infection noted in other species suggests that the sheep is not a natural host or that immunity is quickly acquired. However, since infection is usually limited to a few animals within the flock or group it is advisable to segregate these and administer oral griseofulvin treatment. Also to prevent further spread, contact with the original source of infection should be denied.

ECTOPARASITES

Ectoparasitic infestations of sheep are common, involving a variety of insects, each with its individual life cycle. These can cause serious economic loss through their effects on health, bodily condition and fleece damage if not controlled. Those causing mange are the subject of Chapter 37. Ticks are considered separately because their role as vectors of disease is more important than their direct parasitic effect.

LICE, KEDS AND MITES

Lice : *Damalinea ovis, Linognathus ovillus, L. pedalis*
Ked : *Melophagus ovinus*
Mites: *Trombidiform spp*

Clinical signs

The main symptom of infestation with these ectoparasites is intense irritation causing affected animals to scratch, bite or rub themselves. The biting louse, *D. ovis* feeds on exfoliated epithelium and other skin debris and is present in greatest numbers around the neck and over the back area. *L. ovillus*, the blue louse is found chiefly affecting the face while *L. pedalis* occurs on the lower limbs. These suck blood and tissue fluids and like the sheep ked, *M. ovinus*, when present in significant numbers, can have a marked debilitating effect with signs of anaemia. Forage mites *(Trombidiform spp)* may cause an exudative dermatitis on the face where attacks usually occur.

Diagnosis

When examining sheep exhibiting symptoms of pruritis, skin hypersensitivity or wool damage it is essential to consider mange within the differential diagnosis (see Chapter 37).

Apart from the sheep ked which is readily visible, careful and prolonged examination may be necessary to detect the presence of lice particularly *D. ovis* for which the use of a magnifying glass may be required. Specific identification requires microscopic examination. The detection of forage mites involves microscopic examination of skin scrapings.[6]

Epidemiology and transmission

Louse infestation is essentially a winter disease and can be a considerable problem in housed sheep, particularly hoggs. Although lice can survive for several days after removal from the host, transmission is generally by direct contact with an infected sheep. The eggs attached to the base of hairs or wool

fibres hatch in 7–14 days and develop to egg-laying adults in 14–21 days, so that populations multiply rapidly. Sheep keds spend their entire life on the host, spread by direct contact and like lice they have a short regeneration interval of some 5 weeks. Populations peak also during the winter months.

Forage mite infestation of sheep arising as a clinical problem is not common but can occur when populations of the harvest mite *Trombicula autumnalis* are high in the autumn or when contaminated forage is being fed to housed sheep. The adults generally live in soil with small rodents acting as natural hosts and it is the larval stages which may infest animals and man.

Control

Routine summer and winter dipping of sheep carried out in accordance with the manufacturers instructions ensures effective control of these ectoparasite infestations. Dipping is preferable to spray treatment as it allows complete penetration of the dip at all body sites and supplies residual larvicidal protection to eradicate emerging stages from deposited eggs.

TICKS

Ixodes ricinus
Haemophysalis punctata

Ixodes ricinus is widely distributed throughout Europe including Britain, where it occurs mainly in the high rainfall central and western regions. Although commonly referred to as the sheep or cattle tick it infests a wide range of both birds and mammals including man.[7] *I. ricinus* is a three host tick and all stages in the life cycle, larva, nymph and adult may occur on the same host concurrently. Each stage attaches itself to the host sufficiently long to satisfy its requirement for blood and tissue fluids before detaching and dropping to the ground to continue further development or in the case of the adult female to lay many thousands of eggs. The interval between successive feeds by individual stages may be as short as 4 months and as long as 20 months so that the generation time can vary from 2–6 years.[8]

Tick activity which is highly dependent on climatic conditions is negligible during winter but rapidly reaches a peak in the spring with a second smaller peak occurring during autumn. The duration of the periods of activity is longest in the more mild western areas of Britain. Survival of the unfed stages depends on the persistence of a high relative humidity within the microhabitat which is most likely to exist on uncultivated upland and hill areas where the vegetation mat of grass, bracken or heather is dense.

The predilection sites for tick attachment on sheep are the head, neck, axillary and groin regions. While light infestations are well tolerated, heavy infestations give rise to considerable irritation causing animals to rub and scratch themselves; a condition which is regarded sufficiently note-worthy by shepherds to be named 'tick worry'.

The overriding importance of tick infestation however is in regard to the diseases they may transmit: tick-borne fever, louping-ill and babesiasis (see Chapter 43).

Haemophysalis punctata Another European tick which may occasionally be found infesting sheep in Britain. It appears to be limited in its distribution to the southern and western coastal areas. Although identified as a vector of *Babesia major* and *Theileria mutans* infection of cattle on the European continent its disease transmitting role in Britain is not significant.

The catholic nature of *I. ricinus* in respect of host selection enables the parasite to survive in significant numbers in the complete absence of sheep, therefore any attempt to eradicate this tick even by a rigorous dipping programme would be unsuccessful. Although spring dipping is regularly carried out on many hill farms with the aim of reducing tick numbers it is doubtful whether any significant reduction in the total population is obtained. Such reduction as is achieved is extremely variable since spring tick dipping is generally carried out on a fixed date some weeks before lambing which may or may not coincide with a period of marked tick activity and attachment. Where tick infestation is sufficiently severe to cause the animals irritation then dipping will alleviate 'tick worry'. Even if it were possible to reduce total tick numbers by a significant amount the prevalence of tick-associated diseases would be unaffected since these are transmitted by even very light infestations.

Where dipping can be of significant value is in preventing tick spread into areas in which they have not already become established.

Flock owners farming land known to be free from tick infestation who purchase any sheep which may be carrying ticks should be strongly recommended to dip such stock immediately on arrival.

REFERENCES

1 Austwick P.K.C. (1960) Mycotic dermatitis and the down-grading of wool. pp.39–42. *National Sheep Breeders' Association Year Book,*

2 Lloyd D.H. & Sellers K.C. (Eds) (1976) Dermatophilus infection in animals and man. Proceedings of a Symposium held at the University of Ibadan, Nigeria, and sponsored by the Agricultural Research Council of Nigeria. Academic Press, London.

3 Roberts D.S. (1967) Dermatophilus infection. *Veterinary Bulletin,* **37,** 513–21.

4 Buxton A. & Fraser G. (1977) *Animal Microbiology.* Vol.1. pp.295–7. Blackwell Scientific Publications, Oxford.

5 Ainsworth G.C. & Austwick P.K.C. (1973) *Fungal Diseases of Animals.* 2e, p.30. Commonwealth Agricultural Bureaux, Slough.

6 Lapage G. (1968) *Veterinary Parasitology,* 2e, Oliver & Boyd, Edinburgh.

7 Arthur D.R. (1963) *British Ticks.* Butterworth, London.

8 Donnelly J. (1978) The life cycle of *Ixodes ricinus* L based on recent published findings. In Proceedings of an International Conference. (Ed. Wilde J.K.H.) Tick-borne Diseases and their Vectors. Centre for Tropical Veterinary Medicine, Edinburgh University.

SECTION 7

MISCELLANEOUS DISEASES

43 TICK-ASSOCIATED INFECTIONS

G.R. SCOTT

If the ubiquitous staphylococcus is the common parasite on the skin of sheep then tick-transmitted agents are the common visceral parasites. Three types of agent are incriminated: viruses, rickettsias and protozoal piroplasms.

TICK-BORNE VIRUSES

At least 78 viruses have been isolated from ticks. Fortunately a mere handful infect sheep and only one occurs in the UK viz. louping-ill (see Chapter 14). In the rest of Europe and in temperate Asia members of the tick-borne virus encephalitis complex are endemic in sheep and goats and in the tropics of Africa and Asia infections with members of the Ganjam virus complex are widespread, the best documented being Nairobi sheep disease virus, the causal agent of a fatal febrile gastroenteritis prevalent in flocks in eastern Africa.

TICK-BORNE RICKETTSIAS

Rickettsias are minute, Gram-negative bacteria that multiply by binary fission inside host cells. Unlike other bacteria they cannot be grown on or in inanimate media. Rickettsias are part of the resident flora of most domestic animals including sheep which can harbour at least nine species. In the UK sheep carry in descending order of clinical significance: *Chlamydia psittaci*, the causal agent of enzootic abortion. *Coxiella burnetii*, a rarer source of ovine abortion transmissible by contact to man causing Q fever: *Cytoecetes phagocytophila*, the causal agent of tick-borne fever and *Eperythrozoon ovis*, a benign parasite of red blood cells (Fig. 43.1). The British rickettsias that are transmitted by ticks are *C. phagocytophila* and, on occasion, *C. burnetii*. Overseas in Africa the most serious tick-borne rickettsial infection of sheep and goats is heartwater, a febrile encephalitis complicated by circulatory failure and caused by *Cowdria ruminantium*. The most widely distributed tick-borne rickettsia of sheep is *Anaplasma ovis* which like its newly discovered close relative *A. mesaeterum* is a relatively benign parasite of red blood cells. Although both species occur in Europe neither has been detected in British sheep.

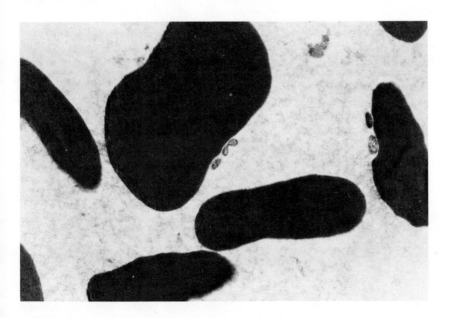

Fig. 43.1 *Eperythrozoon ovis* rickettsias cradled on the surface of a red blood cell (x 20 000).

TICK-BORNE FEVER

One of the first projects commissioned by the Animal Diseases Research Association in the 1920s was a study of 'trembling' the cause of which was identified by Dr J. Russell Greig and his colleagues in 1929 as a virus. A tick *Ixodes ricinus* was suspected as being the vector which led to trials to confirm this hypothesis. Dr J. MacLeod collected unfed nymphs from areas grazed by 'trembling' sheep and allowed them to feed on susceptible sheep all of which developed fevers. Blood from the febrile sheep induced similar fevers when injected into fresh sheep. When these experimental sheep were later challenged with 'trembling' or 'louping-ill' virus they succumbed. Dr MacLeod had inadvertently identified a new disease now called tick-borne fever.[1]

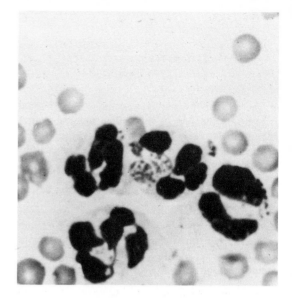

Fig. 43.2 Giemsa-stained smear with neutrophils infected with *Cytoecetes phagocytophila* (x 3000).

Definition

Tick-borne fever is a benign rickettsiosis of domestic and wild ruminants characterized by minimal constitutional disturbance despite a prolonged parasitaemia and a high fever. Defence mechanisms are depressed in affected animals allowing potentiation of other pathogens. Pregnant animals abort.

Cause

The causal rickettsia is pleomorphic and is found typically in membrane-lined vacuoles in the cytoplasm of circulating granulocytes, particularly neutrophils (Figs. 43.2 and 43.3). On occasion it infects monocytes. Its rickettsial nature was suspected by the first observers at the Moredun Institute in the early 1930s and was confirmed beyond doubt by Foggie in 1951 who suggested that the organism be named *Rickettsia phagocytophila*.[2] He, later, proposed that it be renamed *Cytoecetes phagocytophila* because of its close resemblance to *Cytoecetes microti*, a neutrophilic rickettsia of voles.[3]

When stained with Giemsa *C. phagocytophila* organisms appear greyish-blue. They do not react with Macchiavello's stain but fluoresce red with acridine-orange.

All European isolates share common antigens that are detectable either by immuno-fluorescent techniques or by complement-fixation tests. *Cytoecetes ovis* observed in the blood of sheep on the Deccan Plateau of India and *Cytoecetes ondiri* found in

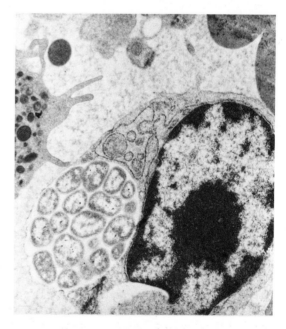

Fig. 43.3 A cluster of *Cytoecetes phagocytophila* rickettsias in a cytoplasmic vacuole of a neutrophil (x 10 000).

bushbuck and cattle in the highlands of East Africa appear to be strains of *C. phagocytophila*.

The rickettsia has been propagated in guinea pigs and mice with limited success. Similarly transient multiplication has been noted in leucocyte cultures.

Clinical signs

The incubation period following exposure to infected ticks ranges from 5–13 days. Injection of infected blood shortens the incubation period to 2–6 days. Most primary infections in the field, however, are not observed because the clinical signs in all but introduced pregnant animals are unobtrusive.

The illness begins with a sudden high fever which fluctuates between 40.5°C (105°F) and 42.0°C (108°F) for 4–12 days. The systemic disturbance, however, is minimal; a close examination reveals a rapid faint pulse, rapid shallow respirations, partial inappetance, low water intake and slight weight loss. There is one published report of transient haemorrhagic colitis. The prolonged fevers are claimed to impair spermatogenesis in rams for at least two months but later work refutes the claim.[4] Most pregnant animals if not previously infected usually abort 2–8 days after the onset of fever while still parasitaemic but in a few the fetus dies and mummifies and is expelled weeks later. Pregnant animals infected before pregnancy do not abort when reinfected.

Death is rare except in aborting ewes. Foggie[2] monitored infections in 447 sheep and concluded that there were only 2 deaths which might be attributed directly to tick-borne fever but he went on to note that a further 13 per cent of the animals died with pneumonia within two months of infection. These late deaths are a manifestation of the ability of tick-borne fever to exacerbate other latent infections with harmful results.

Haematological changes

In sharp contrast to the vague clinical signs the haematological changes are dramatic and diagnostic. At the onset of the fever, the total leucocyte count is unaffected but the relative proportion of the main cell types is changed in that the neutrophils are increased and the lymphocytes decreased. Within 24 hours both leucopenia and thrombocytpoenia are evident. The leucopenia is derived first from a transient lymphocytpenia and an eosinopenia that is often total and persists into convalescence and, then 2–3 days later, from a profound neutropenia. Monocyte numbers increase slightly during the reaction. Lymphocytes return to normal levels shortly after the neutrophil numbers reach their nadir. Neutrophils rise to normal levels 2–3 days after the fever has regressed; there is no reactive neutrophilia.

The lymphocytpenia is associated with a significant decrease in peripheral B-lymphocytes and with only a small reduction in T-lymphocytes.[5]

The number of thrombocytes falls rapidly after the onset of fever reaching its nadir about 3 days later. Thereafter the number increases slowly reaching normal levels at the end of the parasitaemia. The thrombocytopenia is not usually severe enough to produce prolonged bleeding times.

Parasitaemias are patent at the onset of fever and persist overtly throughout the fever. They are manifested by the presence of *C. phagocytophila* organisms in the cytoplasm of granulocytes and, more rarely, monocytes. The organisms are markedly pleomorphic occurring singly or in clusters.

Relapses

Most affected animals become carriers and several relapses of up to 2 day's duration may occur. The relapses are spontaneous and are characterized by sudden leucopenia and overt parasitaemia with or without transient fever.

Sequelae

While unravelling tick-borne fever from louping-ill, MacLeod and Gordon[1] realized that the importance of tick-borne fever lay in its facility to influence adversely the course of other infections including louping-ill. Sheep afflicted with both conditions concurrently sickened visibly and most developed encephalitis. The common and economically troublesome sequel of tick-borne fever is tick pyaemia, a crippling staphylococcal infection of young lambs manifested by the appearance of multiple abcesses in joints and the spinal column (see Chapter 23). Several other pathogens are also potentiated by tick-borne fever. Pasteurellas, for example, are incited to multiply in the nasal passages and are often associated with pneumonic episodes that follow tick-borne fever. Listeriosis is exacerbated. Concurrent *C. phagocytophila* and parainfluenza-3 infections produced severe respiratory illness in lambs killing 20 per cent and prolonging the virus excretion times.[6]

The common sequel in aborting ewes is a fatal post-parturient sepsis. A flare-up of fibrosed epididymitis has been observed to cause seminal degeneration in a ram.[4]

The mechanism underlying the lowered resistance of sheep infected with *C. phagocytophila* is associated with several factors: the neutropenia, the functional impairment of infected neutrophils, B-lymphocyte depression and suppression of the antibody response to other antigens.

Pathology

The pathology, particularly the histopathology, of tick-borne fever in sheep has received scant attention. The prominent gross feature in dead lambs and in sheep killed while febrile, is an enlarged spleen. Sometimes, however, there are no visible macroscopic lesions. Additional less commonly recorded findings are petechiae in the intestinal mucosa, frank haemorrhage into the colon, serosal and subendocardial haemorrhages and slight hydropericarditis.

Microscopic examination of impression smears of cut surfaces of the spleen, liver and lung reveals the presence of *C. phagocytophila* rickettsias in the cytoplasm of neutrophils, Kupffer cells and perhaps alveolar macrophages. The lymph nodes of infected cattle contain noticeably fewer lymphocytes and a large number of large round cells.

Diagnosis

Fever and abortion are the two syndromes that predominate. A provisional diagnosis based on the clinical signs is possible given a knowledge of the flock, the age of the affected animals, the area and the season. Of particular importance is a history of the recent movement of fresh animals on to tick-infested grazings. The clinical signs, nevertheless, are so vague that the provisional diagnosis should be confirmed by finding the rickettsias in thin Giemsa- or Leishman-stained smears of peripheral blood or impression smears of a cut surface of the spleen. The blood smears should be made from animals that are febrile or from ewes as soon as possible after abortion. Because the organism does not invade the fetus, smears of fetal tissues are unhelpful.

Patent parasitaemias can be precipitated in carrier sheep by using immunosuppressive doses of corticosteroids or by splenectomy. Carriers are detected more usually, however, by subinoculation of blood collected in EDTA or heparin into susceptible sheep producing a typical reaction within 14 days. Carriers can also be detected by examining their sera for complement-fixing antibodies.[7]

Epidemiology

Locus Infection with *C. phagocytophila* appears to be limited to the temperate areas of Europe, Africa and Asia. It has been diagnosed in Austria, Finland, Great Britain, Ireland, The Netherlands, Norway and South Africa. *C. ovis* occurs in India and *C. ondiri* in East Africa.

Host range Specific parasitaemias have been observed in sheep, goats, cattle and deer. The two groups of sheep at greatest risk are lambs born on tick-infested grazings and bought-in pregnant ewes. Most of the former develop fevers in the first two weeks of life and most of the latter abort.

Innate resistance Young lambs react less severely than yearlings although the durations of the parasitaemias are similar. Sex has no influence on the reactions but there are significant breed differences not entirely attributable to long, past ancestral association with the parasite; for example, reactions in Blackfaces are shorter and less severe than in Suffolk and Swaledale sheep.

Acquired resistance Carriers are alleged to resist reinfection. The duration of the carrier state, however, is very variable and may be influenced by the strain of *C. phagocytophila* involved; it is measured in months in most sheep and in years in a few. Infection induces the formation of IgM-associated antibodies which persist in carriers. IgG-associated antibodies appear 6 weeks after infection. The antibodies, however, are not protective and there is no passive transfer of resistance from the carrier ewe to its lamb in the colostrum. Reinfections occur more readily with heterologous strains of *C. phagocytophila* than with homologous strains with a waning resistance. The reinfections are usually manifested by a fever, lymphocytopenia, neutropenia and patent parasitaemia but the severity of the reactions and their course are never as severe as a primary reaction. Reinfected pregnant ewes do not abort.

Transmission *C. phagocytophila* is maintained and transmitted by ticks in which it multiplies. The disease, therefore, has a seasonal prevalence and sheep are infected when an infected tick feeds. Transstadial but not trans-ovarian transmission occurs in ticks. The common vector in Europe is *Ixodes ricinus*, in Asia *Rhipicephalus haemaphysaloides* and in South Africa *Rhipicephalus bursa* together with *Hyalomma* spp. It must be emphasized that multiplication in the tick is not essential and accidental transmission can occur through any agency capable of transferring a minute drop of blood from an infected to a susceptible sheep.

Control, prevention and treatment

Most cases are left untreated. When necessary the treatment and drug of choice is a single dose of oxytetracycline administered at the onset of the fever. This abates the fever and clears the parasitaemia but it does not prevent a mild relapse occurring one week later thus ensuring resistance to reinfection. Sulphadimidine is also active but is marginally less efficient.

The most widespread control measure is the practice of purchasing the sheep on the grazing when a farm is bought thus avoiding the introduction of fully susceptible stock to tick infestation. Pregnant animals, in particular, should never be moved from tick-free to tick-prevalent areas. Improving hill pasture lowers the number of ticks, decreases the number of infections and lays the flock open to serious consequences if moved inadvertently on to unimproved tick-infested ground.

Vaccines are not used but experimental attempts to limit the ravages of tick pyaemia by injecting infected blood into newborn lambs show promise (W.B.V. Sinclair, personal communication).

TICK-BORNE PROTOZOA

Protozoal piroplasms are classified into two families, the *Babesiidae* and the *Theileriidae*. Sheep harbour members of both families.

The babesias are piriform parasites of red blood cells. Infections may result in haemolytic anaemia, jaundice and haemoglobinuria but overt disease other than ill-thrift is rare in sheep in the UK where two species have been tentatively identified recently, *B. capreoli* in Argyllshire and *B. motasi* in Wales. The former is a common parasite of red deer being transmitted by the tick, *Ixodes ricinus*.

The theilerias are rod-shaped parasites of red blood cells. Some are very pathogenic while other appear to be innocuous. Fortunately, the only member of the family present in the UK, *Theileria ovis*, is non-pathogenic. It is transmitted by *Haemophysalis punctata* ticks.

REFERENCES

1 MacLeod J. & Gordon W.S. (1932) Studies on louping-ill (an encephalomyelitis of sheep). II. Transmission by the sheep tick, *Ixodes ricinus* L. *Journal of Comparative Pathology and Therapeutics*, **45**, 240–56.
2 Foggie A. (1951) Studies on the infectious agent of tick-borne fever in sheep. *Journal of Pathology and Bacteriology*, **63**, 1–15.
3 Foggie A. (1962) Studies on tick pyaemia and tick-borne fever. Symposia of the Zoological Society of London, **6**, 51–8.
4 Deas D.W. (1976) Project II. Tick-borne fever. Annual Report of the Edinburgh School of Agriculture 1975, p.42.
5 Batungbacal M.R., Scott G.R. & Burrells C. (1982) The lymphocytopenia in tick-borne fever. *Journal of Comparative Pathology*, **92**, 403–7.
6 Batungbacal M.R. & Scott G.R. (1982) Tick-borne fever and concurrent parainfluenza-3 virus infection in sheep. *Journal of Comparative Pathology*, **92**, 415–28.
7 Woldehiwet Z. & Scott G.R. (1982) Immunological studies on tick-borne fever in sheep. *Journal of Comparative Pathology*, **92**, 457–67.

OVINE KERATOCONJUNCTIVITIS 44

G.E. JONES

Synonyms Infectious or contagious ovine ophthalmia, conjunctivitis, keratitis or conjunctivo-keratitis, rickettsial or follicular conjunctivitis, pink-eye, snow or heather blindness.

The term ovine keratoconjunctivitis (OKC) and its synonyms are used to describe a clinical condition which is contagious but apparently not always ascribable to the same agent. OKC may involve, separately or together, two or more micro-organisms whose distribution, prevalence and importance remain largely unknown.

Entropion also causes conjunctivitis and even corneal ulceration in lambs, but is a congenital condition which involves physical irritation of the eye due to inturning of the eyelids, and responds to manual or surgical treatment.

OKC occurs worldwide, and affects all ages of sheep. Morbidity is generally higher in lambs and hoggs, but severity is generally greater in ewes.[1] The disease may reduce twinning in ewes and growth rate in lambs, although the latter effect is nullified following recovery.[2] Any economic effects of OKC largely result, therefore, from the loss through drowning or other misadventure of animals which are bilaterally blinded.

Causal agents

Pathogenicity has been demonstrated for only two of the eight micro-organisms which have been isolated from the eyes of sheep, namely *Chlamydia psittaci ovis* and *Mycoplasma conjunctivae*.

C. psittaci ovis Chlamydia have been isolated from cases of OKC in the USA, UK, Australia and New Zealand. In the American studies 42 per cent of eye swabs and 43 per cent of sera taken from sample animals in six flocks were positive for chlamydia and chlamydia antibodies respectively.[3] Conjunctivitis has been experimentally reproduced by the intra-ocular inoculation of chlamydia isolates,[4] but the isolation and culture techniques employed to obtain the inoculated strains could not exclude the possible concomitant presence of mycoplasmas.

M. conjunctivae This organism has been isolated from cases of OKC in North America, Europe and Australia. In one British study[5] the organism was isolated from 68 per cent of lambs with OKC and from 57 per cent of apparently normal lambs. This reflects two major aspects of OKC involving *M. conjunctivae*, namely ocular infection with the organism continues long after symptoms have disappeared, and *M. conjunctivae* is generally only mildly pathogenic for young lambs. The pathogenicity of *M. conjunctivae* for the eyes of both sheep and goats has been experimentally demonstrated.[5, 6]

Other mycoplasmas Identified isolates from the eyes of sheep with OKC comprise *Mycoplasma arginini*, *Mycoplasma ovipneumoniae* and *Acholeplasma oculi*. The importance of these in the OKC complex is not known, although *A. oculi* has been experimentally shown to reproduce conjunctivitis in goats.

Bacteria *Branhamella ovis* (previously *Neisseria ovis* or *catarrhalis*) is frequently isolated from both normal and diseased eyes, but is probably not pathogenic, even when the cornea is damaged. *Moraxella bovis* and *Listeria monocytogenes* type 04 have been isolated from sheep eyes on rare occasions; their pathogenicity in this site is unknown. Despite this lack of a suitable bacterial candidate, it seems likely that bacteria are implicated in the more severe forms of OKC, particularly corneal ulceration.

Finally there remains the possibility of a rickettsial aetiology, first suggested over 50 years ago on the basis of bodies observed in stained smears of conjunctival scrapings. However, these rickettsiae, named originally *Rickettsia conjunctivae* and later *Colesiota conjunctivae*, have never been cultivated and their appearance in stained conjunctival cells shows a close resemblance to *M. conjunctivae*. It has therefore been suggested that these putative rickettsiae are in fact *M. conjunctivae*.[5]

Clinical signs

Descriptions of the clinical course and findings in OKC attributed to either chlamydia, mycoplasma or 'rickettsia' are largely similar. The disease comprises 3 major stages, namely:

1. A conjunctivitis in which the palpebral conjunctiva becomes hyperaemic, turning brownish in colour, the scleral vessels become congested and there is excessive lachrymation with epiphora and blepharospasm. Follicle development may occur (Plate 7), especially on the third eyelid.
2. The cornea becomes involved, with vascularization and the development of opacity (pannus) in the periphery, particularly of the dorsal aspect. This may spread inwards to involve all corneal tissue (Plate 8).
3. Corneal ulceration.

All stages are generally accompanied by the presence of a muco-purulent discharge in the conjunctival sac and on the cornea, and by the caking of periorbital hair with dried ocular discharges. The stages are progressive, but healing may commence at any time. Stage 1 may occur in over 90 per cent of a flock, and is seen particularly in lambs. The more severe stages generally occur more commonly in ewes and rarely affect more than 5 per cent of the flock.

Descriptions of the chlamydia-associated outbreaks of OKC in the USA emphasize four clinical features, namely the development of lymphoid follicles in the conjunctival sac and on the third eyelid, the occurrence of chemosis (swelling of the conjunctiva), the generally bilateral nature of the disease, and the accompanying occurrence of polyarthritis in 10–25 per cent and sometimes as much as 85 per cent of lambs in feedlots.[3] Of these features, the first three are not pathognomonic for chlamydia-associated OKC, and have been observed in OKC experimentally reproduced with *M. conjunctivae*. The concomitant occurrence of polyarthritis in a proportion of lambs is not a constant feature, and was not observed in outbreaks of chlamydia-associated OKC in the UK and New Zealand.[1] However, polyarthritis and conjunctivitis strains of chlamydia have been shown to be serologically related.

Pathology

In the acute stages of infection with *M. conjunctivae* the cornea is infiltrated by polymorphonuclear and mononuclear cells, and tears contain large numbers of polymorphonuclear cells, some plasma cells and small numbers of lymphocytes. Tears or corneal/conjunctival scrapings stained by May-Grünwald-Giemsa show the presence of large numbers of uni-, bi- or tri-polar bodies of approximately 250–1100 μm in diameter closely associated with epithelial cells (Fig. 44.1). These bodies are specifically stained in

Fig. 44.1 Corneal epithelial cells taken 19 days after experimental infection of a lamb with *Mycoplasma conjunctivae*, and stained by May-Grünwald-Giemsa. (x 1280)

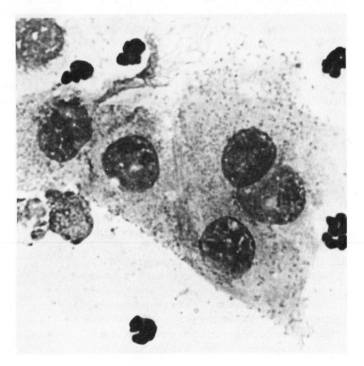

the fluorescent antibody test using antiserum to *M. conjunctivae* (Fig. 44.2)

Histopathological changes in chlamydia-associated OKC have not been described, but tears contain a cell exudate similar to that seen in mycoplasmal OKC. In the American work [3] intracellular clusters of chlamydia bodies, which stain red by a modified Ziehl-Neelsen stain, were observed in about 35 per cent of smears from individual eyes.

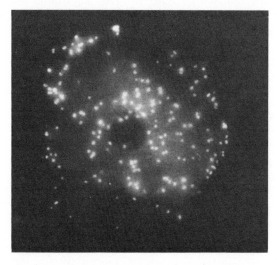

Fig. 44.2 Corneal epithelial cell from an animal infected with *Mycoplasma conjunctivae*, and stained by the indirect fluorescent antibody test using specific antiserum to *M. conjunctivae*. (x 1280)

Diagnosis

Diagnosis may be made on:
1. Isolation of the agent(s) involved. For their isolation mycoplasmas generally require complex, high quality artifical media; chlamydia, require embryonated eggs or specially treated tissue cultures. Moraxellae may be isolated by conventional bacteriological techniques, but *L. monocytogenes* type 04 requires cold enrichment at 4°C in tryptose broth.
2. Demonstration of organisms in tears or conjunctival scrapings by staining or, preferably, by the fluorescent antibody test using specific antiserum. Interpretation of stained preparations may be confused by the presence in epithelial cells of other bodies including melanin granules.
3. Examination of sera for chlamydial antibodies.

Epidemiology and transmission

Development of immunity to *M. conjunctivae* appears to be poor, and the organism may persist in the eyes of sheep for months, long after remission of symptoms. Thus apparently normal animals can act as reservoirs of infection, and outbreaks of OKC frequently occur after the introduction of new animals to a flock. Furthermore, sheep may be repeatedly infected with *M. conjunctivae* if the organism is eliminated by chemotherapeutic agents before each subsequent inoculation, and express signs of OKC with each fresh infection. Transmission does not require vectors and presumably occurs by contact; trough-fed animals are therefore particularly at risk.

The epidemiology and transmission of chlamydia-associated OKC is unknown.

Control, prevention and treatment

Affected eyes should be treated topically with a broad spectrum antibiotic such as oxytetracycline or chlortetracycline. Chloramphenicol, dihydrostreptomycin and penicillin are not recommended.[7] Alternative chemotherapeutic agents are the phenanthridines, in particular ethidium bromide.[7] Relapses following treatment are common. The effectiveness of subconjunctival injection of broad spectrum antibiotics, the treatment of choice in bovine infectious keratoconjunctivitis, does not appear to have been evaluated in sheep.

Since transmission of mycoplasma-associated OKC is probably by contact, increasing trough space per animal and decreasing stocking density should reduce spread of the disease. The only preventive measure available is ocular treatment of new animals with antibiotics or phenanthridines before they are admitted to the flock, irrespective of the clinical condition of their eyes.

REFERENCES

1 Cooper B.S. (1967) Contagious conjunctivo-keratitis (C.C.K.) of sheep in New Zealand. *New Zealand Veterinary Journal*, **15**, 79–84.
2 Axelsen A. (1961) Effect of contagious ophthalmia on multiple lambing and sheep liveweight. *Australian Veterinary Journal*, **37**, 60–2.
3 Hopkins J.B., Stephenson E.H., Storz J. & Pierson R.E. (1973) Conjunctivitis associated with chlamydial polyar-

thritis in lambs. *Journal of the American Veterinary Medical Association,* **163**, 1157–60.

4 Cooper B.S. (1974) Transmission of a *Chlamydia*-like agent isolated from contagious conjunctivo-keratitis of sheep. *New Zealand Veterinary Journal,* **22**, 181–4.

5 Jones G.E., Foggie A., Sutherland A. & Harker D.B. (1976) Mycoplasmas and ovine keratoconjunctivitis. *Veterinary Record,* **99**, 137–41.

6 Trotter S.L., Franklin R.M., Baas E.J. & Barile M.F. (1977) Epidemic caprine keratoconjunctivitis: experimentally induced disease with a pure culture of *Mycoplasma conjunctivae. Infection and Immunity,* **18**, 816–22.

7 Cooper B.S. (1961) Treatment of conjunctivo-keratitis of sheep and cattle with ethidium bromide. II. Contagious conjunctivo-keratitis of sheep (CCK). *Veterinary Record,* **73**, 409–15.

45   TUMOURS OF SHEEP

K.W. ANGUS AND K.W. HEAD

The evidence from published information and the short economic life of the sheep would cause one to believe that tumours in sheep were of little significance. Analysis of numerous surveys and collections of pathological material indicates that neoplasms are uncommon in sheep, compared with some other species.[1] No useful purpose would be served by listing all the various tumour types and attempting to categorize them into some order of importance. A proportion of tumours are symptomless, being early cases found during routine meat inspection; others cause vague ill-health, resulting in culling of animals which are 'going back in condition'. It is useful, however, to be aware of the clinical features of some neoplasms, because a high prevalence in certain circumstances can be related to management factors and hence may be avoidable.

Accurate figures on the comparative prevalence of the various tumours encountered in sheep are difficult to obtain, due to differences in geographical location, breed, or the conditions under which different surveys are undertaken. One survey in the USA mentions that 'lambs represent 95 per cent of sheep admitted for slaughter since the carcass value of most range ewes is not worth the cost of transportation and slaughter'.[2] In contrast, in Britain and New Zealand the age distribution of sheep examined at meat inspection in abattoirs would normally be a mixture of fat lambs (under 1 year) and old cast ewes (5–7 years old), since both categories are economic to slaughter. Ewes in their productive years, i.e. before udder disease or broken mouth leads to culling, are under-represented in such statistics. In this country, there is no requirement to record the exact age of sheep at an abattoir, thus Anderson, Sandison and Jarrett[3] in their extensive abattoir survey, could not report the age distribution of specific tumour types. The other source for tumour statistics is the post-mortem examination reports from Veterinary Investigation Centres or Veterinary Schools. In general, only casualties from major disease outbreaks are submitted to these centres for diagnosis, obviously, another biased sample. Despite these constraints, the available information makes it clear that certain ovine tumours occur much more frequently than others, and it is convenient to think of tumours in lambs and tumours in cast ewes.

TUMOURS IN YOUNG SHEEP

By far the most common and important tumours of young sheep are lymphosarcoma and related tumours of the haemopoetic system. Sheep with lymphosarcoma are unthrifty and in the generalized form there is bilateral enlargement of the superficial lymph nodes, which are readily palpable. By the time these features are evident, the neoplastic process is usually widespread in other lymphoid and non-lymphoid organs, particularly the spleen, liver, kidney, heart, and alimentary tract, all of which are extensively infiltrated by neoplastic cells.[4] Localized forms of the disease are also seen in which the neoplastic tissue may be nodular or diffuse, involving the intestine, liver or kidneys. Splenic, thymic and skin localized forms may occur, but are relatively uncommon.

In Germany, it has been shown that lymphosarcoma can be transmitted to sheep using whole blood from cattle affected with enzootic bovine leucosis (EBL). Paulsen et al[5] noted that in multiple incidence flocks there are many cases of lymphocytosis, and sheep with lymphosarcoma often have leukaemia (28.66–279.5 x 10[9] lymphocytic cells per litre). Because of variation with local breeds and management, it is recommended that a leukaemia be recorded if the lymphocyte figure is above the local normal mean plus three times the standard deviation. Johnstone et al[6] have shown virus-like particles in New Zealand sheep with persistent lymphocytosis following inoculation with material from an ovine malignant lymphoma. Both multiple incidence flocks and sporadic isolated lymphosarcoma cases occur in the UK, but their relationship to EBL has not been clarified.

Adrenal tumours may be found in young sheep at slaughter. These are mainly adenomas or adenocarcinomas of the cortex, medullary tumours are uncommon. Cortical tumours may cause hormonal imbalances, and clinical effects such as androgenism may occur[7], but these have little significance in fattening lambs.

Multiple cystic malformations of the bile ducts in young sheep are non-malignant and have no clinical significance, but their presence at slaughter leads to condemnation of the liver.

Skin tumours occur from time to time in sheep of all ages. Melanomas are found almost exclusively in the Suffolk breed, in areas of pigmentation. Their importance relates to their high degree of malignancy and tendency for widespread dissemination. Neurofibromas are characterized by disorderly proliferation of Schwann cells, and in sheep usually take the form of small, single or multiple nodular skin tumours, which have little clinical importance.

TUMOURS OF ADULT SHEEP

Apart from lymphosarcoma, which can occur in breeding stock, older sheep experience a different spectrum of neoplasia from that found in lambs, one in which prevalence is closely related to age. It is a fact that the longer breeding stock are retained before culling, the greater is the chance of sporadic occurrence of tumours, presumably because of the greater chance of the multiple factors 'causing' tumours occurring simultaneously in a single sheep. This is exemplified by the fibrosarcoma of the jaw found along with other primary tumours (oral and ruminal squamous carcinoma, and intestinal adenocarcinoma) in some flocks in North Yorkshire.[1] There is strong circumstantial evidence that prolonged grazing of bracken may have had a carcinogenic effect in this locality, and experimental feeding of bracken induced bladder tumours and a single case of the jaw tumour in 7 out of 8 sheep in a trial.[8] However, the relatively high incidence of these tumours in some flocks was closely correlated with later culling, up to 8 years old. It should be noted that spontaneous bladder tumours have not been reported in UK sheep even in these 'bracken poisoning' areas.

The most common tumours of adult sheep in the UK are intestinal adenocarcinoma and liver tumours. Pulmonary adenomatosis (jaagsiekte), a transmissible lung tumour with a high incidence in some flocks, is dealt with in a separate chapter.

Whereas an association between bracken, upper alimentary tract viral papillomatosis, and carcinoma of the alimentary tract of cattle is now established, the situation in sheep is less clear. Adenocarcinoma of the intestine does occur in sheep grazing on bracken hills[1] but in other countries where there is a higher incidence of intestinal adenocarcinoma, e.g. New Zealand, Australia or Iceland, bracken can be excluded as a causative factor.

In the late stages of the disease, sheep with intestinal adenocarcinoma are cachectic, anaemic, and there may be distension of the abdomen due to ascites. The usual necropsy finding is one or rarely several, localized annular thickenings (1–3 cm wide) of the intestinal wall, with constriction of the gut at the tumour site and marked dilatation cranial to the constriction. The small bowel is the common site of this tumour, the colon being affected only very rarely. The primary lesion is usually quite small, but secondary serosal involvement is often very extensive, involving up to 100 cm of the gut wall proximal to the tumour. These scirrhous tumour plaques on the serosal surface of the bowel, and similar plaques on the surface of the diaphragm, anteroventral flanks and other peritoneal surfaces, may be mistaken for chronic peritonitis. Metastatic spread via the afferent lymphatics to the regional mesenteric and posterior mediastinal lymph nodes often occurs, but involvement of the liver and lung is rare.

Liver tumours are usually hepatocellular, i.e., derived from liver cells, but may be cholangiocellular, where the tumour arises from the bile-duct lining cells. Theoretically one would expect the liver to be a common site for the growth of secondary tumours, from primaries in the lung or intestine, but in practice liver tumours of sheep tend to be primary.[9]

Hepatocellular tumours are nearly always benign and may not give rise to any clinical signs. They are often single, and may be large, up to 15 cm in size. The tumour is often sharply demarcated from the rest of the liver, and tends to 'shell out' on handling. When incised, the cut surface is yellowish due to central necrosis or greenish due to the presence of bile pigments. In rare cases of malignancy, small secondary tumours may be seen in the liver substance close to the primary, with spread to the portal nodes and occasionally to the lungs. For unknown reasons the dressed carcass in such cases often has a fevered appearance.

Bile duct cell tumours are usually malignant, and affected sheep are anorectic, cachectic and may have ascites. These tumours are multiple, firm, white and nodular, those involving the capsule having a characteristic umbilicated appearance. There is no relationship between either of these tumours and fascioliasis or cirrhosis, but they may be related to ingestion of mycotoxins such as aflatoxin.[10]

Most other tumours fall within the province of the meat inspector rather than the clinician. One fairly common tumour found in the abattoir is papillomatosis of the rumen, a non-malignant, often multiple neoplasm found in all age groups on the ruminal pillars as 2–20 mm sessile or pedunculate nodules. Sporadic tumours with clinical manifestations include chondrosarcoma, which may arise, causing deformities, in the sternum, ribs, scapulae or occasionally in vertebrae, where they may give rise to locomotor disturbances. Squamous-cell carcinoma is very rare in the UK, but sporadic cases involving the ear pinnae, jaw, muzzle or eye may occur. This tumour is of considerable economic importance in Australia. A vulval form, once common in parts of Australia as a result of the Mules' operation in docked Merino sheep, does not occur in the UK. Similarly, the squamous cell carcinoma of grass seed awn cysts described in a strain of Merinos does not occur in Britain. It is possible that ovine facial and aural carcinomas have a similar complex aetiology involving unpigmented skin, exposure to sunlight and infection with a papilloma virus.[11]

Although the tumour is seldom recorded in British sheep, mention should be made of nasal adenopapilloma or adenocarcinoma, an intranasal tumour of the ethmoturbinate olfactory mucosa. Lewis et al [10]

reported a case in sheep fed on aflatoxin in an attempt to induce liver tumours, and one of us (KWH) has seen a single case in a series of over 800 sheep tumours. Multiple cases of this tumour have been reported in Europe, North America, Nigeria and Japan, amongst other countries, with a tendency for clustering in related flocks to occur. In one study[12] viral particles morphologically similar to maedi-visna virus were seen in the tumours, and tumour cells cultured in vitro had greater RNA-dependent DNA polymerase activity when compared with control cultures. The free movement of sheep between countries may be a factor in the spread of this and perhaps other sheep tumours.

Finally, a word of caution: not all cellular masses are tumours. For example, the so-called 'loops and fingers' of fibrous tissue occasionally seen in the udder region are not fibrous tissue tumours but represent excessive granulation tissue around blood vessels after sloughing of the udder following acute necrotizing mastitis.

REFERENCES

1 McCrea C.T. & Head K.W. (1978) Sheep tumours in north east Yorkshire. I. Prevalence on seven moorland farms. *British Veterinary Journal*, **134**, 456–61.

2 Brandly P.J. & Migaki G. (1963) Types of tumours found by federal meat inspectors in an eight-year survey. *Annals of the New York Academy of Sciences*, **108**, 872–9.

3 Anderson L.J., Sandison A.T. & Jarrett W.F.H. (1969) A British abattoir survey of tumours in cattle, sheep and pigs. *Veterinary Record*, **84**, 547–51.

4 Johnstone A.C. & Manktelow B.W. (1978) The pathology of spontaneously occurring malignant lymphoma in sheep. *Veterinary Pathology*, **15**, 301–12.

5 Paulsen J., Best E., Frese K. & Rudolph R. (1971) Enzootic lymphatic leucosis in sheep-lymphocytosis, pathological anatomy and histology. *Zentralblatt für Veterinärmedizin*, **18B**, 33–43.

6 Johnstone A.C., Manktelow B.W., Jolly R.D & Belton D.J. (1979) Persistent lymphocytosis and virus-like particles in lymphocytes of sheep inoculated with cell-free extracts derived from ovine malignant lymphomas. *Journal of Pathology*, **128**, 183–91.

7 Appleby E.C. & Sohrabi I. (1978) Pathology of the adrenal glands and paraganglia. *Veterinary Record*, **102**, 76–8.

8 McCrea C.T. & Head K.W. (1981) Sheep tumours in East Yorkshire. II. Experimental production of tumours. *British Veterinary Journal*, **137**, 21–30.

9 Jubb K.V.F. & Kennedy P.C. (1963) *Pathology of Domestic Animals.* 1e. Vol.2. pp.213–14. Academic Press, London & New York.

10 Lewis G., Markson L.M. & Allcroft K. (1967) The effect of feeding toxic groundnut meat to sheep over a period of five years. *Veterinary Record,* **80**, 312–14.

11 Vanselow B.A., Spradbrow P.B. & Jackson A.R.B. (1982) Papillomaviruses, papillomas and squamous cell carcinomas in sheep. *Veterinary Record,* **110**, 561–2.

12 Yonemichi H., Ohgi T., Fujimoto Y., Okada K., Onuma M. & Mikami T. (1978) Intranasal tumour of the ethmoid olfactory mucosa in sheep. *American Journal of Veterinary Research,* **39**, 1599–606.

FURTHER READING

International Histological Classification of Tumours of Domestic Animals. Bulletin of the World Health Organization Vol.50. (No.1–2, 1974) and Vol.53 (No.2–3, 1976). Geneva.

Moulton J.E. (1978) *Tumours in Domestic Animals,* University of California Press, Berkeley, London.

INTERNATIONAL IMPORTANCE OF SOME EXOTIC VIRUS INFECTIONS IN SHEEP

46

R.F. SELLERS

In every country there are endemic diseases which persist from one year to the next. However, if diseases exotic to the country are introduced, heavy losses may occur among the sheep population. In addition, export of sheep to countries where such diseases are endemic may lead to illness and death of the sheep if they have not been suitably protected.

The virus diseases discussed in this chapter include foot-and-mouth disease, bluetongue, Akabane disease, Rift Valley fever, rabies, peste des petits ruminants, rinderpest and sheep pox as well as two of lesser importance, Nairobi sheep disease and Wesselsbron disease. The viruses responsible also cause disease in other ruminants beside sheep. With rabies other animals and man can be affected. Sheep are not or may not be the main host in foot-and-mouth disease, Akabane, rabies and rinderpest; with bluetongue, sheep are the indicator host, the main reservoir being cattle. Nevertheless, with all these virus diseases damage could be extensive among susceptible sheep flocks.

The viruses responsible are spread by movement of animals or their products, by arthropods such as biting flies, mosquitoes, midges or ticks, by the bite of animals, by the airborne route and through contact with people, animals, vehicles and materials contaminated with virus. The diseases are closely connected with the pattern of husbandry and trade in and between countries, and it is by national and international control that spread of disease can best be avoided.

FOOT-AND-MOUTH DISEASE

Foot-and-mouth disease is a highly contagious vesicular disease of cattle, sheep, goats, pigs and wild ruminants.[1] It is caused by a virus belonging to the *Aphthovirus* genus of the family Picornaviridae. Seven immunological types: O, A, C, SAT 1, SAT 2, SAT 3 and Asia 1, as well as over 60 subtypes have been recorded.

Clinical signs

The first sign observed in a flock is lameness in some sheep, with affected animals resting or lying away from the others. On closer examination, salivation may be seen. High temperature is found in affected animals. Vesicles may be present in the mouth, on the hard palate, on the dental pad, where the incisors touch, on the gums, lips and tongue. Occasionally they may be found in the nostrils. In older cases the vesicles may be ruptured and necrotic tissue with an offensive smell may be found.

Affected feet are warm to the touch and vesicles may be found in the interdigital space, at the bulbs of the heel and along the coronary band. Rupture of the vesicles may lead to secondary infection and foot rot is often present at the same time. Occasionally vesicles may be seen on the teats, prepuce and vulva.

The severity of the disease varies considerably with the breed of sheep, the strain of foot-and-mouth disease virus responsible and the environmental conditions. Sometimes there may be little to see, especially in sheep with dark pigment. On mixed farms the disease may be seen first in the cattle or in the pigs.

Some strains of virus cause sudden death in lambs due to the virus affecting the heart muscle; this may occur before vesicular lesions are apparent.

Distribution

Foot-and-mouth disease occurs world-wide. At present North and Central America, Australia, New Zealand, Japan and Scandinavia are free of the disease. Types O, A and C occur in Europe and South America, O, A, C and Asia 1 in the Middle East, the Indian subcontinent and the Far East; SAT 1 has also occurred in the Middle East. In Africa, O, A, C, SAT 1, SAT 2 and SAT 3 have been found.

Transmission and epidemiology

Spread occurs in a number of ways: by direct contact between livestock, by contact with or feeding on infected meat products and milk, by the airborne route, by mechanical carriage on people, domestic or wild animals, birds, vehicles and fomites.[2]

Sheep are more likely to be infected by direct contact, by the airborne route or through contact with contaminated people or carriage in contaminated vehicles.

Sheep are infective for other sheep by all routes both before lesions are apparent and in the early stage of lesions. Later, the amount of virus given out diminishes with recovery and production of anti-

body. However, virus can still be found in the tonsillar area of the throat for 9 weeks in 50 per cent of sheep and in some animals for as long as 9 months. Such virus is unlikely to spread unless a variant virus develops.

The importance of sheep in the epidemiology of disease varies with the types of husbandry and relative numbers of livestock. For example, in the Middle East, where cattle are kept enclosed and where sheep are moved from place to place, virus is maintained in the sheep. In the UK, on the other hand, disease was usually first seen in an area in pigs and cattle and subsequently spread among the sheep population. Such disease was usually introduced by infective bones or offal or by airborne spread from northern France and the Low Countries. However, it must be pointed out that, because of the nature of lesions in sheep, disease has not always been recognized immediately and several cycles of infection, with spread, may have occurred.

Diagnosis

Diagnosis is based on clinical signs, together with laboratory examination to establish the type involved. Because of its highly infectious nature and the danger of spread in many countries, it is a notifiable disease and diagnosis after the first reports is carried out by the official veterinary services.

Epithelium is collected from vesicles in the mouth and on the feet.[3, 4] As far as possible, samples should come from animals showing early lesions and as much material as possible should be collected. If cattle are present on the farm, they should be examined and epithelium sent from any affected. In certain instances blood for serum examination and oesophageal-pharyngeal samples for virus isolation are collected.

The epithelium is tested for the presence of antigen to the seven types by complement fixation tests. If there is insufficient antigen present, the virus is grown in tissue cultures and further tests carried out. Results from the first test take about three hours after receipt, but it may be 48 hours or longer before virus has been grown in sufficient quantities in tissue culture.

It is important that, if there is any suspicion of foot-and-mouth disease, it should be reported as soon as possible to the appropriate authorities; otherwise, damaging spread could occur.

BLUETONGUE

Bluetongue is an infectious non-contagious disease of sheep, cattle, goats and wild ruminants such as deer.[5, 6] It is caused by a virus (*Orbivirus* of the family Reoviridae) and is transmitted by midges, species of the genus *Culicoides*. Bluetongue virus has been divided serologically into 20 types at least and there may be more. In addition, the bluetongue complex, as it is termed, contains epizootic haemorrhagic disease of deer virus, Ibaraki virus and three viruses occurring in Australia and central Africa.

Clinical signs

In a flock affected with bluetongue, the first signs may be a sheep looking sick and depressed, standing or lying down away from the other animals.

On closer examination the animal is found to have a fever, oedema of the muzzle and hyperaemia and oedema of the lips, buccal and nasal mucosae.[5] There is a discharge from the nose and from the eyes, at first serous but later mucopurulent. Frothing saliva drips from the lips and the muzzle may become encrusted with discharge from the nose and mouth. Oedema is sometimes found in the ears and over the brisket. In some sheep, especially English breeds, however, oedema may not be such a prominent feature but nasal discharge is apparent. Others may only show a swelling of the lips and reddening of the buccal mucosa.

The feet are often affected with coronitis, pain in the hooves and lameness. Reddening and haemorrhages may be found along the coronary band. Other complications include torticollis. Some animals may die at the time of acute signs. Others may develop bronchopneumonia due to *Pasteurella* infection and die in the later stages.

The morbidity and mortality rate varies with the strain of virus and the breed of animal. For example, in the 1943 outbreak in Cyprus 60–70 per cent of the sheep in the flocks were affected; in an outbreak in 1977 about 14 per cent of sheep in the flocks were affected. Breeds of sheep where the disease is endemic are often less affected than sheep in epidemic or sporadic areas. Exotic breeds of sheep introduced into endemic areas often show disease.

If sheep are infected during pregnancy, abortion, malformation or the birth of stunted lambs may occur.

Distribution

One must distinguish between distribution of the disease and distribution of the virus.[6]

Virus has been found to occur in America approximately from the USA–Canada border in the north to Paraguay in the south and in Africa from the Mediterranean coast to Cape Province. In Europe it was in the Iberian peninsula in 1956–60 (but not since) and in Greece (Lesbos) in 1979. It is found in Turkey, Cyprus, Israel and other countries of the Middle East, the Indian subcontinent, Japan, China, South-East Asia, Indonesia and parts of Australia.

Disease, however, has only been found in 1. exotic sheep introduced into endemic areas, i.e. Australian Merino sheep into India, 2. sheep in fringe areas such as higher altitudes in endemic areas, e.g. Kenya, or at higher latitudes on the fringe of the distribution of the virus, as in Cyprus and western Turkey or Spain and Portugal. In some areas such as South-East Asia and Australia the disease has not been seen in sheep, although the virus has been isolated and antibodies are present in sheep and cattle.

Transmission and epidemiology

The virus is transmitted by midges of the genus *Culicoides*.[5, 6] The main ones involved are *C. variipennis* in North America, *C. imicola* in Africa and the Middle East, and *C. brevitarsis* in Australia. Other species are involved in Africa and Australia, and *C. insignis* is suspected to be responsible for transmission in Central and South America. As far as Europe is concerned, a vector *C. obsoletus*, which was found to contain bluetongue virus in Cyprus, occurs in the UK as well, and there are species in this country which might transmit bluetongue if they should become infected.

In many areas of the world the main hosts for the virus are cattle. However, the virus does not always cause disease in cattle but multiplies to high titre and can persist in the animal for at least 100 days. In addition, it may be transmitted through the semen of the bull, or the dam may be infected during pregnancy and the virus pass to the fetus; calves may then be born with virus in their blood and thus may transmit infection.

The *Culicoides* bite cattle as well as sheep and virus is thus maintained in a cycle. However, it is the sheep that normally develop disease. Virus persists in sheep blood, but not so long as in cattle blood; virus is also transmitted across the placenta but, as the gestation period in sheep is shorter and the amounts of virus smaller, the chances of transmission this way are considerably less than in cattle.

Spread of disease from one area to another or from one country to another is by movement of the animal or carriage of midges on the wind or possibly in vehicles, trains, aircraft and ships.

The time of appearance of the disease or infection with bluetongue varies. In countries with Mediterranean climate it usually appears in late summer and autumn and dies away with the first cold weather. In tropical countries, infection is usually associated with the movements of winds at the beginning and end of the rainy season.

Diagnosis

Diagnosis is based on the clinical picture in sheep. The time of year it occurs and the location are also taken into consideration.

Confirmation is obtained by:
1. transmission of virus to sheep, with isolation of virus or at least confirmation of an antibody rise;
2. demonstration of a group-specific antibody response in the agar gel precipitation test;
3. isolation of virus in the laboratory in chick embryos or tissue culture, together with subsequent determination of virus type.[3, 4]

The samples to take are whole blood and serum from animals in the early stages of infection or from unaffected animals in the flock, and spleen and lymph nodes from dead animals.

Other diseases which must be taken into account in differential diagnosis include contagious pustular dermatitis (orf), foot-and-mouth disease, sheep pox and peste des petits ruminants. Other possibilities include photosensitization, *Oestrus ovis*, mycotic stomatitis and pneumonia. Where deformed lambs are seen, Akabane virus should be considered.

AKABANE DISEASE

Akabane disease is an epidemic virus disease giving rise to abortion, mummified fetuses, premature births, stillbirths and congenital arthrogryposis-hydranencephaly in cattle, sheep and goats.[7] It is caused by Akabane virus, a member of the Simbu group of the family Bunyaviridae.[8] The disease is transmitted by insects.

Clinical signs

A number of clinical signs may be presented, depending on the breeding season of the sheep and length of time the sheep has been pregnant at the time of infection.[9]

If infection occurs early in pregnancy, abortions may occur; later, mummified fetuses may be expelled. Infection later in pregnancy leads to premature births. At parturition the lambs may be born dead with various degrees of decomposition or malformation. Among lambs born alive, some may be dummy lambs unable to stand or suck, others may exhibit arthrogryposis. The dummy lambs die later and the prognosis for lambs with arthrogryposis is poor. No signs are seen among older animals in the flock.

Distribution

The disease in sheep has been seen in Israel and Australia.[9, 10] It has also been seen in cattle in these countries and in Japan and Turkey. Evidence of infection with Akabane virus has been found in Australia, South-East Asia, Japan, southern Africa, Kenya, the Near East, Israel, Cyprus and Turkey.

Transmission and epidemiology

Akabane virus is transmitted in a cycle between cattle, sheep and goats as hosts and *Culicoides* midges and mosquitoes as vectors.[7] It is not known whether other animals are involved in the cycle but antibodies to Akabane virus have been found in buffaloes, camels, pigs, horses and monkeys.

Three areas can be identified: a free area where the vector species involved and the virus are not found, an endemic area where virus, vectors and hosts are present all through the year, and an epidemic area containing hosts, into which vectors infected with Akabane virus are carried at irregular intervals. Disease is not seen in the free or endemic areas. In the endemic area, the young, after birth, are protected by maternal antibody. When maternal antibody wanes, the animals are bitten by infected insects and develop antibodies. Hence, when the females become pregnant, the antibodies already present protect against multiplication of Akabane virus and its transplacental transfer.

The disease is seen in the epidemic areas. Virus is carried into the area by infected insects blown by the wind or brought by infected animals. Since it may be several years since the previous introduction, there may be a substantial number of susceptible animals. In those animals that are pregnant and do not have antibodies, the virus after multiplying in the dam may cross the placenta. The point in the gestation period when this occurs determines the outcome. Infection after 50 days of pregnancy in the sheep has no effect on the lambs born. After infection between 30th and 50th days of pregnancy malformations may develop. Infection before and around 30 days of pregnancy results in abortions, mummified fetuses and apparent barrenness. The virus, however, does not persist in the fetus.[7,9]

Diagnosis

Diagnosis is based on the clinical picture of abortions, mummified fetuses, malformation, hydranencephaly and arthrogryposis. Confirmation is difficult to obtain since the virus does not persist in the lambs born. Aborted material may contain virus which could be isolated in tissue culture. In lambs with hydranencephaly and arthrogryposis it may be possible to demonstrate antibody, preferably in serum from a bleeding before colostrum ingestion. Ewes and other animals in the flock may be bled to establish the antibody picture. Serum may also be obtained from fetuses for testing.

At the same time, arthrogryposis and hydranencephaly may be noticed in cattle and goats in the area.

RIFT VALLEY FEVER

Rift Valley fever is an acute virus disease of ruminants.[11] It is caused by a virus which, it is suggested, belongs to the *Phlebovirus* genus of the family Bunyaviridae. The virus is transmitted by insects, especially various species of mosquitoes and possibly also midges and black flies.

Clinical signs

The disease can be peracute, acute or subacute or there may be infection without appearance of the disease. In the peracute form there may be high mortality among lambs, with up to 80 per cent of the lamb crop affected. The lamb becomes listless, develops a fever, collapses and dies. Bloody diarrhoea and urine may be seen and on post-mortem

examination there may be widespread haemorrhages in the organs and in the subcutaneous tissues.[11]

The acute form is mainly found in adults. The signs are fever, sickness, nasal discharge, haemorrhagic diarrhoea and an unsteady gait. Pregnant ewes often abort. In the subacute form there is a transient fever and in suckling ewes the milk often dries up.

Where cattle and goats are kept on the same farm or in the district, similar signs of peracute and acute disease may be seen in the calves and kids and in the adults.

Distribution, transmission and epidemiology

The disease has only been found in Africa, although evidence of infection has been found in Sinai.

There are epidemic and endemic forms of Rift Valley fever. The endemic form is found in West Africa, Uganda, Kenya, central Africa and southern Africa and involves the circulation of a less virulent virus, the presence of which is detected only by the antibody in ruminants and other animals.

The epidemic form is found from time to time in Kenya (1968, 1978/9), South Africa (1950/51, 1974/75), Zambia (1974) and Zimbabwe (1978). Epidemics have also occurred in Sudan (1973) and Egypt (1977 onwards), with sickness and death in man, cattle, sheep and goats.[12, 13]

The virus circulates in a cycle between hosts—sheep, goats, cattle, camels, wild ruminants, monkeys and vectors—mosquitoes of various species, black flies *(Simulium)* and midges *(Culicoides)*.[12, 13]

The factors involved in the development of epidemics are not understood. However, they take place in areas that have not been previously affected or have not been affected for some years. It is possible that Rift Valley fever virus is brought into the area by infected animals or by insects carried on the wind. The incursion coincides with the presence of a large susceptible ruminant population and an abundance of breeding sites for insects provided by unusual rains or through extensive irrigation.[13]

The virus is transmitted to man by insect bites or through the handling of infected animals or carcasses.

Diagnosis

Diagnosis is based on the clinical signs and on post-mortem examination. In addition, a number of laboratory tests can be carried out.[4]

1. Virus isolation may be attempted from blood, serum or spleen in tissue cultures or mice.
2. Complement fixation and gel precipitin tests on tissues.
3. Detection of inclusion bodies in the liver.
4. Demonstration of antibody rise by neutralization and haemagglutination inhibition tests.

It must be pointed out that this virus is dangerous to man and care must be taken in examining affected animals and in carrying out post-mortem examinations.

The sudden death of lambs has to be distinguished from enterotoxaemia and Wesselsbron disease. Other causes of abortion should also be borne in mind.

RABIES

Rabies is a virus infection of the central nervous system affecting mainly dogs, wolves, foxes, jackals, coyotes, the cat family, skunks, mongooses, raccoons and bats. Infection spills over from these animals to affect man and domestic animals, including sheep.

Clinical signs

Usually there is a history of sheep being bitten or attacked by one or possibly more dogs, wolves, foxes, stoats, coyotes, mongooses or bats, followed by a change in behaviour 2–4 weeks later. Some of the wounds or bites may still be evident.

The clinical signs are variable.[14] The sheep may be restless, become excited and paw the ground. Abnormal sexual behaviour may be seen, with sheep mounting each other. Some sheep may appear to be choking as though there was an obstruction in the throat. Butting of other animals, tearing out wool and aimless trotting to and fro may be seen. This restless stage may be absent and all that may be seen is muscular incoordination, dullness, depression and somnolence. There may be an excessive flow of saliva and nasal discharge. Death usually occurs. Lambs born to ewes after infection but before the appearance of clinical signs do not appear to be affected.

Distribution

Rabies is found in all continents except Australasia. Some countries are free, e.g. Great Britain, Eire and Scandinavian countries.

Transmission and epidemiology

Essentially, the virus is transmitted from one animal to another by biting, virus being present in the saliva. Occasionally infection by aerosols from bats may occur.

Two main cycles can be distinguished—urban and in the wild. In the urban cycle the virus is transmitted among dogs and cats in towns; occasionally man and other domestic animals are bitten. In the wild the cycle involves different species in various parts of the world; for example, in Europe the fox and wolf; North America the skunk, wolf and fox; South Africa and Asia the jackal and mongoose; South America the vampire bats; and West Indies the mongoose. In some situations there can be overlapping between the urban and wild cycles; for example, where foxes live on the outskirts of towns and cats become involved, or urban dogs attack sheep kept in parks or farms near cities.[14]

Diagnosis

Diagnosis is based on the clinical signs in the affected animal and is confirmed by laboratory examination of samples from dead or killed sheep.[3, 4, 15] Tests are carried out on the brain or other tissues to detect the presence of rabies. Such tests include the fluorescent antibody test, the mouse inoculation test, histological examination of the brain for Negri bodies. Identification of virus isolated after mouse inoculation is confirmed by a neutralization test.

PESTE DES PETITS RUMINANTS AND RINDERPEST

Peste des petits ruminants (PPR) is a contagious disease of sheep and goats and rinderpest one of cattle, sheep, goats and pigs. They are caused by viruses of the *Morbillivirus* genus of the family Paramyxoviridae.[8, 16, 17] Originally PPR was considered a variant of rinderpest virus but now the two viruses are distinguished from each other on serological grounds.

Clinical signs

The clinical signs in sheep, whether caused by PPR or by rinderpest, are essentially similar.

Some of the flock are noticed to be depressed,

listless, not feeding and standing away from the others. On closer examination, a rise in temperature is found. Nasal and ocular discharges are present and in the mouth, on the tongue and in the nasal cavity erosions of the epithelium and skin are seen. Frequently a profuse diarrhoea is observed. Later a cough develops and bronchopneumonia may be present. Death occurs in severe cases.

However, in some instances the disease is milder, especially with rinderpest in West Africa although not in India, where the disease is more severe. The severity of PPR for sheep varies. Sometimes it is as described before; at other times all that will be found is a slight rise in temperature, depression and some nasal and ocular discharge.

With PPR, goats in the same area are also affected, often with severe clinical signs similar to those seen in the sheep. However, cattle are not affected. With rinderpest in an area where vaccination is not practised, disease is also be observed in cattle and possibly in goats. If vaccination has been carried out, cattle are resistant.

Distribution

PPR is found mainly in West Africa. In the southern parts of Nigeria it is termed 'kata'. It has also occurred in the Sudan. Evidence of infection has been found in the Arabian peninsula but no disease has been reported from that area.

Rinderpest in sheep has been reported from West Africa, Sudan, the Middle East, the Indian subcontinent and South-East Asia. Undoubtedly, the vaccination campaigns in cattle have reduced the incidence of rinderpest in sheep in Africa but rinderpest is still a problem among sheep and goats in parts of Asia.

Transmission and epidemiology

Virus is present in the sick animal and in its secretions and excretions. Thus, these diseases can be transmitted by direct contact between animals or by indirect contact through man, animals, vehicles or fomites contaminated by the infective discharges. The route of infection is respiratory and animals may be infected through droplets from others coughing.[17]

Rinderpest may be passed from cattle to sheep, goats and wild ruminants; with PPR, a cycle of virus is maintained with sheep and goats. Spread over long distances is through movement of animals or carriage in contaminated vehicles. Spread through feeding of infected animal products is less likely.

Diagnosis

Diagnosis is based on the clinical signs and, in dead animals or animals killed for post-mortem examination, on the presence of intense inflammation in the alimentary tract, the so-called 'zebra striping', as well as erosions and necrosis of Peyer's patches.

Laboratory confirmation [3, 4] is obtained by:
1. agar gel precipitin tests on mesenteric lymph nodes or mucosal scrapings from the large intestine (this will not distinguish between PPR and rinderpest);
2. virus isolation from lung or large intestine samples in primary lamb or sheep kidney cell cultures (PPR) or primary calf kidney cell cultures (rinderpest);
3. demonstration of a rise in antibody in serum taken from animals at the time of disease and 14 days later.

The diseases must be distinguished by differential diagnosis from bluetongue, sheep pox, contagious pustular dermatitis and foot-and-mouth disease.

SHEEP POX

Sheep pox is a highly infectious disease of sheep characterized by fever and the development of generalized pocks. It is caused by a virus, which is a member of the *Capripoxvirus* genus of the family Poxviridae.[8, 18]

Clinical signs

Three forms of the disease are defined: peracute, acute and subacute.

In the peracute form deaths among lambs can be expected. In addition, if there are exotic or imported sheep in the flock, deaths may be found among them. Closer examination of the animal reveals haemorrhagic pocks, especially over those areas of the skin not covered by wool, e.g. around the mouth, udder, teats, scrotum and vulva.

In acute cases, affected sheep appear sick. Fever, nasal and lachrymal discharges and salivation can be seen. On examination, pocks at various stages of development are observed on areas not covered by wool, although vesicles are not a prominent feature.

Pocks may also be visible in the mouth and nostrils. Some animals die and on post-mortem examination pocks may be found in the alimentary and respiratory tracts. In sheep that survive, healing of the skin may be slow and permanent scars are left.

In subacute cases there may be no obvious clinical signs. Some sheep appear to be off-colour but on closer examination pocks may be found, especially around the head and about the tail.

Secondary infection of the sheep may occur, leading to bacterial and other infections. In general, the course of the disease is about three to four weeks. Morbidity can be up to 70 per cent in a flock, with mortality of 60–100 per cent in lambs and 5–50 per cent in adult sheep.

With some strains of sheep pox, goats in the flock may also be affected; similarly, some strains of goat pox may also affect sheep. However, the clinical signs do not differ.

Distribution

Sheep pox is found in north Africa, Sudan, east Africa, Turkey and the Middle East, the Indian subcontinent and other parts of Asia.[18] It was eradicated from the UK in 1866 and was last seen in France in 1967, Spain in 1968 and Portugal in 1969.

Transmission and epidemiology

Sheep pox is divided into two types; one type affects sheep only, the other type affects sheep and goats and is related to lumpy skin disease of cattle.

The route of infection of sheep is by contact, through the respiratory tract or through skin abrasions. The virus spreads from the primary site of multiplication to other sites on the skin and mucous surfaces, where pocks are formed. Virus is excreted from the pocks, in the nasal secretions, milk and possibly urine.

Spread occurs by contact between animals, by animals coming into contact with contaminated wool or feeding stuffs. At certain times of the year transmission may also occur through flies.

In some countries the disease is noted more frequently during the winter months, when animals are housed; elsewhere, the disease has been reported during the late summer, when biting insects are numerous.

Diagnosis

Diagnosis is based on clinical signs and the discovery of pocks.

Laboratory diagnosis is based on electron microscopic examination, when the typical rectangular virus (195 x 115 nm) can be recognized.[3,4] Elementary bodies can be seen in impression smears under the light microscope. Tissues can be examined histologically for the typical 'cellules claveleuses', large stellate cells with oval nuclei containing enlarged nucleoli.

Isolation of virus can be attempted in sheep and goats or in tissue cultures of sheep testes or sheep thyroids. Viral antigen may be detected in gel diffusion or complement fixation test on material from skin lesions or on blood serum. A rise in antibody may be demonstrated by gel diffusion or by neutralization tests.

The disease has to be differentiated from contagious pustular dermatitis (orf) and from bluetongue.

MISCELLANEOUS DISEASES

Two other virus diseases of sheep have been described: Nairobi sheep disease, which is found in Kenya and Uganda, and Wesselsbron disease in southern Africa.[8,19]

The virus of Nairobi sheep disease is a member of the Bunyaviridae family of viruses, to which Dugbe (Congo, Nigeria, Senegal) and Ganjam (India) viruses belong. It is transmitted by *Rhipicephalus appendiculatus* ticks and causes a haemorrhagic gastroenteritis in sheep, with high fever and pulmonary oedema.

Wesselsbron virus is a *Flavivirus* of the Togaviridae family of viruses and its transmitted by *Aedes* mosquitoes. Evidence of infection by the virus has been found in South, Central and West Africa, Madagascar and Thailand. It resembles Rift Valley fever in that it gives rise to abortions in pregnant ewes and deaths of newborn lambs with jaundice. Other sheep in the flock are not affected.

With both diseases, diagnosis is based on clinical signs together with attempts to isolate virus from blood and tissues in sheep, mice and tissue cultures and, in the case of Wesselsbron virus, fertile hens' eggs. A rise in antibody level may be demonstrated in blood serum.

REFERENCES

1 Special Feature (1978) Foot-and-mouth disease. *Veterinary Record,* **102**, 184–98.

2 Sellers R.F. (1971) Quantitative aspects of the spread of foot-and-mouth disease. *Veterinary Bulletin,* **41**, 431–9.

3 Commission of the European Communities (1976) Information on Agriculture, No.16. Methods for the detection of the viruses of certain diseases in animals and animal products. CEC, Brussels.

4 French E.L., Geering W.A. (Eds.) (1978) *Exotic Diseases of Animals.* A manual for diagnosis. 2e, Australian Government Publishing Service, Canberra.

5 Symposium on Bluetongue (1975) Contributions by various authors. *Australian Veterinary Journal,* **51**, 165–232.

6 Bluetongue (1980) Contributions by various authors. *Bulletin de l'Office international des Epizooties,* **92**, 461–600.

7 Inaba Y. (1979) Akabane disease: an epizootic congenital arthrogryposis- hydranencephaly syndrome in cattle, sheep and goats caused by Akabane virus. *Japan Agricultural Research Quarterly,* **13**, 123–33.

8 Andrewes C.H., Pereira H.G. & Wildy P. (1978) *Viruses of Vertebrates.* 4e. Bailliere Tindall, London.

9 Parsonson I.M., Della-Porta A.J. & Snowdon W.A. (1977) Congenital abnormalities in newborn lambs after infection of pregnant sheep with Akabane virus. *Infection and Immunity,* **15**, 254–62.

10 Shimshony A. (1980) An epizootic of Akabane disease in bovines, ovines and caprines in Israel, 1969–70: epidemiological assessment. *Acta Morphologica Academiae Scientiarum Hungaricae,* **28**, 197–9.

11 Easterday B.C. (1965) Rift Valley fever. *Advances in Veterinary Science,* **10**, 65–127.

12 Meegan J.M., Hoogstraal H. & Moussa M.I. (1979) An epizootic of Rift Valley fever in Egypt in 1977. *Veterinary Record,* **105**, 124–5.

13 Sellers R.F., Pedgley D.E. & Tucker M.R. (1982) Rift Valley fever, Egypt 1977. Disease spread by windborne insect vectors? *Veterinary Record,* **110**, 73–7.

14 Kaplan C. (1977) *Rabies: the Facts.* Oxford University Press, Oxford.

15 Kaplan M. & Koprowski H. (1973) *Laboratory Techniques in Rabies.* 3e. World Health Organization, Geneva.

16 Scott G.R. (1964) Rinderpest. *Advances in Veterinary Science ,* **9** ,113–24.

17 Plowright W. (1968) Rinderpest virus. *Virology Monographs,* **3**, 25–110. Springer-Verlag, Vienna, New York.

18 Singh I.P., Pandey R. & Srivastava R.N. (1979) Sheep pox: a review. *Veterinary Bulletin,* **49**, 145–54.

19 Berge T.O. (1975) *International Catalog of Arboviruses.* 2e, US Department of Health, Education and Welfare.

SECTION 8

POISONS AND POISONING

47 PLANT AND INORGANIC POISONS

G.A.M. SHARMAN

Plants may be poisonous in different degrees. Animals in their wild habitat avoid the highly dangerous ones but may with impunity eat those that are less poisonous, because they are part of a mixed flora and there is plenty of choice elsewhere. By contrast today's domestic livestock have little choice of diet. For example if a flock of sheep is enclosed in a field containing a crop composed of a single plant species which contains harmful toxins the sheep have no means of avoiding them.

The plant kingdom contains a great variety of phytotoxins some of which have been identified chemically but many of which are still recognized only by their effects. For example plants may contain one or more of the following known or potential poisons: cyanogenetic glycosides, alkaloids, copper, selenium, nitrate, fluoride, oxalate; agents which cause abortion or photosensitization; agents which are goitrogens, oestrogens or teratogens. These and many others may act directly or they may be present in the plant as harmless pro-toxins requiring further elaboration. Conversion to the toxin may take place during feed storage, in the feed or water trough, in the alimentary canal or after absorption from the gut. Saprophytic or parasitic fungi growing on the plants may produce toxic metabolites (mycotoxins) detrimental to animals and their role may be attributed to the host plant falsely. Plant surfaces may become contaminated with fall-out from atmospheric pollution arising from industrial or natural processes and thus poison grazing herbivores.

It is noteworthy that in parts of the world where sheep flocks are taken to the mountains or herded through them in summer, poisonous plants form the greatest hazard, requiring great skill from the herder in their recognition and avoidance.

Inorganic poisoning usually results indirectly from industrial processes, if not by fall-out from the atmosphere, then by injudicious disposal of waste or accidental contamination of processed animal food or of pasture. Occasionally it arises from natural deposits of minerals but usually after disturbance by man for industrial exploitation.

This chapter deals briefly with the poisons of greatest economic significance to the sheep industry.

BRASSICA POISONING (KALE ANAEMIA, REDWATER)

The invention of the seed drill by Jethro Tull during the 'agricultural revolution' in Britain, permitted the turnip to emerge as a key crop for livestock rearing in the 18th century. Continuing agricultural improvement then produced many species of the genus *Brassica* as feed for stock, either by direct grazing or after harvesting, among them the Swede turnip, kale, rape and increasingly in the 20th century, the residues and surpluses from vegetable crops grown for human consumption such as Brussels sprouts, cabbages and cauliflowers.

During World War II, the first reports came from Germany and Canada of cattle poisoned by feeding too much kale and rape, and from Britain in 1953 of deaths from kale and rape poisoning in grazing sheep.[1] Haemolytic anaemia with Heinz-Ehrlich bodies (HB) in the circulating erythrocytes are the prominent features of the disease which in the acute form progresses rapidly to give haemoglobinuria and death but more commonly is characterized by lowered production.

Brassicas provide soluble carbohydrate, good quality protein, vitamins and minerals in proportions that should theoretically give high yields in the animals to which they are fed. These are achieved when brassicas form part only of the ration but when they are fed exclusively, the potential is seldom reached. The deficit may result from the sub-clinical effects of toxic substances in the brassicas, from suppression of appetite or from the effects of copper deficiency if the crop is fed for a long time.

There are 3 potentially toxic substances in the brassicas, all of which require conversion to produce their toxic effects. These are S-methylcysteine sulphoxide (SMCO) which is the precursor of the haemolytic anaemia factor, the glucosinolates which after conversion to thiocyanates cause goitre, and nitrate, which if reduced to nitrite, causes methaemoglobinaemia. Sheep grazing brassica crops occasionally develop photosensitization but the precipitating cause has not been identified.

The haemolytic anaemia factor

SMCO is a sulphur-containing aminoacid occurring in the brassicas and *Allium spp* (garlic and onions). In experiments at the Rowett Research Institute in 1973,[2] a fraction of kale juice that caused haemolysis in a goat was found to be rich in SMCO. The coincidence of this finding with the detection of volatile sulphur compounds in the rumen of a kale-poisoned goat, led to the hypothesis which was subsequently proved, that bacterial fermentation in the rumen converted SMCO to dimethyl disulphide which in the blood stream caused HB formation and haemolysis. Dimethyl disulphide, oxidises reduced glutathione which is essential within the erythrocyte for the maintenance of haemoglobin in its reduced state. Continuous consumption of brassicas by a ruminant animal therefore subjects the erythrocytes to severe oxidative stress and causes irreversible changes in the haemoglobin, part of which is precipitated as HBs. The affected cells then lyse and the HBs are removed from circulation by the spleen.

The severity of the disease relates directly ιo the amount of SMCO ingested but may be modulated by the ability of the rumen flora to convert SMCO and also by the availability in the rumen of fermentable substrate; an ample supply of sugars and carbohydrates may accelerate the production of dimethyl disulphide and hence the onset of acute haemolytic disease. However, analysis of the crop for SMCO gives a useful indication of the potential hazard. Daily intakes of 10–18 g SMCO/100 kg liveweight (LW) produce low-grade anaemia and higher intakes acute haemolytic anaemia. A 33 kg LW weaned lamb ingesting 1 kg dry matter (DM) daily of a brassica crop containing 0.4 per cent of its DM as SMCO would be in the low grade anaemia category (12 g SMCO/100 kg LW). A crop with 1.4 per cent of its DM as SMCO would provide the lamb with 42 g/100 kg, an amount which, even if the appetite were drastically reduced, would still be dangerous.

The SMCO content increases with the age of the crop. In Maris Kestrel kale it may rise from 0.4 per cent DM in August to 1.4 per cent in January. Brussels sprouts, cabbage and kale contain similar relatively high levels of SMCO, whereas rape, stubble turnips and Swede turnips have considerably less. Experiments in New Zealand have shown that SMCO has an appetite-limiting effect in sheep.[3] While this may protect against the acute haemolytic disease by preventing ingestion of too much SMCO it also contributes to the lower than expected growth rate. The SMCO content of the crop responds to manurial nitrogen. In the north of Scotland as the nitrogen application to the soil before sowing was increased from zero to 220 kg/ha the SMCO content rose, reaching a maximum at 160 kg/ha. There was no response to the application of potassium or phosphorus.

At times of increased metabolic demand, as in lactation, the voluntary intake may increase greatly. Experience with dairy cattle suggests that lactating ewes ought to be rationed to no more than 30 per cent of their diet as kale. The coincidence of lactation in sheep with available brassica crops, however, is unusual.

Clinical signs and pathology

The clinical signs vary from clinically undetectable shortening of the erythrocyte life-span, to severe anaemia. The subclinical signs are depression of growth or production; for example weaned lambs grazing Lair (giant) rape, grow slowly but steadily at 50–150 g/day. Haematological examination reveals moderate anaemia with HB production. Chronologically the changes are first a rise in HBs followed by depression of haemoglobin and packed cell volume both of which then return to normal levels while HBs decline again to stabilize at about 30 per cent of the erythrocytes affected. Minor secondary cycles of anaemia and partial recovery follow, caused by the continuing oxidative stress and a compensatory increase in erythropoesis. The reduced glutathione level in blood also falls and recovers cyclically but cholesterol levels decline progressively.

The gross pathological features in the acute disease are haemolysis, icterus, haemoglobinuria, anaemia, congestion of kidneys, spleen and liver and the presence of active healthy red bone marrow. Histologically there are zones of necrosis in the liver and widespread haemosiderosis which are secondary to the haemolytic anaemia. Haemorrhages are not a feature.

Diagnosis and treatment

Iι is important to differentiate acute haemolytic anaemia from nitrite poisoning so that immediate treatment of the latter may be given if necessary. If a freshly dead sheep is not available for post-mortem examination blood samples should be withdrawn from sick animals for assessment of colour and consistency. In acute haemolytic anaemia the blood is dark red and watery, while in nitrite poisoning it is chocolate brown and opaque. This is sufficient evidence for determining the course of action. For laboratory confirmation the samples required are; crop for SMCO and nitrate plus nitrite determination and blood from sick sheep for haematology. An additional sample may be diluted with 20 volumes of phosphate buffer of pH 6.6 to stablize methaemoglobin which in acute haemolytic anaemia seldom exceeds 12 per cent of the haemoglobin, but in nitrite poisoning exceeds 20 per cent and may be as high as 80 per cent.

Treatment of sheep affected with acute haemolytic anaemia requires removal from the crop and careful substitution of highly nutritious food. It may not be necessary to administer iron, as much of the iron released during haemolysis is recycled.

Low growth-rate may be counteracted by the partial substitution of other appropriate foods thus lowering the SMCO intake. Differentiation from the debilitating effect of copper deficiency should be considered.

Copper deficiency

Brassicas are relatively rich in sulphur and contain molybdenum. As both elements limit copper absorption by ruminants prolonged brassica feeding depletes copper reserves. When the feeding of brassicas is to be continued and growth-rate is low, diagnostic blood samples should be sent to a laboratory for copper determination and if the values are very low, treatment with copper should be given. If treatment is by injection, a site, such as high in the neck should be chosen, to avoid blemishes that might otherwise spoil the carcass. Withdrawal times, according to the manufacturer's instructions must be observed.

Glucosinolates

The glucosinolates present in the brassicas, are hydrolysed to thiocyanates which have a goitrogenic effect. After 2–3 months of rape feeding, prime lambs at slaughter have had thyroid glands ranging in fresh weight from normal to ten times normal. No association has been found between thyroid size and growth rate and no detriment noticed in carcass quality.

Clinically, enlarged thyroid glands can be palpated. The goitrogenic effect can be prevented or treated, by the intramuscular administration to a weaned lamb of 1 ml of 40 per cent iodine in poppyseed oil repeated as required.

Nitrate

Nitrate can account for up to 1.5 per cent of the dry matter of kale, but is relatively harmless unless reduced to nitrite by bacterial action in the crop before ingestion or intra-ruminally. Poisoning occurs only rarely, but when it does many sheep may become very ill simultaneously. Nitrite is absorbed and combines with haemoglobin to form methaemoglobin which is unable to transport oxygen. Cyanosis and all the signs of progressive anoxia develop.

Differential diagnosis from acute haemolytic anaemia is made by venepuncture and demonstration of chocolate brown coloured blood.

Treatment is with a reducing agent such as a 4 per cent aqueous solution of methylene blue given intravenously at 9 mg/kg LW and repeated as necessary. Laboratory diagnosis is by analysis of blood for methaemoglobin (see above).

RHODODENDRON POISONING

Rhododendrons are attractive flowering shrubs grown in gardens throughout Britain. They also grow wild in the west Highlands of Scotland, frequently on land accessible to sheep. Poisoning from ingestion of the evergreen leaves is rare but when it occurs mortality may be high. A white-out of the pasture with snow is often the stimulus for sheep to shelter among the shrubs which commonly border driveways; the opportunity to browse green leaves may be taken, with fatal results. Garden clippings, thoughtlessly discarded into a sheep pasture, are another common cause of poisoning.

Rhododendron leaves contain andromedotoxin which causes drooling and distressing attempts to vomit, abdominal pain, respiratory distress, staggering, collapse and death after several days of illness. The toxin appears to act directly on the vagal nerve endings in the stomach.

Post-mortem examination is negative apart from the presence of portions of the leaves in the rumen.

Treatment is symptomatic. Emptying of the rumen by rumenotomy is indicated.

YEW POISONING

Taxus baccata is the commonest species of yew tree in Britain; all species are very poisonous. They have been grown near places of worship since pre-

Christian times and are still to be found in church-yards.

The poisonous principle is the alkaloid taxine which is thought to act directly on the heart, slowing its rate, causing cyanosis and heart failure. Death is sudden; the symptoms which are rarely seen, are trembling, dyspnoea and death within about 5 minutes of onset, with no struggling.

At post-mortem examination, fragments of yew leaves are present in the rumen, the abomasal mucosa may be inflamed and liver, spleen and lungs are congested. The heart is reported to stop in diastole.

Treatment is seldom possible. Rumen emptying by rumenotomy is the logical procedure when the opportunity is presented.

BRACKEN-FERN (*PTERIDIUM AQUILINIUM*) POISONING

Bracken is distributed world-wide. In the British Isles it is most abundant in the moister west, covering large areas of woodland, thickets and open hillsides. Its success as a colonizer arises from a rhizome system with enormous reserves of energy and an ability to tolerate wide ranges of soil types and pH. In the UK farming depression years of the 1930s, when marginal land was abandoned from cultivation, bracken gradually descended the hillsides and displaced large areas that might otherwise have been useful pasture. It has great potential to kill or maim livestock in several ways through the toxins it contains. All parts of the plant, whether fresh or air dried, are poisonous.

By feeding bracken experimentally to the horse, neurological and other signs classical of vitamin B$_1$ (thiamine) deficiency were produced; bracken contains a thiaminase enzyme which destroys the dietary thiamine. Cattle and sheep grazing bracken do not become thiamine deficient because bacteria in their rumens synthetize an excess of thiamine, but bracken contains other, as yet unidentified, substances which cause diseases of considerable economic importance and shorten the life-span of sheep which have unrestricted access to it. These toxins are cumulative poisons, that is their effects are additive and usually pass unnoticed until the terminal phase of the disease is reached, hence the misapplication of the term 'acute' to the syndrome to be described.

Acute bracken poisoning (Haemorrhagic Fever)

Prolonged exposure of the red bone marrow to a bracken-derived toxin causes aplasia particularly of the granulocytic and thrombocytic series of cells. This weakens the body defences, increases blood clotting time and leads to microbial invasion, haemorrhage and anaemia.

The clinical signs are well-documented for cattle but not for sheep. Usually the first sign in a group is the passage of blood in the urine and faeces: a most useful signal to this may be an individual being followed by the others, presumably attracted to the source of the blood by smell. Other signs are high fever (rectal temperature 40.3°C/104.5°F and above), anaemia, epistaxis, melena, haematuria and very fast respiratory and heart rates.

The main pathological features are severe anaemia indicated by paleness, watery blood and sparseness of blood clot in the heart chambers, blood clots in bladder and lumen of gut, mucosal haemorrhages and infarctions in liver, kidneys, lungs or rumen wall. The important primary histological feature is aplasia of red bone marrow.

Progressive retinal degeneration (Bright Blindness) of sheep

Watson, Barlow and Barnett[4] described in 1965 a blindness of ewes grazing the upland bracken-infested pastures of West Yorkshire which was known to local farmers for a long time but becoming commoner. It was reproduced experimentally by feeding ewes for many months 1 kg/day of pelleted ration containing 50 per cent bracken.

The clinical signs are progressive bilateral blindness of sheep over 2 years old, identified usually in the autumn by inadvertent separation of individuals from the flock, the high-stepping gait and alert high-held head characteristic of blind sheep. The eyes shine abnormally brightly in semi-darkness, the pupils become circular and react poorly to light. There is no inflammation and no opacity of the lens. Through an opthalmoscope the arteries and veins appear narrower and the main vessels more widely separated than normal; subclinical cases may be identified by such examination.

The pathological changes are confined to the eye and are characterized by degeneration of the neuro-epithelium of the retina. Early, the rods and cones become fragmented and later, completely destroyed together with the outer, and portions of the inner, nuclear layers.

Neoplasia (tumours, carcinoma)

After flockmasters in North Yorkshire were convinced in the late 1960s that bracken was killing their ewes, the flocks spent less time on the infested pastures. Subsequently 'acute' poisoning diminished but tumours of the jaw or alimentary tract emerged as the most common remaining cause of wastage.[5] Ingestion of smaller amounts of bracken over a long time together with late culling of the ewes may have been important predisposing factors in the incidence of these tumours.

An association between bracken and a form of cancer in cattle was made by Jarrett[6] who found a high incidence of carcinoma of the intestines and bladder in cattle from upland bracken-infested areas. The carcinomas were preceded by benign papillomas caused by a certain type of papilloma virus. Transformation to malignancy results from ingestion of a carcinogen contained in bracken. However, there is no conclusive evidence that a similar association exists in sheep on bracken uplands (see Chapter 45).

Diagnosis

Seek evidence that bracken was available for long periods and appears to have been grazed; look for bracken in the rumen. Determine that clinical and gross post-mortem findings are appropriate. On-the-spot tentative diagnosis is then usually possible and immediate action justified. Sick animals should be separated for treatment; they also provide material for laboratory confirmation. Blood in anticoagulant is required from live sheep for haematology, including differential white cell and thrombocyte counts, which will show leucocytopaenia, thrombocytopaenia and severe anaemia. Faeces should be taken for detection of blood if clots are not obvious, and urine to confirm haematuria. Red marrow smears may be taken using anaesthesia and biopsy equipment but only if necessary and by special arrangement with the laboratory. The appropriate post-mortem specimens are rumen contents for microscopical identification of bracken if necessary, red bone marrow fixed immediately after death and lesions (eye, tumours) fixed for histology.

In an 'acute' outbreak, the flock should be excluded from bracken-infested pastures for as long as possible and inspected immediately and daily thereafter for detection of new cases. Sick sheep should be provided with shelter and treated symptomatically with vitamin K and iron and introduced gradually to a highly-nutritious diet based, during convalescence, on the lactation requirements of ewes.[7]

In the long term, it is important to reduce as much as possible the exposure of the flock to bracken. Consideration should be given to the possibility of chemical or other methods of controlling bracken growth, bearing in mind that the wilted, dried or grubbed up plant may be palatable and toxic.

COPPER POISONING

Chronic copper (Cu) poisoning

Although the ovine has a remarkable capacity to store Cu in its liver, signs of centrilobular necrosis accompanied by increased plasma amino-aspartate transaminase (AAT) concentrations appear when liver concentrations exceed 750 mg Cu/kg DM. At higher levels, unknown factors cause a sudden release of Cu into the bloodstream and precipitate a haemolytic crisis. Kidney function then declines dramatically causing the accumulation of metabolites such as urea in the bloodstream and spongiform degeneration occurs in the brain. Characteristic post-mortem features are widespread jaundice, a discoloured (yellow/orange) liver, distended and discoloured (bronze/black) kidneys and haemoglobinuria.

Chronic Cu poisoning is often associated with prolonged ingestion of diets from which Cu is efficiently absorbed (around 10 per cent) rather than high dietary Cu concentrations. Suffolk and Texel-cross lambs absorbed Cu twice as efficiently as Blackface lambs and are particularly at risk from poisoning: subclinical toxicity has been recorded in such lambs on diets containing 12 mg Cu/kg DM, a level often exceeded in commercial foodstuffs. Chronic Cu poisoning is therefore a perpetual risk in intensive sheep rearing. The risk bears a curvilinear relationship to dietary Cu concentration because at high Cu levels, a smaller proportion of the dietary Cu is absorbed and more of the absorbed Cu is excreted via the faeces. Some breeds, notably the Blackface, seem more able to adapt to Cu loading than others. Ingestion of hepatotoxic alkaloids can precipitate Cu toxicity, presumably by impairing the ability to excrete excess Cu.

Prevention

Every effort should be made to reduce the Cu content of rations for intensively reared sheep. Use of silage as a maintenance feed may lower the absorbability of Cu in the concentrate part of the ration. Where cases of Cu toxicity have occurred, oral dosing with ammonium molybdate and sodium sulphate has successfully lowered the liver Cu reserves of surviving lambs. Continuous medication of the drinking water with ammonium molybdate (20 mg/l) and reliance on sulphur in the diet to potentiate the inhibition of Cu absorption, may be equally effective.

Diagnosis

Other hepatotoxic factors such as carbon tetrachloride produce liver lesions similar to those in Cu toxicity and other haemolytic factors may also give rise to some of the post-mortem features characteristic of Cu toxicity. The full combination of features with elevated concentration of Cu in the liver (> 1000 mg/kg DM) and kidney (> 50 mg/kg DM) confirm the diagnosis. The risk of losing further animals during an outbreak of Cu toxicity should be assessed by looking for elevated AAT and blood Cu concentrations in the surviving animals.

Acute Cu poisoning

Acute toxicity may occur within 2–3 days of the subcutaneous administration of Cu for prevention purposes. The features at necropsy are quite distinct from those of chronic toxicity in that signs of jaundice and haemolysis are absent. The common features are pulmonary oedema and inflammation of the gastric mucosa. Tissue Cu concentrations are much lower than those associated with Cu poisoning, particularly in the liver where levels may not have exceeded 300 mg/kg DM. It has been suggested that stress factors such as withdrawal of food and the simultaneous administration of anthelmintics may render some sheep more susceptible to acute Cu toxicity.

LEAD POISONING

Acute lead poisoning is rare but clinically characteristic and is usually diagnosed correctly. Chronic poisoning is commoner than generally recognized, but is insidious because it is asymptomatic during the long, cumulative phase and generally terminates fatally. In individual sheep, deaths of this kind are often not investigated and the existence of a problem is not suspected until several have died.

The most common sources of acute poisoning are flakes of old lead paint, red lead, putty, old car batteries, linoleum, roofing felt and other such lead-containing material which, though often carried to sheep pastures by flood, more frequently arrive there by people dumping rubbish without due care. The forms usually associated with livestock poisoning are lead oxide, sulphate, sulphide, carbonate and acetate. Swallowed metallic lead is retained indefinitely in the reticulum, only insignificantly small amounts being absorbed.

Affected animals have fits within a day or two of ingesting the toxic material. They roll their eyes, stagger in distress and collapse, with intermittent convulsive seizures before death. Any survivor is blind and between fits unresponsive to external stimuli, often standing with its head pressed against a wall.

At post-mortem examination the musculature is dirty grey, there is gastroenteritis, degeneration of the liver and the kidneys are pale with haemorrhagic patches on the surface.

Chronic lead poisoning results from continuous exposure to lead by the oral route. Fits may occur without the owner appreciating their significance. The syndrome is generally characterized by anorexia, constipation, recumbency and death, while pregnant ewes may abort. At post-mortem examination there is emaciation and other features which resemble those of acute poisoning but gastroenteritis may be absent.

Outbreaks have been reported of stiffness in suckled lambs around disused lead mines in south Scotland,[8] the northern Pennines and Derbyshire, associated with high levels of lead in tissues, soil and pasture. The lambs develop osteoporosis with brittle bones, resulting in fractures and bone malformations. They take short steps, walk on their toes or may become paraplegic. Cappi, a Scottish term applied to a softness of the bony plate covering the frontal sinus, may develop. The syndromes differ slightly and may be complicated by concomitant excess or deficiency of other elements. At necropsy the essential feature is rarefaction of cancellous bone, fractures and their sequelae.

The administration of lead compounds to cattle and sheep experimentally[9] provided invaluable information on lead metabolism and yielded diagnostic criteria for lead levels which have withstood the test of time. Only 1–2 per cent of ingested lead is absorbed, most being eliminated in the faeces whether given as the water soluble lead acetate or the insoluble carbonate as in paint. Most of the absorbed lead is re-excreted in bile and a small amount in urine. In the ewe, significant amounts pass through the placenta to the fetus and through the milk to suckled lambs. In acute poisoning highest values are found in kidneys. The lethal dose for young ruminants is 0.2–0.4 g/kg LW and is somewhat higher for the mature animal. After continuous oral administration the largest amounts are found in the bones.

Following a single oral administration the blood level may rise from a normal of < 0.25 mg/l to 4 mg/l and death may supervene. If the animal survives the level declines very slowly and even 6 months later may still be elevated. The faecal lead level on the other hand rises steeply to > 1000 parts/10^6 dry matter but declines to normal (10–30 parts/10^6) in a few weeks. This is of considerable value diagnostically: if both blood and faecal levels are high it can be assumed that ingestion occurred recently. If blood lead is high but faecal lead is normal, ingestion probably occurred some time before and the animal is likely to survive.

The pathogenesis of lead poisoning is not fully understood. Accumulation of lead in kidneys, liver or brain gives rise to cellular degeneration typified sometimes by intranuclear inclusions and glomerular lesions in the kidneys. In free-ranging lambs in plumbiferous areas it appears that lead acquired *in utero*, in the milk and possibly also directly ingested from the pasture or soil, causes failure of osteoid synthesis. Stress factors, such as virus infections, are thought in some cases to precipitate the acute syndrome during an asymptomatic cumulative phase.

Diagnosis

Consideration must be given to the symptoms, the availability of lead, lead levels in blood and faeces of survivors and the post-mortem findings. Kidney cortex should be sent for lead analysis. A value for fresh tissue of > 40 parts/10^6 is definitely positive; > 10 parts/10^6 can be regarded as positive with collateral evidence and highly suggestive without such evidence.

Treatment

The agent of choice is sodium calcium edetate (CaEDTA: calcium disodium versenate) which is often dramatically successful. It is given at the rate 75 mg/kg daily by slow intravenous injection, preferably during the first 48 hours as a divided dose at intervals of a few hours. It is a chelating agent which combines with lead and is excreted in the urine and bile. It should be used sparingly as it may cause toxic nephrosis or death from too rapid mobilization of lead.

LUPIN POISONING AND LUPINOSIS

These two subjects were reviewed extensively in 1967[10] and in 1974, after confirmation that lupinosis is caused by a mycotoxin, it was reviewed again.[11]

Lupins (*Lupinus*) are annual legumes of the temperate and warm regions, with palmate leaves and tall terminal racemes of white or brightly coloured blue, yellow or red pea-like flowers. Of approximately 100 existing species most are native to the western USA where the wild plants may be grazed by sheep on the mountain pastures. A few species grow naturally in the east USA, Mexico, Central and South America and around the Mediterranean littoral. They have been spread around the world by man as garden flowers and crops.

Crops of lupins are grown as green manure or as animal feed. On certain poor soils lupins yield considerably more dry matter, energy, and protein than pasture grass. They are grown as forage mainly for sheep particularly in West Australia and South Africa.

The term lupin poisoning is reserved for an acute CNS disorder caused by the alkaloids contained in bitter lupins. By a programme of selective breeding the alkaloid content has been reduced to negligible amounts, so that the plants now known as sweet lupins, have since 1934 been cultivated as animal feed. However, this programme has not eliminated the condition known as lupinosis which is a disease, characterized by hepatitis and its sequelae, caused by a mycotoxin produced by the fungus *Phomopsis leptostromiformis* which parasitizes the growing plant and also grows on the straw and litter.

Lupin poisoning

In the bitter lupins the alkaloids are concentrated mainly in the seeds and seed pods with less in the leaves and still less in the stalks. The alkaloids first stimulate and then paralyse the nerve centres of the medulla and cord causing initial excitement followed by prostration, lowered blood pressure and heart rate and dilation of the pupils. Death results from asphyxia and associated convulsions. Experimental inoculation of the alkaloids into laboratory animals and sheep has shown that there is a critical level above which clinical signs and death occur, but below which the alkaloids are detoxified so that there is no cumulative effect. In the field, the onset is quick and the course rapid. Symptoms and death usually occur within 7 hours of exposure but occasionally there may be delay until 24 hours after ingestion of the lupins. When recovery occurs it appears to be complete and recovered animals remain fully susceptible to further occasional exposures. However, by gradually increasing the amount of bitter lupin seed offered, tolerance of up to 0.5 kg/day may be acquired.

Diagnosis depends upon a history of bitter lupins which have reached the pod stage having been recently available and the characteristic clinical signs. The post-mortem findings are associated with convulsive struggling and of asphyxia. Fragments of lupin may be found among the rumen contents. Laboratory confirmation is available only in certain specialized research centres.

Unaffected sheep should be removed immediately from the lupin pastures disturbing any prostrate survivors as little as possible, because stimulation of the survivors may precipitate death but if they are left alone they may recover.

On mountain pastures, if a route through lupins cannot be avoided, it is safer to take the flock slowly through the stands of lupins to give the sheep the opportunity to graze the intervening sward. In a hurried traverse some are likely to pull off and eat lupin pods in passing.

Lupinosis

Sweet lupins become toxic, generally after summer rain. In 1880 this association between moisture and lupinosis in sheep led Kühn in Germany to suspect fungal contamination as the cause but it was not until 90 years later that *Phomopsis leptostromiformis* (syn. *Phomopsis rossiana, Cryptosporium leptostromiforme*) growing on the lupins was found to be the cause of the progessive hepatitis which is the primary lesion of the disease.

Whether the fungus parasitizes the living plant or grows in the dead straw and litter it presents a threat to sheep grazing lupin pastures. The severity and the form of the disease depend on the amount of the fungus or its mycotoxin ingested. At high levels it causes acute illness with anorexia, depression and stupor. The sheep stand with the head lowered and drawn to the side or backwards. They tend to sway and if they fall over are unable to regain their feet unassisted. Pulse and respiration are accelerated and temperature elevated (40–40.6°C/104–105°F). There are clinical signs of icterus, constipation, dark yellow urine and sometimes swelling of the ears. Anorexia becomes complete in 24–48 hours and appetite returns only very slowly. Severe acute cases die in 3–11 days and mortality may be high. At lower levels anorexia is partial with voluntary food intake insufficient to sustain growth and body condition. Anaemia and slight icterus may develop. The cachexia of chronic lupinosis mimics that of some trace element and other deficiency diseases.

At necropsy, in acute lupinosis, there is generalized icterus associated with an enlarged friable, bright-yellow, fatty liver. Subacute and chronic cases are characterized by atrophy, fatty degeneration and cirrhosis of the liver which may be shaped like a boxing glove. Other necropsy findings include enlargement of the gall bladder, splenomegaly, nephrosis, impaction of the large intestine, ascites, hydrothorax and hydropericardium.

Histological examination of the liver in the acute disease shows fat metamorphosis, eosinophilic globules in the cytoplasm of the hepatocytes, karyorrhexis, and vesiculation of the nuclei. In the subacute and chronic forms there are, accumulation of pigment in the hepatocytes, megalocytosis, proliferation of the bile ducts and their epithelium and centrilobular fibrosis. Haemosiderosis occurs in other organs. In both the acute and chronic forms there is spongy transformation of the brain commensurate with the degree of liver damage, and vacuolation of the white matter of brain stem but no malacia, neuronal damage or gliosis.[12]

Evidence that lupins may have been grazed, along with the appropriate clinical signs and post-mortem findings are sufficient to warrant immediate action. Blood samples may be taken from affected sheep for supportive laboratory evidence of liver dysfunction. Liver should be preserved in fixative for histology following post-mortem examination.

The sheep should be withdrawn from the affected pastures and the clinically affected should be grouped for special care and attention. Treatment is symptomatic. Ambulatory cases should be inspected at least once daily as they are liable to misadventure.

Basic slag poisoning

Basic slag, a by-product of steel production, is used as a source of phosphate for grass and other crops. It may be delivered in bags or when in bulk may be dumped in a heap in the field in which it is to be spread. Livestock should be excluded from the pasture during and after spreading until sufficient rain has fallen to wash it into the soil. When this has not been done, cases of poisoning have been reported in sheep and cattle.[13]

Basic slag appears to be attractive to lambs, but all ages may be poisoned. The sheep may die quickly or develop dark greenish diarrhoea and hind limb incoordination. Particles of slag may be seen in the liquid faeces. The lesions are those of an irritant poison namely inflammation of the abomasal and intestinal mucosae usually with particles of the slag present.

When deciding, after fertilizer has been spread, when to readmit the sheep, factors to consider are that long grass is safer than short, because of its dilution effect and the area where bulk slag was dumped may be so heavily contaminated that it requires to be excluded with a temporary fence.

The blood inorganic phosphorus levels of affected survivors may be high. Following post-mortem examination the stomach and its contents should be sent for laboratory examination.

REFERENCES

1 Stamp J.T. & Stewart J. (1953) Haemolytica anaemia with jaundice in sheep. *Journal of Comparative Pathology and Therapeutics*, **63**, 48–52.

2 Smith R.H. (1974) Kale poisoning. Annual Report of the Rowett Research Institute (Studies in Animal Nutrition and Allied Sciences), **30**, 112– 31.

3 Barry T.N., Manley T.R. & Millar K.R. (1982) *Journal of Agricultural Science, Cambridge*, **99**, 1–12.

4 Watson W.A., Barlow R.M. & Barnett K.C. (1965) Bright blindness: a condition prevalent in Yorkshire hill sheep. *Veterinary Record*, **77**, 1060–9.

5 McCrea C.T. & Head K.W. (1978) Sheep tumours in north-east Yorkshire. I. Prevalence on seven moorland farms. *British Veterinary Journal*, **134**, 454–61.

6 Jarrett W.F.H. (1980) Bracken fern and papilloma virus in bovine alimentary cancer. *British Medical Bulletin*, **36**, 79–81.

7 Robinson J.J. (1980) Energy requirements of ewes during late pregnancy and early lactation. *Veterinary Record*, **106**, 282–4.

8 Butler E.J., Nisbett D.I. & Robertson J.M. (1957) Ostoeporosis in lambs in a lead mining area. 1. A study of the naturally occurring disease. *Journal of Comparative Pathology and Therapeutics*, **67**, 378–96.

9 Allcroft R. (1951) Lead poisoning in cattle and sheep. *Veterinary Record*, **63**, 583–90.

10 Gardiner M.R. (1967) Lupinosis. *Advances in Veterinary Science*, **11**, 85–138.

11 Marasas W.F.O. (1974) In *Mycotoxins*. (Ed. Purchase I.F.H.) pp.111–27. Elsevier Scientific Publishing Co., Amsterdam.

12 Allen J.G. & Nottle F.K. (1979) Spongy transformation of the brain in sheep with lupinosis. *Veterinary Record*, **104**, 31–3.

13 Crowley J.P. & Murphy M.A. (1962) Basic slag toxicity in cattle and sheep. *Veterinary Record*, **74**, 1177–8.

48   POISONING BY PHENOLIC COMPOUNDS

K.W. ANGUS

Synonym Carbolic acid poisoning

Poisoning by phenolic compounds can occur not only by ingestion but also as a result of skin absorption and the use of phenolic preparations for topical application, for example in sheep dips, must take account of this proclivity. Poisoning can occur in three different circumstances. First, accidental poisoning by toxic industrial gases or effluents may result from inhalation or from ingestion of contaminated herbage, or both. Secondly, phenol-containing products commonly used around the farm, for example disinfectants or wood preservatives, may be accidentally ingested if spilt, washed into puddles, or left in open containers which could be knocked over. Thirdly, licenced preparations containing phenols for topical application, such as carbolic dips, may be toxic in certain circumstances or if used incorrectly.

Sources of phenols The main sources on the farm are disinfectants, creosote, coal-tar and coal-tar pitch.[1] Many common disinfectants e.g. Lysol, Jeyes Fluid, contain coal-tar derivatives including phenols and there is usually little difficulty in establishing an association between availability of the product and deaths or illness in the stock. The same may be true of some cases of creosote poisoning, but chronic creosote toxicity can result from exposure to sawdust impregnated with creosote and used as bedding.[2] Creosote is a distillate of coal-tar produced by high temperature carbonization of coal. It consists of liquid and solid aromatic hydrocarbons with appreciable amounts of tar acids and tar bases. Creosotes from different sources vary widely in composition, depending on the type of coal and the temperature used for carbonization. Sheep may be poisoned by ingesting the raw preserving fluid carelessly exposed, by licking newly treated fences or other timber, or most commonly by ingesting herbage contaminated by spillage. Sheep have also been poisoned when rain washed creosote off galvanized sheeting into their lying area (T.W.H. Jones, personal communication). Chewing of dry impregnated timber is not likely to cause toxicity. Pentachlorophenol (PCP) is considered less toxic than creosote for sheep.

Carbolic sheep dips, once fashionable but now used almost solely as bloom dips for show sheep, contain cresylic acid and other tar acids, derived mainly from the low-temperature carbonization of coal in the manufacture of smokeless fuels. These dips suffer from the same lack of standardization as creosotes, and consequently different dips, or even different batches of the same dip, may vary somewhat in their constitution. Toxicity generally occurs only when the manufacturers' directions are ignored or misunderstood. A common factor in many episodes is inaccurate knowledge of the capacity of the dipper, leading to dipping at too high a concentration. Farmers are inclined to err on the side of greater concentration rather than lesser and instances of combining different dips, or prewashing with carbolic sheep washes, are not uncommon. Additionally, there is strong circumstantial evidence that some carbolic dips can be toxic even when correctly prepared.[3] Phenol-containing antiseptic balms must also be used circumspectly; poisoning occurred when

a young lamb rejected by its mother was liberally coated, together with the dam's nose, with a proprietary disinfectant normally used for dressing navels and small cuts.[4] Finally, contamination of pasture by poisonous phenolic fumes from an asphalt boiling plant caused deaths, with high levels of phenols in stomach contents and other organs, in sheep and cattle.[5] Contamination of silage by phenols has also been reported.

Clinical signs

Sudden death, with no premonitory clinical signs, is often the first warning of an episode of phenol poisoning. In live cases, the main clinical signs are muscular convulsions or fasciculations, nervous system depression, diarrhoea due to irritation of the gut, and poor thermoregulatory ability.[1, 4] Experimental acute creosote poisoning (6 g/kg) caused dullness, inactivity, scouring and passage of dark urine with a strong tarry odour, followed by death in a few days, while inappetance, loss of condition, reluctance to move and diarrhoea were seen in chronic cases.[2]

Severe respiratory distress accompanied by peculiar grunting respirations has been reported in carbolic dip toxicity, usually within 1–3 days after dipping. Affected sheep do not eat, stand away from the rest of the flock and death usually ensues within 12 hours of the onset of clinical signs.[3] Visible membranes may be cyanotic, a brown mucopurulent fluid discharges from the nostrils, often accompanied by marked salivation, and the breath smells strongly of carbolic. The rectal temperature may rise slightly initially, but soon becomes subnormal as prostration supervenes. The pulse is strong and rapid initially, but later becomes weak and almost imperceptible.[6] More variably excessive thirst, clonic spasms and opisthotonos may occur.

Clinical signs in sheep poisoned by fumes from an asphalt boiling plant included fever, rapid pulse, inappetance, vomiting and tympany, with the passage of dark, viscous faeces with a pungent smell.[5]

Pathology

Carcases of sheep dying of poisoning from all sources often have a strong odour of carbolic. Phenols are irritant poisons when ingested, and features commonly found at necropsy in such cases include rumenitis, abomasitis, and patchy enteritis in both small and large intestines. Less commonly, liver enlargement, cystitis, and petechiation of the renal cortex, or the epicardium along the course of the coronary grooves, may be present. The lungs may be congested. Histological examination does not provide additional information in these circumstances.

In cases of poisoning by carbolic dips, the main feature is pneumonia with severe consolidation of both lungs. Lungs are usually dark red or dark blue, and heavy, firm or rubbery in consistency. Cervical, mediastinal, pharyngeal and prescapular lymph nodes are usually enlarged and oedematous and petechiation of many tissues, e.g. subcutis, tonsils, larynx, trachea, thymus, epicardium, kidneys and intestines, may be found. These necropsy findings have been verified experimentally (Fig. 48.1) and have been shown to be due to skin absorption, rather than accidental swallowing or inhalation of the dip during dipping.[4, 7] The microscopic changes are characteristic and do not resemble other sheep pneumonias. There is striking engorgement of vasculature, particularly alveolar capillaries. Alveolar walls are swollen by fluid leakage from dilated capillaries and by infiltrates of cells, mainly neutrophils and macrophages. Alveolar spaces contain serofibrinous material and variable numbers of neutrophils, macrophages and necrotic alveolar lining cells. The most prominent feature is diffuse lining of alveolar spaces by continuous sheets of cuboidal cells, though these may be separated from underlying structures by autolytic processes, so that they lie loosely in the alveoli (Fig. 48.2). Mitotic figures can be seen in some of the cuboidal cells and ultrastructural studies in experimental poisoning cases have shown that these cells are Type II pneumocytes. Hyaline membranes and focal necrosis of bronchio-

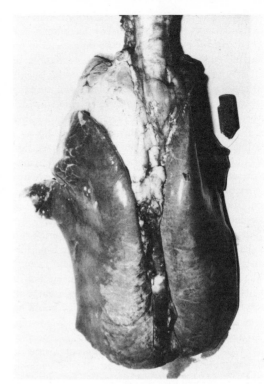

Fig. 48.1 Consolidation of lungs in experimental carbolic dip poisoning.

lar epithelium are occasionally found. Microscopic changes were absent from extrapulmonary tissues in cases of experimental poisoning.[3]

Diagnosis

Where accidental ingestion of phenols is resposible for deaths or illness, a cause–effect relationship can

Fig. 48.2 Histological appearance of the lung of a sheep which died after dipping in a carbolic dip. Note the fluid and cellular exudate, and the sheets of cuboidal cells lining many alveoli. (arrows). H & E x 48. (This Figure is reproduced by kind permission of Dr. K.A. Linklater, *et al*, and the Editor of the Veterinary Record).

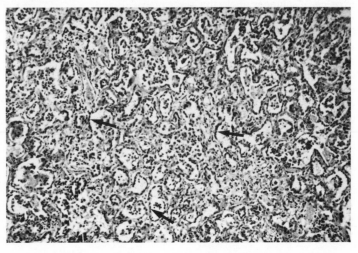

usually be established from the history. In deaths apparently due to pneumonia following dipping with carbolic dips it is essential to take lung material for bacterial culture and histological examination, as well as stomach contents and other organs for phenol and cresol analysis. Failure to isolate common lung pathogens, particularly *Pasteurella haemolytica*, together with the characteristic lung histopathology are good evidence for the implication of the dip as a possible cause of death. It must be emphasized that only traces of phenols absorbed through the skin may be present in organs at the time of death, partly due to metabolism and partly to elimination, thus assessment of the lung pathology is of great importance in differential diagnosis.

Samples from all cases of suspected poisoning can be analysed for phenols and cresols by high-performance liquid chromatography. They should include stomach contents, lung, liver, kidneys, urine and if possible, dipping fluid or other suspected source of phenols, together with blood. All samples should be frozen at $-20°C$ and transported as soon as possible to an analytical laboratory. Samples for transport by post or rail should be packed in leakproof plastic, or other suitable containers, in solid CO_2 in vacuum flasks or expanded polystyrene boxes and in the case of rail transport, an arrangement should be made with the laboratory to have the samples picked up at the depot on arrival.

Control and treatment

Prognosis in cases of phenolic poisoning is usually poor, and treatment is symptomatic. The use of demulcents is recommended for ingested phenols, with anti-diarrhoeal medicine and parenteral administration of solutions of electrolytes, aminoacids and dextrose as fluid support therapy in cases of prostration. Antibiotics and steroids have little or no mitigating effect on the respiratory form of poisoning, other than protection against secondary infections, but some improvement has been claimed with the use of tylosin (Tylan 200, Elanco Products Ltd.). Removal of excess dip from the fleece and skin by washing in soap and water is recommended.

Most measures are designed to prevent stock from gaining access to areas where phenolic compounds are stored or likely to be used. These should include avoiding housing or penning sheep where newly-treated timber is available. Sawdust for bedding should be obtained from untreated timber if possible, pasture contaminated by spillages or industrial wastes should be fenced off and silage should not be made from grassland contaminated by phenolic fumes. When carbolic dips are used, it is essential to adhere strictly to the label directions, and the practices of combining different dips or pre-washing in carbolic soap solutions, must be avoided. Finally, it is essential to know the exact capacity of the dipper, so that the dip is used at the correct dilution. These factors are of particular importance to breeders of Suffolk sheep, since recent episodes of dip toxicity have occurred almost exclusively in this breed.

REFERENCES

1 Humphreys D.J. (1978) A review of recent trends in animal poisoning. *British Veterinary Journal*, **134**, 128–45.

2 Harrison D.L. (1959) The toxicity of wood preservatives to stock. 2. Coal-tar creosote. *New Zealand Veterinary Journal*, **7**, 94–8.

3 Linklater K.A., Angus K.W., Mitchell B., Spence J.A., Rowland A.C. & Hunter A.R. (1982). Pneumonia in sheep associated with dipping in carbolic dips. *Veterinary Record*, **110**, 33–6.

4 Eales F.A., Small J., Oliver J.S. & Quigley C. (1981) Phenol poisoning in a newborn lamb. *Veterinary Record*, **108**, 421.

5 Hogstad J. (1965) Smoke pollution from an asphalt boiling plant as a cause of poisoning in cattle and sheep. *Nordisk Veterinaermedicin*, **17**, 220–4.

6 Steyn D.G. (1929) A note on the symptomatology of phenol poisoning in sheep induced by certain dips. 15th Annual Report of the Director of Veterinary Services, pp.657–8. Union of South Africa.

7 de Kock G. (1929) Pathology of phenol poisoning in sheep induced by certain dips. 15th Annual Report of the Director of Veterinary Services, pp.643–56. Union of South Africa.

SECTION 9

FLOCK HEALTH

49

FLOCK MANAGEMENT
FOR HEALTH

A.O. MATHIESON

In Britain, a closely integrated and interdependant stratified sheep production system exists, within which are many breeds and crosses and widely differing systems of management. In broad terms flocks can be categorized as hill, upland and lowland in the latter of which the greatest degree of intensification is practised. However, not all hill flocks are maintained on a completely extensive system and housing of sheep for varying periods may be a feature of the management system at all levels of sheep production for differing reasons. On the hill and exposed upland areas some degree of in-wintering may be carried out because the stock carrying capacity of the hill is limited and to protect certain classes of stock from adverse weather and reduce labour requirements, while in the upland and lowland areas a main consideration is to reduce land poaching. Not all sheep owners maintain breeding flocks, some purchase lambs and cast ewes solely for fattening while others purchase ewe lambs to be sold the following year as gimmers to breeding flock owners requiring female replacement stock.

The variations in sheep management which exist therefore make it difficult to do more than generalize on flock health programmes and in practice these require to be tailored to suit the specific type of husbandry being pursued on individual farms.

HILL FLOCKS

Apart from ram purchases the majority of hill flocks are self-contained. This feature coupled with the low stocking densities associated with extensive hill grazing systems generally results in a low incidence of infectious disease. The disease pattern on individual farms is influenced considerably by geographical location with the wet western regions of Britain having a higher incidence of fascioliasis while sheep grazing tick-infested pastures are exposed to tickborne fever infection, the possibility of tick pyaemia in lambs and in some areas louping-ill infection. The factor most influencing productivity in hill areas however is the adequacy of nutrition.[1]

In hill flocks vaccinations and other health care programmes should be planned, for economic utilization of labour, to coincide whenever possible with normal main flock gatherings in the management routine; tupping (mid November), lambing (mid April), marking and castration (mid May), shearing (mid July), weaning (mid September) and dipping (mid October). The exact timing of these operations varies in different localities throughout the country and the number of gatherings also varies between farms, depending on local conditions and the degree of intensification which is being imposed. For individual farms therefore, a calendar of events should be devised and displayed for the information of the shepherds, showing month by month procedures to be implemented for the different categories of stock; ewes, rams, ewe hoggs and gimmers. Sheep handling facilities suited to the size of flock are essential for routine health care procedures.

Vaccinations

Protection against clostridial diseases is an essential component of any sheep vaccination programme. Lambs should receive their first sensitizing injection at the marking gathering and the second immunizing dose when the flock is gathered for clipping some six weeks later. Ewe hoggs retained for breeding stock should receive their first booster dose at the beginning of April and in subsequent years as pregnant ewes, a booster injection at least two weeks before lambing to ensure maximum colostral antibody levels.

On farms where protection against louping-ill infection is required stock ewe hoggs should be vaccinated in mid-February in advance of the spring tick activity. Where ewe hoggs are wintered away this vaccination is best carried out before they are transported to the wintering area.

Internal parasite control

In hill flocks the main emphasis in helminth control should be placed on minimizing the spring egg rise. This can be achieved most effectively where the ewes are brought down to lowground or in-bye areas for lambing. Anthelmintic treatment is administered as the ewes are returned to the hill with their lambs or, in the case of ewes with twins being kept on in-bye fields, as they are transferred to these grazing areas. Stock ewe hoggs should also be treated in the spring. A second anthelmintic treatment of all breeding stock during October will help to ensure that these are in good bodily condition prior to tupping. When ewe hoggs are being winter housed anthelmintic treatment is best administered some ten days after housing.

Frequency of flukicide therapy is governed by the degree of risk and guidance should be taken from annual forecasts of likely disease prevalence. The basic strategy should incorporate treatment of all breeding stock in mid-October, during January and at the May gathering, the latter designed to reduce the level of contamination of snail habitats with fluke eggs. Additional treatments may be necessary in areas where liver fluke disease is a serious problem (see Chapter 12). The implementation on one hill farm of a strategic dosing programme designed to eliminate adult fluke stages throughout the year, thereby preventing the passage of eggs in the faeces, has demonstrated that this means of control is a practical proposition.[2]

Ectoparasite control

To ensure freedom from lice and ked infestation during the winter a late autumn dipping is essential. A dip providing control of mycotic dermatitis is recommended at this time. Summer dipping may not be necessary on farms where there is little risk of blowfly myiasis but even then consideration should be given to any need to prevent mycotic dermatitis.

Foot care

At each flock gathering careful observation should be made for lame animals and appropriate treatment carried out. Before returning sheep to the hill all should pass through a footbath containing 5 per cent formalin as a routine measure.

Nutrition

Apart from the need to provide winter supplementary feeding as dictated by weather conditions, particular times of the year can be identified when additional nutrition or the administration of trace elements is necessary to maintain hill sheep in a healthy condition. At weaning time the bodily condition of the ewe flock should be assessed and any lean ewes segregated for supplementary feeding to achieve a target condition score between 2 and 3 at mating. During late pregnancy any ewe with a body condition score less than 2 should be given extra feed, additional to any supplementary feeding which may be provided to the flock at this critical period. Unless this is done, lambs born to these ewes may be undersized, undernourished and the ewe will not produce an adequate supply of milk.

In circumstances where there is a risk of swayback lambs being born, particularly on areas of improved hill pasture where liming has been carried out, the administration of a copper preparation to ewes during mid-pregnancy is advocated.

Improved areas of pasture may result in an induced copper deficiency condition in lambs during the summer grazing period which can be counteracted by copper therapy.[3] In many hill areas a more likely problem is that of cobalt deficiency affecting lamb growth for which the administration of a cobalt intraruminal bullet is recommended. On farms where this deficiency has been diagnosed an additional cobalt bullet should be given to all stock ewe lambs in October to maintain adequate vitamin B_{12} levels throughout their continuing growth period.

Perinatal care

Detailed investigations of perinatal lamb losses in hill sheep flocks have consistently demonstrated that the majority of deaths result not from infectious causes but are due to a combination of exposure and starvation. In addition to the need to provide additional temporary shelter for the lambing flock on many hill farms there is obviously considerable advantage to be gained from applying the techniques developed to counteract hypothermia (see Chapter 51). To provide such a system on many hill farms even where the flock is gathered into the lower lying areas would probably require the development of a portable rewarming container.

UPLAND FLOCKS

In most upland farming areas sheep production is a main enterprise with large flock sizes, often between 300 and 600 ewes, regularly under the care of one full-time shepherd with additional labour provided during the lambing season. In the main, flocks consist of ewes of one of the hill breeds which are mated with rams of a longwool breed to produce ewe lambs or gimmers as replacement breeding stock for lowground flocks. Whether lambs and ewe lambs not selected for breeding are sold off for further fattening or retained to be sold for slaughter. Breeding flock replacements originate generally from one of three sources. Sufficient ewes may be bred pure for this purpose but, more commonly, ewe lambs are purchased from hill flocks. Some upland flock owners purchase ewes drafted from hill flocks after they have produced three or four lamb crops. In the more grass-productive upland areas, flocks of first cross ewes are kept for mating with rams of a Down breed, principally the Suffolk.

The now regular practice of sire marking of ewes at mating enables the breeding flock to be subdivided within two weeks of lambing for optimum feeding management and to avoid overcrowding in the lambing fields.

The requirements for lambing areas and proper care of the upland flock at this time have been described.[4]

Vaccinations

Due to the greater degree of intensification practised in upland flocks a rigid policy of full clostridial vaccination must be pursued. It has to be assumed that female and male breeding flock replacements have not been previously vaccinated. These should be given a sensitizing dose of a multi-component vaccine in September followed by an immunizing dose six weeks later, prior to tupping which generally commences in mid-October. Both gimmers and ewes require a booster dose before lambing to provide their lambs with passive protection for 12–16 weeks. Lambs destined for slaughter after the age of 16 weeks should be immunized against pulpy kidney disease.

In areas where enzootic abortion is prevalent a single vaccination of female replacements pre-mating provides protection for three years.

Internal parasite control

The practice of closely confining ewes during lambing enables anthelmintic treatment of ewes to be carried out as a routine measure as they are transferred from the lambing field. On those upland farms on which a large part of the grazing is permanent pasture and has therefore carried sheep

stock in previous years regular anthelmintic treatment of lambs may be necessary by mid-May and monthly thereafter throughout the grazing season. Increasingly, however, upland farms have significant acreages of shorter term pastures which allow the adoption of a clean grazing system.[5]

A control scheme for liver fluke disease in upland areas is likely to involve a combination of improved drainage or fencing off of high risk areas as well as flukicide treatment of the sheep flock; the frequency of treatment depending on the disease risk forecast. Serious consideration should be given to the adoption of the dosing strategy referred to previously, to reduce the contamination of snail habitats with fluke eggs.

Ectoparasite control

Both summer dipping for myiasis and mycotic dermatitis control and autumn dipping for mycotic dermatitis and lice control should be regarded as essential procedures in the calendar of flock care.

Foot care

While every shepherd's aim should be to minimize the number of lame sheep at all times, particular attention to feet condition is necessary before mating and during mid-pregnancy. The elimination of foot rot is scarcely a practical proposition in large commercial flocks where control is more likely dependent on treatment of individual cases of lameness, regular use of the footbath and the use of foot rot vaccine (see Chapter 22). Scald, affecting lambs principally, is most prevalent during early summer when the whole flock should be walked through a bath containing 5 per cent formalin as soon as early cases are observed.

Nutrition

With the majority of ewes in the upland flock expected to be twin bearing, matching the feed requirement to physiological needs is an essential part of good management, with assessment of bodily condition score a prerequisite several weeks before mating and during mid-pregnancy. A high energy concentrate should be fed to lean ewes identified at these critical periods. The provision of a high protein concentrate in late pregnancy and throughout early lactation results in well-developed lambs and ensures satisfactory milk yield from the ewes.[6] Adequate trough space to enable shy feeders to obtain their ration and special attention to ewes that are in any way incapacitated further helps to avoid the occurrence of pregnancy toxaemia cases (see Chapter 32).

Perinatal care

The greatest potential for improving productivity in many flocks lies in increasing the lamb survival rate when it is considered that the average perinatal loss in upland flocks is about 12 per cent, and that between individual flocks a range from 3–30 per cent has been recorded.[7] Two major factors influencing these figures are the extent to which abortion is a problem and the weather conditions prevailing during the lambing season. The various infectious causes of abortion along with control measures which can be taken are discussed elsewhere. The adverse effect of inclement weather can be modified by the provision of ample shelter devices but it has to be accepted that ensuring high lamb survival rates demands a high labour input over this critical period. This is essential if continuous flock supervision is to be maintained, cases of dystokia dealt with promptly, and lambs suffering from varying degrees of hypothermia and starvation identified and given appropriate nursing care. Obviously losses are unavoidable but where apparent still-births and deaths attributed to exposure and starvation account for 60–70 per cent of neonatal loss, investment in reducing this figure significantly must be worthwhile. To say that investigations of neonatal loss show on average that only some 15 per cent of these are due to infectious causes is not to infer that these are not insignificant. The most common amongst these is enteritis which invariably becomes more prevalent as the lambing season progresses and the area in use becomes increasingly soiled. To check the course of an outbreak it may be necessary to move the remaining in-lamb ewes to an alternative lambing field. Rubber ring castration and docking, which cause temporary discomfort and immobility in lambs, should not be carried out when climatic conditions are particularly adverse. Lambing pens should always be freshly bedded for each ewe and as a routine hygienic measure each lamb's navel dressed with iodine, primarily to prevent necrobacillosis.

LOWLAND FLOCKS

Such a wide variety of flock management systems exist on lowground farms that within the scope of this book only general recommendations can be made on health care. Lowland flocks may consist of pure Down and longwool breeds maintained essentially for ram production, Down cross ewes for early fat lamb production, Dorset Horns kept for twice yearly lambing or draft hill and upland ewes producing fourth and more lamb crops. Additionally hormone controlled breeding techniques have resulted in a situation within Britain where lambing may be occurring at any time of the year although the majority of lowground flocks are managed to lamb within the December–March period. The degree of intensification practised in many of these flocks is such that much dependence needs to be placed on vaccination and parasite control programmes to maintain the animals in a healthy state, particularly where full-time experienced shepherding may not be available.

Forward planning should include the preparation of a time-table of vaccinations, pasture use, external and internal parasite control and feeding regimes for the individual flock.

Vaccinations

A comprehensive vaccination programme covering the infectious diseases commonly encountered in lowland flocks is shown in Table 49.1.

In fat lamb producing flocks the majority of lambs will be sold for slaughter by 16 weeks of age and only those being retained for breeding need be actively immunized against clostridial diseases.

Internal parasite control

The availability of new short-term pastures on all lowground farms facilitates the operation of a clean grazing system in which all ewes are treated with an anthelmintic on introduction to the grazing area, with no necessity to dose lambs until mid-summer when they are weaned and moved to silage or hay aftermaths.

Housing

On many farms housing for a variable period has become an integral part of lowland flock systems.[8]

Table 49.1 Vaccination schedule

	Pre-mating	Mid pregnancy	Late pregnancy	Early lactation	Weaning	Post weaning
Flock replacements	Cl–P E	O	Cl–P E			
Ewes or gimmers	EAE 2 × F			F		
Stock ewes	F	O	Cl–P	F		
Rams	Cl–P F		Cl–P	F		
Lambs					Cl–P	Cl–P

Key: Cl–P = combined Clostridia and Pasteurella; EAE = Enzootic abortion of ewes; F = footrot; E = Erysipelas; O = Orf.

When sheep are being housed for a prolonged period a number of basic procedures can be recommended. Good ventilation of the sheep house is essential to avoid respiratory disease, trough space must be sufficient to allow all the animals within a group to feed simultaneously and fresh clean drinking water should be provided.

Fleece damage due to ectoparasites, principally lice infestation, is avoided by dipping several days prior to housing and the flock housed in a dry state.

Immediately prior to housing, treatment of lame ewes and foot paring should be carried out followed by walking the flock through a footbath containing 5 per cent formalin. Providing the floor area is kept reasonably dry and ewes are subjected to footbath treatment at regular 3 week intervals further foot problems should be minimal.

Introducing sheep to the concentrate ration to be fed indoors, while they are still outside, helps to prevent cases of acidosis arising from overeating during the initial confinement period. Anthelmintic treatment should be carried out about 10 days after housing. In formulating rations for housed sheep the mineral content should be balanced to reduce the risk of urolithiasis in males and castrates and particular attention should be paid to the copper level in the concentrate ration which should not exceed 15 mg/kg; sheep being susceptible to chronic copper poisoning. This hazard can be alleviated by the addition of ammonium molybdate to the ration.[9]

When ewes are to be kept inside continuously over the lambing season they are best penned in groups of 30 or less so that individual animals can be easily observed for signs of ill-health, to reduce bullying and to control infectious diseases for which no vaccines are available such as salmonellosis, ovine pulmonary adenomatosis and scrapie. The close confinement of stock indoors facilitates a build-up of pathogenic microbes particularly those causing en-

teric disorders. Thorough disinfection of lambing pens and fostering units between ewes is recommended. Any animals showing evidence of contagious conditions like orf and periorbital dermatitis should be segregated. Divisions between pens should be constructed so that young lambs are prevented from straying between groups. In lowland flocks where high prolificacy is sought, special provisions should be made for the artificial rearing of triplets, orphaned and weakly lambs. While indoors the risk of chilling from cold exposure is much reduced, close observation during the neonatal period is vital to detect cases of colostrum deprivation and starvation. Reserve supplies of colostrum easily obtained from lambed ewes can be held at 4°C for rewarming before use. The maintenance of healthy and productive flocks indoors demands a high degree of good stockmanship.

REFERENCES

1 Eadie J. (1981) Science in hill sheep management. *The Veterinary Annual*, **21**, 122–5.

2 Whitelaw A. & Fawcett A.R. (1977) A study of a strategic dosing programme against ovine fascioliasis on a hill farm. *Veterinary Record*, **100**, 443–7.

3 Whitelaw A., Armstrong R.H., Evans C.C. & Fawcett A.R. (1979) A study of the effects of copper deficiency in Scottish blackface lambs on improved hill pasture. *Veterinary Record*, **104**, 455–60.

4 Mitchell B. (1979) Management and disease control in lambing ewes. *The Management and Diseases of Sheep*. pp.257–64. Commonwealth Agricultural Bureaux, Slough.

5 Rutter W. (1975) Sheep from grass. Bulletin No.13. East of Scotland College of Agriculture, Edinburgh.

6 Speedy A.W. (1980) *Sheep Production: Science into Practice*. Longman, London.

7 East of Scotland College of Agriculture (1977) Perinatal losses in lambs. A collection of papers from a symposium at Stirling University in February, 1975. High Output Lamb Production Group of the Scottish Agricultural Colleges.

8 Fell H. (1979) *Intensive Sheep Management*. Farming Press, Ipswich.

9 Harker D.B. (1976) The use of molybdenum for the prevention of nutritional copper poisoning in housed sheep. *Veterinary Record*, **99**, 78–81.

CONTROL OF GASTROINTESTINAL HELMINTHIASIS

50

J. ARMOUR

The important gastrointestinal helminth diseases of sheep are:
1. nematodiriasis in lambs;
2. parasitic gastroenteritis in lambs and occasionally older sheep;
3. fascioliasis, usually in sheep over 4 months.

Cestode infections are also common in lambs but are generally not considered to be a problem.

The successful control of any helminth disease is based on a sound knowledge of its epidemiology. The latter often varies according to the helminth species involved, so it is proposed to consider the control of each of the above diseases separately.

NEMATODIRIASIS

Although several species of *Nematodirus* occur in British sheep it is *Nematodirus battus* which is the species responsible for the severe outbreaks seen in lambs in late spring. Epidemiology of this disease is based on three main factors.
1. The egg which contains the infective third stage larva (L3) has a high resistance to freezing or desiccation and can survive on pasture for up to 2 years.
2. Hatching, with release of the L3, requires special stimuli in the form of a period of chill followed by a mean day/night temperature of more than 10°C.
3. Adult sheep are resistant.

Because of the great capacity of the eggs for survival the infection can be continued from lamb crop to lamb crop, without the necessity for any intervening passage. Accumulation of infection on pasture therefore takes place over a period of years and not in a single season as in parasitic gastroenteritis. Disease never occurs on first year grass, is rare on second year, but by the third year of grazing by lambs, contamination may be at pathogenic levels.

As a result of the critical hatching requirements there can be an almost simultaneous appearance or 'flush' of large numbers of L3 on the pasture. Though the flush happens each year, disease does not always follow even on heavily contaminated grazing for if the flush is very early (April) many young lambs are ingesting insufficient grass to take in large numbers of L3, and if it is late (June) they are able to resist the larval challenge since age resistance appears by about three months of age and is high by six months.

Clearly, due to the annual hatching of *N. battus* eggs in spring the disease can only occur on fields grazed by young sheep in the previous year. Control can therefore be successfully achieved by avoiding the grazing of successive lamb crops on the same pasture and this is the ideal method. Where such alternative grazing is not available each year then control can be achieved by anthelmintic prophylaxis, the timing of treatments being based on the knowledge that the peak time for the appearance of *N. battus* L3 is May to early June. Ideally, dosing should be timed at three-weekly intervals over May and June and it is unwise to await the appearance of clinical signs of diarrhoea before administering the drugs.

The Ministry of Agriculture have developed a forecasting system for Britain based previously on soil temperature in the early spring which can predict the likely severity of nematodiriasis.[1] In years when the forecast predicts severe disease then three treatments are recommended during May and June, in other years two treatments in May should suffice. Several drugs are to be recommended, particularly levamisole or one of the modern benzimidiazoles: levamisole (Nemicide, ICI, Macclesfield, Cheshire, England) at 7.5 mg/kg, fenbendazole (Panacur, Hoechst Ltd., Milton Keynes, Bucks, England) at 5 mg/kg., oxfendazole (Systamex, Wellcome, Berkhamstead, Herts, England) at 5 mg/kg, albendazole (Valbazen, Smith Kline & French, Welwyn Garden City, Herts, England) at 5 mg/kg.

PARASITIC GASTROENTERITIS

Parasitic gastroenteritis in Britain is primarily a disease of lambs and occasionally older sheep. The principle genera present are *Ostertagia* and *Trichostrongylus*; less frequently *Haemonchus, Strongyloides, Cooperia, Nematodirus* spp. other than *N. battus, Bunostomum, Chabertia, Trichuris* and *Oesphagostomum* are involved.

As mentioned previously (see Chapter 11), the herbage numbers of L3 of the above species increase markedly from mid-summer onwards and this is when disease problems arise. Usually *Ostertagia* and *Haemonchus* larvae appear first, followed later in the summer and early autumn by the other species. The source of these larval infections are two-fold.[2]

1. Strongylate eggs passed in the faeces of ewes during the peri-parturient relaxation of immunity. The duration of the egg output by the ewes is from about 2 weeks prior to lambing until 6 weeks post-lambing, i.e. 8 weeks in total.
2. Strongylate eggs passed by lambs and resulting from the ingestion of overwintered L3; the latter can overwinter in fairly high numbers but decline in number rapidly during April and May and few survive beyond June.

It is important to realize that it is the eggs deposited in the first half of the recognized grazing season, i.e. April–June which are responsible for the potentially dangerous populations of infective larvae which accumulate in the second half, i.e. July–September. If ingested prior to October, the majority of these larvae mature in a few weeks; thereafter until the following spring many of the larvae ingested become arrested in development for up to several months.

The object of any control programme is to prevent contact between the host and the infective stage of the parasitic. In sheep this may be achieved by using drugs to prevent or limit the contamination of pasture by ewes and lambs prior to June or, by avoiding the grazing of pastures after June where contamination has occurred and larval populations are likely to be high. In some instances a combination of both methods may be applied. The method of prophylaxis applied depends on whether or not alternative grazing is available for the sheep either on an annual basis or in mid-season.

CONTROL ON FARMS WITH LIMITED ALTERNATIVE GRAZING

These farms consist primarily of permanent pasture which must be grazed all year round and therefore includes many upland and most hill farms. In these farms control may be achieved in two ways, namely:
a. by anthelmintic prophylaxis, or
b. by alternate grazing on an annual basis with cattle.

Clearly, the former is the only method where the farm is primarily stocked with sheep while the latter is to be recommended where cattle and sheep are both present in reasonable proportions.

Prophylaxis by anthelmintics

The most important source of infection for the lamb crop is undoubtedly the peri-parturient increase in strongylate eggs in ewe faeces. Prophylaxis is only efficient if this is kept to a minimum. Effective anthelmintic therapy of ewes during the fourth month of pregnancy should eliminate most of the worm burdens present at this time including arrested larval stages and in the case of hill ewes where nutritional status is frequently low this treatment often results in improved general body condition. However, during late prenancy and early lactation such treated ewes become reinfected from the ingestion of overwintered larvae on the pasture. It is therefore recommended that for optimal prophylaxis a further treatment be given within one month of lambing although in practice this is often delayed until 6–7 weeks post-lambing; by this time the treatment is less effective, as the egg counts of the ewes will have declined.

An alternative to the gathering of ewes for these treatments is to provide anthelmintic incorporated in a feed or energy block during the peri-parturient period. The results obtained with the latter system appear to be best when the ewes are contained in small paddocks or fields and uptake of drug is less efficient under extensive grazing systems. Rumen boluses containing anthelmintics and designed for the slow release of drug over a prolonged period are under development and indeed one is available for cattle (Paratect, Pfizer Limited, Sandwich, Kent, England). Such a delivery system seems an ideal way of controlling the peri-parturient increase in strongylate egg output and would avoid the problems of regular handling for individual treatments.

Apart from specific treatment for *Nematodirus* infection, as mentioned previously, lambs should be treated at weaning, and if possible moved to safe pasture, i.e. pasture not grazed by sheep since the previous year. Where such grazing is not available then prophylactic treatments should be repeated through until autumn or marketing. Hoggs and tups should be treated in the same manner as the ewes. The number of treatments vary depending on the stocking rate, one treatment in September sufficing for hill lambs and two under upland conditions.

The following anthelmintics are recommended for the above prophylactic treatments and the list includes only drugs effective against arrested larvae which accumulate during autumn and winter:

levamisole (Nemicide, ICI, Macclesfield, Cheshire, England) at 7.5 mg/kg.

The following benzimidazoles:

  ivermectin (Ivomec, Merck, Sharp & Dohme Ltd., Hoddesdon, Herts, England) 200 mcg/kg.
  fenbendazole (Panacur, Hoechst Ltd., Milton Keynes, Bucks, England) at 5 mg/kg;
  oxfendazole (Systamex, Wellcome, Berkhamstead, Herts, England) at 5 mg/kg;
  albendazole (Valbazen, Smith Kline & French, Welwyn Garden City, Herts, England) at 5 mg/kg.
  the pro-benzimidazole, thiophanate (Nemafax, May & Baker Ltd) at 50 mg/kg.

Several other benzimidazoles and pro-benzimidizoles are marketed but these are only fully effective against arrested larvae at dosage rates in excess of that routinely recommended.

For low-level administration in feed blocks the following have proved useful:

thiophanate (Sheep Energy Wormer Block, Colborn-Dawes, Canterbury, Kent, England);
fenbendazole (Wormablock, Rumenco, Burton-on-Trent, England).

The prophylactive programmes outlined above are costly in terms of drug and labour handling but unfortunately they are currently the only methods available in upland and hill farm where the enterprise is heavily dependent on one animal species.

Prophylaxis by alternate grazing

On farms where sheep and cattle are both present in significant numbers, good control of ovine parasitic gastroenteritis can be achieved by alternating the grazing of fields on an annual basis with the different host species. The basis for this control is two-fold:

1. the host specificity of the different nematode species. Theoretically, only *Haemonchus contortus* and *Trichostrongylus axei* can develop to maturity in both sheep and cattle though some of the *Ostertagia* spp. e.g. *O. leptospicularis* are also adapted to both hosts.
2. the annual mortality of overwintered L3. But recent evidence, particularly in cattle, suggests that it may require 2 years for the larval population on the herbage and upper soil layers to completely die out.[3]

However, despite these two qualifications good control is possible by simply exchanging in the spring, the pastures grazed by sheep and cattle over the previous year, preferably, combined with an anthelmintic treatment at the time of exchange. Care should be taken to use one of the efficient anthelmintics mentioned previously.

CONTROL ON FARMS WITH A PLENTIFUL SUPPLY OF ALTERNATIVE GRAZING

In these farms which are mostly situated in the lowlands, rotation of crops and grass are often a feature and therefore new leys and aftermaths are available each year. In such a situation control should be based on a combination of anthelmintic prophylaxis and grazing management.

Control by grazing management and anthelmintics

Good control is obtained with only one anthelmintic treatment of ewes and this should be carried out when the ewes leave the lambing field.[4] This treatment terminates the peri-parturient increase in nematode eggs prior to moving the ewes and lambs to a safe pasture, such as a new ley. At weaning, the lambs should be moved to another safe pasture. Although an anthelmintic treatment of the lambs at this time is good policy it may be unnecessary if the new pasture is really safe, i.e. not grazed by sheep since the previous season.

An excellent control system [5] which is not costly in terms of labour or drugs has been devised for farms with arable crops, sheep and cattle as equal components of the total enterprise. The system can be adapted to suit farms where sheep and cattle dominate the livestock component. A 3 year rotation is recommended in Table 50.1.

Table 50.1 A three year rotation is recommended

	Field 1	Field 2	Field 3
Year 1	Cattle	Sheep	Crop
Year 2	Sheep	Crop	Cattle
Year 3	Crop	Cattle	Sheep

Where aftermath grazing is available after cropping, this can be used for weaned calves and sheep but lambs or hoggs must not be allowed access to the cattle area since this is intended as the next year's safe grazing. It has been suggested that anthelmintic prophylaxis can be disposed of completely under this system but clinical parasitic gastroenteritis has sometimes occurred when such a recommendation has been adopted. It is worth remembering that even the high quality drugs currently available do not necessarily remove all the worms present, that some cattle nematodes can infect sheep and vice versa and that a few infective larvae on the pasture can survive for beyond 2 years. So, it is advisable to use at least an annual spring treatment within the 3 year rotation outlined above. This treatment should be given at the time of moving to new pastures.

Control by grazing management

Many schemes have been devised to control the acquisition of L3 purely based on grazing management. One recommendation was to rotate grazing through paddocks but since it involved the return of the sheep to their previously grazed paddocks in the same season, it was of little value since we now know that contamination in the spring results in larval infections in the herbage during summer. Two other

methods, namely strip grazing, in which sheep are confined to a narrow strip across the field by fences which are moved every few days, and creep grazing, in which a single fence confines the ewes but allows the lambs to graze forward, do not involve a return to the original pasture. These systems can be highly effective in preventing parasitic gastroenteritis but are costly in fencing and labour.

FASCIOLIASIS

Severe outbreaks of fascioliasis or liver fluke disease only occur following wet springs and summers. The British Ministry of Agriculture have therefore been able to develop formulae for forecasting the incidence and severity of fascioliasis based mainly on rainfall data from the preceding month. In the case of the summer infection of snails responsible for outbreaks of acute ovine fascioliasis accurate predictions can be made by the end of the summer; however, and 'early warning' can also be issued if May, June and July have been unduly wet.[6]

Control of fascioliasis whether on a long term or on a short term basis in conjunction with the forecast system may be approached in two ways; firstly, by reducing population of the snail intermediate host *Lymnaea truncatula* or secondly by using anthelmintics to limit the availability of *Fasciola hepatica* eggs and therefore miracidia to surviving snail populations. Readers should refer to Chapter 12 for details of the methods of control.

TAENIASIS

Although tapeworms of several genera, namely, *Moniezia, Thysanosoma, Avitellina* and *Stilesia* have been reported as causing disease in sheep in various parts of the world only *Moniezia* occurs in Britain. The pathogenicity of *Moniezia* spp. in lambs in Britain has yet to be conclusively demonstrated and it is almost certainly the ease by which segments are recognized in the faeces that induces farmers to treat lambs for these tapeworms.

Drugs with a special activity against tapeworms in sheep are available, e.g. niclosamide (Mansonil, Bayer) at 100 mg/kg and more recently praziquantel (Droncit, Bayer) at 5 mg/kg which is active against adult and larval stages. However, treatment with a specific drug is usually unnecessary and, since several of the benzimidazole group, e.g. fenbendazole, oxfendazole and albendazole routinely used to prevent nematodiriasis and parasitic gastroenteritis, possess good efficiency against *Moniezia* spp, treatment for tapeworms is usually incorporated with prophylaxis of these roundworm diseases. Tapeworm burdens are highest in lambs in spring and early summer and if considered a problem one of the above benzimidazoles should be selected for routine prophylaxis at this time.

Combined anthelmintics for roundworms and fluke

The indiscriminate use of a combined treatment of gastrointestinal nematodes and liver flukes can be wasteful. The optimal times for anthelmintic prophylactic treatments of these two helminthiases do not usually coincide except in the spring when a product containing a roundworm and fluke drug, e.g. Ranizole (Merck, Sharp & Dohme), which contains the benzimidazole, thiabendazole and rafoxanide or Nilzan (ICI) which contains levamisole and oxyclozanide can be used. The benzimidazole, albendazole has high activity against roundworms, cestodes and adult flukes and at 7.5 mg/kg is a good choice for the spring treatment although it would be preferable to use a drug with more activity against immature flukes. So, as a general rule it is better to use separate drugs for the two groups of helminths to ensure the best control.

Drug resistance

Resistance to anthelmintics has been recorded mainly from the tropics where *Haemonchus* spp predominate and where the number of annual generations of nematodes and the number of annual treatments are more numerous than in temperate zones such as western Europe. This resistance or 'tolerance' as it is frequently called has been reported principally with the benzimidazoles or pro-benzimidazoles and refers to gastrointestinal nematodes of sheep.

Side-resistance has been reported between different benzimidazoles and cross-resistance to drugs with different modes of action such as levamisole and morantel tartrate.

In Britain, reports of anthelmintic resistance in the field have only recently been substantiated and involve both the benzimidazoles and levamisole and the nematodes *O. circumcincta* and *H. contortus.*[7, 8, 9]

Australian workers have observed that nematodes have sufficient genetic variation, to develop resistance to different anthelmintic groups to which they are exposed i.e. multiple resistance, at rates at least equal to that which develops to a single drug.[10] They suggested that in order to reduce the chances of multiple resistance developing it would be sound policy to alternate the anthelmintics used on a farm provided that the alteration occurs between different generations of worms and that the alternative drugs are from different chemical groupings. It also appears that if the use of a particular anthelmintic is suspended for a few generations, some reversion to susceptibility to that drug is likely to occur. In practice, since the number of generations of worms per annum in Britain is no more than two, an annual change of drug is recommended.

However, it should be borne in mind that while resistance in Australia is associated with frequency of dosing other selection pressures may exist under British conditions. For example where anthelmintic treatments are combined with a move to safe pastures selection for resistance may be accelerated since only the progeny of worms which survive the treatment will be present on the safe pasture. This aspect is currently under investigation.

REFERENCES

1 Smith L.P. & Thomas R.J. (1972) Forecasting the spring hatch of *Nematodirus battus* by use of soil temperature data. *Veterinary Record,* **90**, 388–92.

2 Thomas R.J. & Boag B. (1972) Epidemiological studies on gastro-intestinal nematode parasites of sheep. Infection patterns on clean and summer-contaminated pasture. *Research in Veterinary Science,* **13**, 61–9.

3 Armour J., Al Saqur I.M., Bairden K., Duncan J.L. & Urquhart G.M. (1980) Parasitic bronchitis and ostertagiasis on aftermath grazing. *Veterinary Record,* **106**, 184–5.

4 Gibson T.E. (1973) Recent advances in the epidemiology and control of parasitic gastro-enteritis in sheep. *Veterinary Record,* **92**, 469–73.

5 Rutter W. (1975) Sheep from grass. Bulletin No. 13, East of Scotland Agricultural College, Edinburgh.

6 Ross J.G. (1977) A five-year study of the epidemiology of fascioliasis in the North, East and West of Scotland. *British Veterinary Journal,* **133**, 263–72.

7 Britt D.P. (1982). Anthelmintic resistance in sheep nematodes. *Veterinary Record,***110**, 343.

8 Cawthorne R.J.G. & Whitehead J.D. (1983). The isolation of Benzimidazole resistant strains of *Ostertagia circumcincta* from British sheep. *Veterinary Record,***112**, (in press).

9 Cawthorne R.J.G. (personal communication).

10 Pritchard R.K., Hall C.A., Kelly J.D., Martin I.C.A. & Donald A.D. (1980) The problem of anthelmintic resistance in nematodes. *Australian Veterinary Journal,* **56**, 239–51.

51 HYPOTHERMIA IN NEWBORN LAMBS

F. A. EALES

Hypothermia means a lower than normal body temperature which in a lamb is 39.2±0.4°C (about 102°F). Each year in the UK 10–15 per cent of all lambs born die in the first few days of life. At least half of these losses (about 1 million lambs) are caused by hypothermia. Hypothermia is progressive in nature; once body temperature has fallen more than a few degrees Centigrade it continues to fall producing a depression of all body functions. Ultimately the lamb will die unless the decline in body temperature is halted and reversed by the shepherd.

The maintenance of a normal body temperature depends on a precise balance between the heat lost from the lamb to the environment and the heat produced by the lamb. It is thus necessary when discussing possible causes of hypothermia to consider what factors can affect both heat loss and heat production.

Heat loss

Factors which affect heat loss from a lamb can be divided into 'lamb' and environmental factors. There are two important lamb factors which affect heat loss. The first is the body surface area to which the rate of heat loss from a lamb is directly proportional. The body surface area of a small lamb is proportionately greater than that of a large lamb, i.e. a small lamb has a high surface area:body weight ratio. This means that a small lamb loses heat faster than a large lamb in proportion to its body weight. The second lamb factor affecting heat loss is the insulation value of the coat. The short fine birth coat of the Merino lamb has a low insulation value when compared with the long coarse coat of the Scottish Blackface lamb.[1] Wetness of the coat, the normal state at birth, considerably reduces the insulation value especially in lambs with short fine coats.

The major environmental factors affecting heat loss are wind speed and temperature (Fig. 51.1). Heat loss is low in warm still conditions but is increased by either a decrease in environmental temperature or an increase in wind speed. The affects of wind speed are most pronounced when the lamb's coat is wet (Fig. 51.1). The ewe can considerably reduce heat loss from the newborn lamb by licking it dry immediately after birth and providing it with shelter.

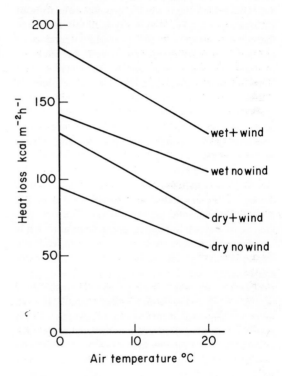

Fig. 51.1 The effects of wind, environmental temperature and coat wetness on the rate of heat loss from newborn lambs.
From Alexander, 1962.

Heat production

When a newborn lamb is exposed to cold, heat production is increased by two mechanisms.[2] The first is non-shivering thermogenesis, mostly if not all attributable to brown fat metabolism, and the second is shivering thermogenesis. In the newborn lamb these two mechanisms are of equal importance. Heat production, by whatever means, is dependent on a supply of energy rich substrate and oxygen. Any restriction on the availability of either of these two metabolites depresses heat production. It is convenient to consider factors which can affect heat production in chronological sequence starting with those that occcur in pregnancy and ending with events in the first day or so of life.

Three important characteristics of the newborn lamb are determined during pregnancy. These are size, quantity of energy reserves and the state of maturity at birth. Size determines the rate of heat loss from the lamb and thus the requirement for heat production. The quantity of energy reserves determines how long the lamb can produce heat without sucking and the state of maturity is closely related to the maximum rate at which heat can be produced [3] and also the ability of the lamb to stand and suck, i.e. its viability. These three factors are related to two characteristics of fetal life; the size of the placenta through which nutrients and oxygen gain access to the fetus and the nutrition of the ewe which determines how much nutrient is available. The desirable state is a well-developed placenta and a well-nourished ewe.

The size of the placenta depends on the maturity of the ewe, the number of fetuses and the nutrition of the ewe in the first half of pregnancy for it is in this period that placental growth occurs. The first two factors are beyond immediate control but nutrition is not. Poor nutrition in early pregnancy can lead to a small placenta[4] which can have a number of serious consequences for the growing fetus later in pregnancy. A restriction of nutrient supply to the fetus results in poor growth and a small lamb. Of even more consequence is restriction of oxygen[5] which may result in either fetal death or the premature birth of small immature lambs with a low heat production capacity and thus a high susceptibility to hypothermia.

Good nutrition in late pregnancy is essential if the growth potential of the fetus is to be fully exploited.[6] Poor nutrition results in a small lamb with low energy reserves even if the lamb has a large placenta. Such a lamb has a high heat production requirement but has little fuel with which to produce this heat. In addition it is likely to be short of milk since poor nutrition in late pregnancy depresses milk production. Poor placental development and poor nutrition are likely to have the most serious consequences in twins and especially in triplets.

Birth presents the lamb with a number of hazards. Significant in the context of heat production is an acute shortage of oxygen (hypoxia).[7] Hypoxia, if severe, can cause death and stillbirth but many lambs survive this insult and are born alive. However, for the first few hours of life these lambs are very lethargic and can produce little heat, so hypothermia is inevitable. This depression of heat production is temporary in nature and normal thermoregulatory ability is regained within 12 h of birth. Providing hypothermia is prevented in this period these lambs survive and develop normally.

Colostrum provides the lamb with two essential ingredients for survival; immunoglobulin which helps to prevent disease and energy which is essential for the maintenance of heat production. Starvation of the newborn lamb can lead to an exhaustion of energy reserves within six hours of birth which results in depressed heat production and hypothermia.[8, 9] The early ingestion of colostrum has an additional benefit to the newborn lamb in that even if body energy reserves are still replete colostrum ingestion leads to an increase in heat production of about 17 per cent and thus an increased resistance to hypothermia.[10]

There are a number of disease conditions occurring in early post-natal life which can directly or indirectly lead to depressed heat production and hypothermia. Pneumonia, often caused by inhalation of milk during bottle feeding, leads to poor respiratory function, tissue hypoxia and a direct depression of heat production whereas enteritis leads to a physiological starvation with the consequences already outlined.

Causes of hypothermia.

In the past, most cases of hypothermia have been attributed to the 'exposure/starvation' syndrome. Recent work has demonstrated that this is an inappropriate term and that there are two major causes of the condition.[11]

The first cause, (Table 51.1) which affects lambs aged between birth and 5 hours is a high rate of heat loss normally attributable to the combination of a wet birth coat, inclement climatic conditions and a ewe which is slow to lick her lambs dry. Examination of blood from such lambs shows high levels of glucose (Fig. 51.2). The second major cause of hypothermia is depressed heat production due to starvation in lambs aged 6 h or more but commonly 12–48 h. These lambs have very low levels of blood glucose (hypoglycaemia) (Fig. 51.2). The elucidation of these two distinct causes of hypothermia which are

Table 51.1 The causes of hypothermia in newborn lambs

Cause	Age at which hypothermia occurs (h)	Predisposing factors	High risk lambs
High rate of heat loss	0–5	Hypoxia during birth: depressed heat production	Twins and triplets
		Immaturity: depressed heat production	Lambs from young ewes: take longer to lick lambs dry
Low rate of heat production due to starvation	12–48	Immaturity: lambs not strong enough to suck	Twins and triplets
		Acquired disease	Lambs from old ewes: often in poor condition and short of milk

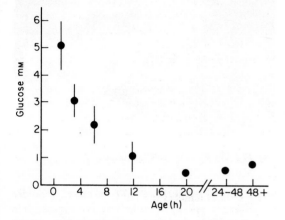

Fig. 51.2 The plasma concentration of glucose (mean ± SEM) in 79 hypothermic lambs related to the ages at which they became hypothermic. The concentration in a healthy lamb should be 4–8mM.

of equal significance, demonstrates the importance of considering both heat production and heat loss. Lambs which become hypothermic in the first five hours of life have a heat loss problem whereas lambs becoming hypothermic later in life have a heat production problem. Immature lambs and lambs which have been subjected to hypoxia during birth have subnormal capacities for heat production and are especially prone to hypothermia (Table 51.1).

Twins and triplets are considerably more susceptible to hypothermia than single lambs. After lambing the ewe takes longer to lick twins or triplets dry than she would a single lamb and thus heat loss is greater. Twins and triplets take longer to suck than single lambs and the benefits that early sucking confers will be delayed. Twins and triplets must produce more heat per unit body weight than the larger single lamb and energy reserves are thus exhausted faster. Twins or triplets require more milk than a single lamb and starvation is more likely. Immaturity and thus low heat production capacity is more common in twins and triplets than in single lambs. Finally a twin or triplet lamb is more likely to be rejected by its ewe.

Prevention

Prevention of hypothermia is a very desirable objective not only because hypothermia will be avoided but it is highly likely that other problems such as those associated with infections and lack of colostrum will also be avoided. Prevention is basically common sense based on understanding. Some idea as to the major cause of hypothermia in a particular situation can be obtained from examination of records from previous years. If most cases have occurred in the first few hours of life then a high rate of heat loss is likely to be the major cause and more attention should be paid to shelter. However if most cases have occurred from 12 h of age onwards starvation is likely to be the major cause and the nutrition of the ewes during pregnancy and after lambing should receive attention. At lambing many problems can be avoided by spotting susceptible lambs before hypothermia arises, e.g. a set of small triplets out of an old lean ewe.

Pathology

Hypothermia can only be positively diagnosed by recording the rectal temperature of the live lamb.

Post-mortem examination together with a clinical history may give some indication of the cause of the condition.

Gross pathology Lambs which die from hypothermia due to a high rate of heat loss soon after birth are often still wet with fetal fluids. The stomach is normally empty. Fat reserves are not exhausted. Lambs which die from hypothermia due to starvation have empty stomachs and fat reserves are markedly depleted. Signs of dehydration may be seen if the period of starvation was prolonged. A low body weight may suggest immaturity as a predisposing cause.

Histopathology Liver glycogen reserves are always depleted following starvation but are only partially so if death occurs soon after birth due to heat loss hypothermia. An absence of fat globules in the mucosal cells of the small intestine indicates that the lamb has not sucked. The retention of fetal lung characteristics or the presence of non-myelinated areas in the central nervous system suggests that the lamb was immature.[11]

Chemical pathology An absence of immunoglobulin in fresh plasma or serum indicates that the lamb has not sucked.

Other conditions Evidence of post-natal disease or injury, which can be factors predisposing to hypothermia, may be found.

Diagnosis

The behaviour and appearance of the hypothermic lamb are described in Table 51.2. It can be seen that the presence of hypoglycaemia has a profound effect on the appearance of the lamb. Death from hypothermia can occur within 90 min of the onset of the condition and thus early detection is essential. The

only reliable way to detect hypothermia, especially in the early stages, is to use a thermometer. The temperature of any lamb that appears at all weak should be checked. Four instruments are currently available; the ordinary clinical thermometer which is of limited use since it only reads from 35°C, the subnormal clinical thermometer which reads from 25°C, the electronic thermometer which reads 0–100°C and the Moredun Lamb Thermometer designed for shepherd use (Fig. 51.3) which indicates whether the lamb

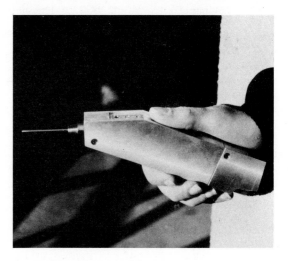

Fig. 51.3 The Moredun lamb thermometer.

has a normal temperature (39°C or more), is mildly hypothermic (37–39°C) or is severely hypothermic (less than 37°C). Each instrument has its uses and the choice depends on the information required, the conditions of use and the ability of the user.

Treatment

Mild hypothermia Mild hypothermia (37–39°C/ 99–102°F) may be treated conservatively. The affected lamb should be removed to shelter, dried if wet

Table 51.2 The appearance and behaviour of hypothermic newborn lambs

Age (h)	Blood glucose	Appearance and behaviour				
		35°C	30°C	25°C	20°C	<20°C
0–5	Normal or high	Weak but can stand	Recumbent	Coma	Deep coma	Death
12+	Low	Recumbent	Coma and death			

and fed colostrum by stomach tube. There is no need to separate the lamb from its ewe providing she also can be removed to shelter. Care should be taken to ensure that the lamb's temperature returns to normal and that future nutrition is assured.

Severe hypothermia Severe hypothermia (less than 37°C/99°F) requires active treatment.[12] There are three components of this treatment. The first is the reversal of hypoglycaemia. Lambs older than five hours are likely to be hypoglycaemic (Fig. 51.2) and if they are rewarmed in this state death from cerebral hypoglycaemia is a likely sequel. Reversal is effected by an intra-peritoneal injection of glucose solution (10 ml/kg of a 20 per cent solution) (Fig. 51.4). The second component is warming. This is best achieved in moving air at 37–40°C. This may be created by using a 'bale' warmer (Fig. 51.5) or by use of the Moredun Lamb Warming Box (Fig. 51.6). Whatever means is used it is essential to monitor the air temperature in the warmer. Depending on the initial temperature of the lamb, warming takes between one and three hours. The lamb's temperature should be checked every 30 min and the lamb removed from the warmer when it has reached 37°C/99°F. It is essential to avoid hyperthermia which is rapidly fatal. After warming the lamb should be fed colostrum by stomach tube (50 ml/kg) and providing that it can stand and suck well it should be returned to its ewe in a small sheltered pen. If the lamb is still weak it should be maintained in a small isolation pen warmed by an infra-red lamp (Fig. 51.7) and fed colostrum thrice daily by stomach tube (50 ml/kg per feed) till stronger. Few problems are encountered on

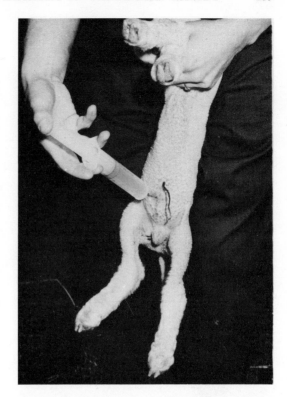

Fig. 51.4 The intra-peritoneal injection of glucose solution

returning the resuscitated lamb to its ewe or to a foster ewe providing good shepherding skills are employed. After resuscitation it is essential that good nutrition is ensured and that any other disease condition is appropriately treated.

Fig. 51.5 The Moredun 'bale' warmer

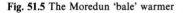

Fig. 51.6 The Moredun lamb warming box. (Courtesy of Macam Ltd., Livingstone, EH54 5DG Scotland.)

Fig. 51.7 Isolation pens used to house weak lambs after warming.

REFERENCES

1 Alexander G. (1962) Temperature regulation in the new-born lamb. IV. The effect of wind and evaporation of water from the coat on metabolic rate and body temperature. *Australian Journal of Agricultural Research,* **13** , 82–99.
2 Alexander G. & Williams D. (1968) Shivering and non-shivering thermogenesis during summit metabolism in young lambs. *Journal of Physiology,* **198,** 251–76.
3 Alexander G., Thorburn G., Nicol D. & Bell A.W. (1972) Survival, growth and the metabolic response to cold in prematurely delivered lambs. *Biology of the Neonate,* **20,** 1–8.

4 Mellor D.J. & Murray L. (1982) Effects of long-term undernutrition of the ewe on the growth rates of individual fetuses during late pregnancy. *Research in Veterinary Science,* **32,** 177–80.
5 Mellor D.J. & Pearson R.A. (1977) Some changes in the composition of blood during the first 24 hours after birth in normal and growth retarded lambs. *Annales de Recherches Veterinaires,* **8,** 460–7.
6 Mellor D.J. & Murray L. (1981) Effects of placental weight and maternal nutrition on the growth rates of individual fetuses in single and twin bearing ewes during late pregnancy. *Research in Veterinary Science,* **30,** 198–204.
7 Eales F.A. & Small J. (1980) Summit metabolism in newborn lambs. *Research in Veterinary Science,* **29,** 211–18.
8 Alexander G. (1962) Energy metabolism in the starved new-born lamb. *Australian Journal of Agricultural Research,* **13,** 144–64.
9 Eales F.A., Small J. & Armstrong R.H. (1980) Plasma composition in hypothermic lambs. *Veterinary Record,* **106,** 310.
10 Eales F.A. & Small J. (1981) Effects of colostrum on summit metabolic rate in Scottish Blackface lambs at five hours old. *Research in Veterinary Science,* **30,** 266–9.
11 Eales F.A., Gilmour J.S., Barlow R.M. & Small J. (1982) Causes of hypothermia in 89 lambs. *Veterinary Record,* **110,** 118–20.
12 Eales F.A., Small J. & Gilmour J.S. (1982) Resuscitation of hypothermic lambs. *Veterinary Record,* **110,** 121–3.

52 SYNCHRONIZATION OF OESTRUS

B. MITCHELL

Ewes breed during only part of the year having a summer anoestrus period. The onset of the breeding season is dictated predominantly by breed type, Dorset and Suffolk crosses commencing in July–August, Border Leicester crosses in September and the hill breeds Cheviot and Blackface from October. Ewes if non-pregnant cycle regularly till January but anoestrus supervenes from February–March. Synchronization of oestrus provides a management tool whereby groups of ewes can be mated at predetermined dates, useful for procedures like artificial insemination and/or embryo transfer, and after approximately 143–149 days gestation ewes lamb in groups to maximize the use of labour and the potential for cross-fostering etc.

Normal oestrus cycle

The oestrus cycle is variable between 14 and 18 days with the majority of ewes returning to service every 16 days. In the ensuing discussion oestrus is identified as being on days 0 and 16. Hormonal control of the oestrus cycle is humoral and depends on the interrelationship of secretions from the ovary (follicles and corpus luteum), the uterus and the anterior pituitary.

Following ovulation a corpus luteum forms in the ovarian site of ovum release and secretes progesterone from about day 4 till day 13. The persistence of the corpus luteum and its progesterone secretion is controlled directly by the uterus. If no pregnancy ensues prostaglandin $F_2\alpha$ is evident in the uterine vein about day 13 and is exchanged to the ovarian artery of the ipsilateral side by a special anatomical configuration between the blood vessels. Prostaglandin causes luteolysis of the corpus luteum and a precipitous fall in progesterone levels on day 13–14.

From about day 9 developing follicles become evident in the ovary. By days 14–15 they have matured. Coincident with this there is a sharp peak in blood oestrogen levels and this immediately precedes a surge of luteinizing hormone (LH) secretion from the pituitary. Follicle stimulating hormone (FSH) is a further gonadotrophic hormone secreted by the pituitary during the maturation of the ovarian follicles. On day 16 there are two phenomena: oestrus and ovulation. Oestrus and psychological receptivity to the ram are precipitated by the high levels of oestrogen acting on tissues sensitized by progesterone over the preceeding number of days. Ovulation, an independent activity, is triggered by the surge of LH and occurs 4–20 hours after the onset of oestrus.

Manipulation of oestrus cycle

Synthetic preparations to control the oestrus cycle are of two types; progestagens and prostaglandins.

The progestagens mimic the persistence of endogenous progesterone from the corpus luteum and block the maturation of follicles and the sensitivity of the pituitary to the gonadotrophin releasing hormones. In field practice, ewes are most conveniently treated by the intravaginal insertion of progestagen impregnated sponges left *in situ* for 12–14 days. During this time follicle maturation is arrested so that the oestrus cycles of all ewes are synchronized. Removal of the progestagen depot simulates day 13–14 of the natural cycle and oestrus follows two days after sponge withdrawal. Ovulation in ewes already cycling normally should be coincident with, or 4–20 hours, after the onset of oestrus and the manufacturers of some commercial products recommend that rams be withheld from ewes until 48 hours after sponge withdrawal to achieve optimal synchronization of service with ovulation even though some ewes exhibit oestrus the day before this. Ovulation can be more actively induced to coincide with oestrus by the intramuscular (i/m) injection of pregnant mares serum gonadotrophin (PMSG) at the time of sponge withdrawal. Where attempts are being made to advance the breeding season of ewes which are in anoestrus PMSG *must* be injected when the sponges are removed. The dose of PMSG required to initiate ovulation depends on the degree of activity of the ewe's ovaries and varies from 750 iu for non-cycling ewes to 375 iu for ewes whose ovaries are already active. A standard dose is 500 iu.

Commercially available sponges are impregnated with (1) 60 mg medroxyprogesterone acetate ('Veramix', Upjohn Ltd.) or (2) 30 mg fluorogestone acetate—cronolone ('Chronogest', Intervet Laboratories Ltd). For advancing the breeding season the 'Veramix' sponges are marketed with serum gonadotrophin BP (Vet) (750 iu i/m) and 'Chronogest' with PMSG 'Folligon' (500 iu i/m).

Prostaglandins require to be injected twice at about 10-day intervals to synchronize all the ewes whose natural oestrus cycles randomly span 16 days. The first injection of prostaglandin synchronizes all ewes having corpora lutea (days 5–13 of natural cycle) to the equivalent of day 14 in the cycle. At the second injection of prostaglandin, 10 days later, ewes range from days 8–14 in their cycle and are synchronized to the 14 day stage. Oestrus and ovulation follow 2 days after this second injection.

Cloprostenol ('Estrumate', Imperial Chemical Industries, Ltd) is marketed for use in cattle and doses suitable for ewes are 0.5 ml containing 131.5 μg cloprostenol sodium BP (Vet) equivalent to 125 μg cloprostenol. Tiaprost ('Iliren', Hoechst UK, Ltd) similarly can be used at a dose of 1–1.5 ml (each ml containing 0.196 mg tiaprost-trometaol salt corresponding to 0.150 mg tiaprost). It is noteworthy to realize that inadvertent injection of prostaglandins to pregnant animals results in abortion because a functional corpus luteum is necessary to maintain pregnancy. The progestagen sponges have no effect on pregnancy if inserted into animals already pregnant.

Mating schedules

The management decisions relating to synchronization of oestrus fall into three areas. First, adequate numbers of fit, fertile and mature rams must be available. One ram per ten ewes is needed and if there is any doubt about the adequacy of rams their effort should be spread over two or three days by staggering the sponge withdrawal or injections over days 12, 13 and 14. A second decision pertains to the size of mating groups. Preferably each ram should have its own group of 10 ewes and be rotated between groups every 6 hours with one shift being a rest period. If the number of individual paddocks is limited up to 4 rams can be set-stocked with 40 ewes.

The final decision is the choice between mating '+ 2 days' after synchronization or '+ 18 days' (2 days plus normal cycle). Several years of experience using progestagen sponges indicate that with '+ 2 day' mating fertility to first service is 60–65 per cent, with aggregate fertility after 18 days of about 90 per cent. Mating at '+ 18 day' results in first service fertility of about 80 per cent. Prolificacy is also better at the '+ 18 day' than at the '+ 2 day' mating.

The timetable of the hormonal treatments is derived by first deciding the date on which ewes are to lamb. Mating date is calculated 146 days before this. The 'treatments' are then calculated to *end* 2 or 18 days before this nominated mating date. The trigger date for *starting* treatment is then identified to allow either 14 days for progesterone sponge control or 10 days for the first prostaglandin injection. Early in the breeding season the author's preference is to mate ewes at '+ 2 days' but when normal cycling is established the '+ 18 day' choice is utilized to maximize fertility and prolificacy.

53

APPROACH TO THE INVESTIGATION OF A DISEASE OUTBREAK

J.A.A. WATT

The basic objective of all investigations into frank 'disease' or indeed mere failure to thrive, is accurate diagnosis, not merely of the immediate cause of death or illness but an assessment of the factor or factors which have led to the development of the condition. We are fortunate in Britain that 'infectious' as opposed to 'infective' diseases are comparatively rare though the 'infectious' group includes a number of diseases with a long incubation or latent period which can make an unravelling of the origin and development of the disease difficult, if not impossible of achievement, e.g. scrapie, Johne's disease, enzootic abortion and more recently, maedi. Elucidation of the problems, however, are probably of more interest to the epidemiologist than the clinician and though of long-term significance in future prevention of disease, are not of great importance in the short term, where immediate remedial measures are required.

In the subject we are considering, therefore, we are confronted first with the decision as to whether or not we are dealing with a specific, recognized disease entity which can be identified and confirmed. A diagnosis in this group, therefore, obviously requires a knowledge of the clinical signs, clinical pathology and confirmatory tests for such recognized conditions. These are dealt with individually in this book and so can be dismissed here rather briefly. The general principles of any investigation apply, however, for the solution of a problem does not end with the diagnosis and the method of control recommended must be based on the management of the flock so it is rarely possible to outline a regime which would cover all eventualities.

The investigator must concern himself primarily with accumulation of the basic facts. It is unfortunate but understandable that the facts presented by the farmer or shepherd are not necessarily those which are of most importance; even a carcass or a sick sheep submitted for examination and autopsy may not be typical or representative of the major condition. Furthermore, the initial request for help is often delivered by someone with little knowledge of the problem. Thus the first principle is that considerable caution should be exercised before the investigator commits himself on the evidence of such single specimens even with the evidence of supportive tests. A good example is provided by an outbreak of abortion which not uncommonly may involve more than a single pathogen, e.g. enzootic abortion and toxoplasmosis, either of which may be the more significant while with toxoplasmosis a 'positive' blood test cannot be regarded as diagnostic.

Secondly, the system affected must be determined to reduce the list of possibilities to manageable proportions.

A comprehensive clinical assessment, together with 'laboratory' examination of specimens or post-mortem dissection and specific serological tests, both immunological and biochemical, in most cases, leads to identification of the immediate cause for which the appropriate remedial measures may be applied when the problem presented is, as postulated, a definite disease entity. However, as suggested earlier, this may not provide the whole answer to the problem and to assess the relative significance of other factors and devise methods for their elimination, a study of the history and the management system is necessary. Solution of the problem may subsequently require consultation with husbandry and nutritional experts so it is essential that an early collection of factual information be made. Unfortunately, sheep are very rarely individually identifiable which limits information largely to a flock basis.

A detailed and accurate account of the origins of the outbreak is essential with some account of the signs exhibited by the affected animals. This is not always as simple as it sounds, and requires careful questioning to elicit all the relevant facts. The shepherd is generally the most reliable source of information. Diagnosis of previous illnesses or deaths are not infrequently accompanied by a diagnostic opinion but these should be treated with considerable reserve, especially such diagnoses as 'staggers', 'moss ill', 'inflammation', etc., but additionally more specific diagnoses such as louping-ill, lambing sickness, 'braxy' etc., may mean something quite different to the shepherd than to the veterinarian. The post-mortem examination of fresh casualties after the history is usually the most important and crucial part of the investigation.

The basic information which should be acquired apart from those itemized above are:

1. Number of sheep, age groups and origins of stock
2. Area grazed
3. Types and age of pasture
4. Is the problem confined to a particular group or age? What is the previous management history of the affected group?
5. Is supplementary feeding provided? If so, what and in what quantity and for how long?
6. Vaccinations, dosings etc.
7. Date first case noted
8. Morbidity
9. Mortality
10. Any previous history of disease

With this basic information at hand, lines of enquiry can be identified and remedial and preventative measures suggested, especially by the other disciplines which may be consulted. No such list can, of course, be generally applicable and the investigator must be guided by the information supplied which may well indicate the most promising line of investigation.

While most problems can be identified as manifestations of a specific and recognized cause, a considerable amount of detective work may be required before the diagnosis can be established. Fortunately over recent years considerable advances have been made in the field of serological and biochemical tests. Such tests are of particular value when the more vague problems are presented, e.g. copper, selenium and cobalt deficiencies, confirmation of nutritional myopathy, etc. Care must be exercised in the interpretation of the results of tests and these should be related to other clinical signs, post-mortem findings and the history presented. The danger, for example, of interpreting 'positive' serological tests for toxoplasmosis as confirming that abortions are due to that organism has already been mentioned and other come to mind, e.g. high levels of ketones in a sample do not justify, *per se*, a diagnosis of pregnancy toxaemia, nor does the isolation of *Clostridium perfringens* type D from the intestine indicate that death was due to enterotoxaemia. Thus the isolation of a pathogen does not invariably justify a firm diagnosis. In many instances it is only when test results are equated with history and clinical findings that a balanced assessment and diagnosis can be reached.

The danger of arriving at a firm diagnosis on the basis of a single carcass is highlighted, for example, in cases where excess barley has been fed or suddenly introduced. Post-mortem examinations in such an incident may show deaths from acidosis, acute *Pasteurella* pneumonia and enterotoxaemia in the same flock. Thus, treatment to control either the pneumonia or the enterotoxaemia would have little effect other than the result of the check given by the gathering and injections. Numerous other instances can be cited related to a sudden increase in the nutritional value of the feed or a reduction in roughage intake.

The incidents quoted are intended to highlight the significance of management and nutritional factors in the elucidation of health in the sheep, factors which are not infrequently relegated to a position lower than the disease itself and this is well illustrated by the emphasis placed on the use of anthelmintics compared to that of pasture management for parasite control. This type of parasite problem highlights the importance of assessing the history of the pasture in the overall picture if satisfactory remedial measures are to be devised. For a time, after the newer and highly efficient anthelmintics were introduced an entity which acquired the name of 'July disease' was quickly found to be the result of returning the lambs to infected pasture after dosing. Coccidial infection was first suspected as being the cause of the continuing scouring as coccidial oocysts were the only pathogens identified in numbers in faecal or post-mortem specimens though equal or greater numbers could be demonstrated in the faeces of normal sheep.

Undoubtedly the most difficult problem with which to deal is that where no specific diagnostic signs are manifest, e.g. poisoning, luckily a rather rare occurrence in sheep in Britain. However, alarming incidents do occur, e.g. rhododendron poisoning after snowfall, basic slag and fertilizer poisoning in lambs, etc. In such cases a careful and exhaustive post-mortem technique must be the major factor in the investigation allied to careful history taking. It is in these obscure cases that the weakness of the standard questionnaire is revealed, as suggestive replies cannot be developed and expanded while other questions are irrelevant. Standard techniques for field and laboratory post-mortem examinations are well documented in the literature [1,2] but history taking is an art which can be acquired only by experience and must be carried out by an experienced veterinarian. A standard questionnaire for the reasons outlined above, being very difficult and even in some cases misleading.

In most cases the clinical picture and history combined suggests a short-list of possibilities and in the absence of dead fresh animals for post-mortem examination, a knowledge of the ancillary tests available and their interpretation in the light of the clinical findings is essential. Most diagnostic laboratories supply a list of available tests, the type of sample required and the method of collection, e.g. clotted or unclotted blood samples and the anticoagulant of choice are relevant. The nature of the specimen and its quality are of paramount importance for meaningful diagnosis and these factors, of course, are in the hands of the clinician in the field. Freshness is all-important, particularly where a post-mortem examination is concerned and a detailed list of the findings should be submitted with relevant specimens. It can also be of great value to the clinical pathologist at the laboratory if diagnostic conclusions and the test particularly requested are appended. Should blood samples be submitted for viral or bacterial antibody assessment it is essential to mark the subject animal so that it can be identified subsequently for sequential samples when rising titres may be diagnostic.

It is usually possible to put the problem into one of several broad groups, nutritional, parasitic, infectious or toxic following which the system or systems most affected can be identified. It is important, however, not to exclude a condition because the classical signs usually associated are absent. An example may be found in severe haemonchosis where the major presenting sign is anaemia, with scour conspicuous by its absence.

When infectious disease has been eliminated and the immediate cause of death or malaise identified the major problem remaining is identification of the precipitating factor, e.g. though the deaths in housed sheep may be typical pneumonia due to *Pasteurella haemolytica* infection, the basic factor may be environmental; faulty ventilation or perhaps nutritional. It is in the identification of these factors and in remedying them that a successful conclusion to the investigation lies.

The veterinary profession in Britain is well served by diagnostic and specialist laboratories and it is important that these facilities be used when it becomes obvious that a problem has arisen. The veterinary investigation centres and research laboratories have immediate contact with specialists in most other disciplines so a wide spectrum of help is readily available through them. While location, communication, etc. governs the laboratory approached, the first approach should generally be to the local veterinary investigation laboratory. Specialist laboratories are geared to their programme of research and cannot immediately disrupt such programmes to undertake *ad hoc* investigations. Where the problem is relevant to research the veterinary investigation officer consults the appropriate worker for his help.

Diagnostic laboratories are not invariably used to the utmost advantage and the most helpful results are not obtained by using them merely to carry out bench tests. The staff are in close touch with other centres and have a library of recent publications which enables them to have an up to date appreciation of the most recent advances in knowledge and of emerging problems. It is therefore re-emphasized that specimens reach the laboratory in the freshest possible condition and that they are the most suitable for the purpose. A prior telephone call can ensure that the best use is made of their expertise and may also determine whether a consultative visit could serve a useful purpose. Most laboratories will, on request, either provide suitable specimen tubes or for more expensive items, advise on the source and type most suitable.

Above all, however, the two fundamental requirements for the solution of health problems on a flock basis are first, a knowledge of the clinical signs, clinical pathology and confirmatory tests for specific diseases and second, an understanding of the effects of feeding, management and physiology of the sheep. Although it may seem superfluous to make the comment, the acquisition of expertise in these fields requires an understanding of the normal healthy animal and its relationship to its environment as well as the scientific knowledge and systematic post-mortem technique.

REFERENCES

1 Belschner H.G. (1965) *Sheep Management and Diseases,* 8e. pp. 396, Angus & Robertson, Sydney, London.
2 McFarlane D. (1965) Perinatal lamb losses. 1. An autopsy method for the investigation of perinatal losses. *New Zealand Veterinary Journal,* **13** , 116–35.

SAMPLES AND SAMPLING FOR THE DIAGNOSIS OF VIRUS DISEASES

54

P.F. NETTLETON

Clinicians faced with a disease outbreak in a flock must be wholly confident of their diagnosis before being able to institute effective control measures. The clinical signs and epidemiology of some virus diseases are readily recognized and laboratory confirmation may not be necessary. There are circumstances, however, when the clinician needs help from a virus laboratory to reach an unequivocal diagnosis, and these would include:

1. Confirming a notifiable or newly imported virus of serious economic importance.
2. Establishing the cause of an unfamiliar clinical syndrome which may have resulted from the introduction of a new virus into an area or the emergence of a more virulent strain of an existing virus.
3. Identifying viral agents in diseases of multifactorial aetiology e.g. pneumonia.
4. Investigating virus strains responsible for a pathognomonic syndrome where there is evidence of a vaccine breakdown.

In the absence of effective chemotherapy the value of diagnostic virology to the practitioner is restricted. It does, however, give the satisfaction of establishing the aetiology of a clinical condition and provide a basis for prognosis, supportive treatment and advice on short and long-term prophylactic measures at the local level. On a national scale samples collected by clinicians for routine diagnosis allow a constant surveillance of the epidemiology and pathogenicity of important virus diseases. New viruses or emerging antigenic variants of established viruses can be recognized and control measures initiated.

Laboratory techniques used for virus diagnosis

All techniques are expensive and the isolation and identification of a virus requires more time than is employed in the recognition of many other disease-producing micro-organisms.

A virus infection can be confirmed in one or more of the following ways:

a. Isolation of virus in cell monolayers or laboratory animals. This is the most sensitive and widely used system for detecting a wide range of viruses since only very low levels of infectious particles are required to produce a recognizable effect. However, some viruses are slow to grow and the final identification of isolates must be made using monospecific antisera against known reference viruses, so that 3–4 weeks may be required to detect and identify a new isolate. Such a retrospective diagnosis is of little immediate value to the clinician and, therefore, rapid methods of demonstrating viral antigens in specimens are continuously being sought.

b. Direct detection of viral antigen either by electron microscopy or by specific immunological identification.

   The electron microscope may provide a rapid diagnosis of conditions in which high concentrations of virus are present in the specimen, e.g. skin and teat lesions, and neonatal faeces samples.

   The identification of viral antigen or viral induced antigen in clinical material by specific immunological identification encompasses many techniques such as gel diffusion, haemagglutination, immunofluorescence and enzyme immunoassays, all of which are rapid and have the advantage of detecting and identifying viral antigen at the same time. Again, these methods are only effective when high concentrations of antigen are present, but the sensitivity of immunofluorescence and enzyme immunoassay systems is continuously being improved, and the development of monoclonal antibodies for diagnostic use will further enhance the sensitivity and specificity of such tests.

c. Identification of characteristic cellular pathology by light microscopy. This is not a sensitive method but histological examination of the brain is useful in the diagnosis of louping-ill, and slow virus infections of the CNS. Similarly examination of affected lungs helps to confirm slow virus infections of the respiratory tract.

d. Demonstration of an antibody response. In acute viral infections demonstration of antibody conversion by examination of acute phase and convalescent serum collected 2–4 weeks later can provide further evidence that an isolated virus was associated with the disease outbreak. Since antibodies to many viruses are widespread among sheep the collection of paired sera is essential to allow interpretation of the antibody levels at the time of the disease.

The examination of single sera is only justifiable when early antibody of the IgM class is present at the time of clinical symptoms, e.g. louping-ill, or when the distribution of an important disease within a flock is being investigated, e.g. maedi-visna.

e. Particularly slow growing viruses such as maedi-visna are best isolated from explants of affected tissues obtained from live animals by biopsy or from animals immediately after slaughter.

Selection of samples for submission to the laboratory

Whenever it is feasible live animals, freshly dead carcases or aborted fetuses plus placentae should be submitted to the laboratory. However, this is often impossible and, therefore, lists of samples to be submitted from specific disease conditions are shown in Table 54.1. The following general considerations are also important:

i. The amount of virus is usually highest at affected sites and during the early stages of disease. Secondary bacterial contamination is a common sequel of virus infections and can make virus isolation difficult. Samples taken during the later stages of disease or at post-mortem examination are less likely to yield viable virus.

ii. Virus infections usually spread among animals in a group. Although the shepherd presents the sickest animal which may have been ill for some time, a clinical examination of both affected and apparently normal in-contact animals should be made since some of the latter may be showing early clinical signs and provide the best samples. Pyrexia associated with viraemia often precedes

Table 54.1 Specimens to be collected for the diagnosis of viral diseases

Disease	Live Animal	Dead Animal
Acute respiratory/ocular disease	Nasal and ocular swabs. Blood	Tissues from affected areas plus draining lymph nodes in VTM
Slow respiratory diseases (jaagsiekte, maedi)	Submit live animal showing symptoms Blood from asymptomatic animals	Affected lungs in formal saline for histology
Gastroenteritis	10–20 ml faeces. Blood	Tissues from affected areas. Mesenteric lymph nodes in VTM; heart blood
Skin diseases	Fresh moist scabs in a dry bottle Swabs and any fluid from lesions in VTM	Tissues from affected areas plus draining lymph nodes in VTM
Lesions of superficial mucous membranes	Scrapings or swabs from lesions in VTM Blood	Tissues from affected areas plus draining lymph nodes in VTM
Diseases of the CNS a. Acute, e.g. louping-ill	Blood. CSF	Brain stem, part in 50% glycerol saline for virus isolation, part in formal saline for histology.
b. Chronic, e.g. scrapie, visna	Submit live animal	Whole brain and spinal cord in formal saline
Abortion	Blood from dam. Vaginal swabs in VTM	Tissues from placenta, plus several organs and intestinal contents from fetus in VTM; fetal heart blood
Weak lambs, e.g. Border disease	Blood	Spleen, kidney, brain, mesenteric lymph nodes in VTM; heart blood

Table 54.2 Equipment required for collection of specimens

Sterile forceps, scissors and scalpels
Sterile cotton wool swabs*
Bijou bottles containing virus transport medium
Dry sterile universal bottles for collection of faeces and skin scrapings
Bottles containing 50% glycerol saline for large portions of brain
Large bottles containing formal saline for tissues for histology
Tubes for collecting blood. Heparinised for virus isolation; without additive for serum collection
Heavy-duty plastic bags for post-mortem material
Pencil or indelible marker for labelling specimens
Request forms to record details of samples and disease
Insulated container to keep specimens cool

---

* Exogen Limited, Clydebank, Glasgow, U.K.

the onset of other clinical signs and detection and sampling of all such animals gives a very good chance of making a diagnosis.

iii. If in any doubt about what samples to collect, select a wide range from several sheep so that the virologist can choose the most suitable, or telephone the virus laboratory to discuss the best way to proceed with the investigation.

Collection of specimens

The equipment required for the collection of specimens is listed in Table 54.2.

Viruses are killed by drying, high temperatures, light and extremes of pH and, therefore, all specimens for virus isolation should be protected between collection and testing. This is best done by placing specimens in virus transport medium (VTM) which will be supplied by the laboratory to which the samples are to be sent. VTM contains a range of antibiotics and should be stored at −20°C until required; after thawing it should be kept cool, ideally at 4°C, both before and after the specimen has been added. Avoid freezing specimens in VTM since this may result in a loss of infectivity of some viruses. It is also important not to add too large a specimen to VTM, e.g. more than 2 swabs/4 ml medium. At post-mortem examination it is best to aseptically collect several small specimens from affected tissues, particularly the edges of any lesions, keeping the ratio of specimen to VTM about 1:10.

Where samples are to be examined by direct methods, e.g. skin scabs and faeces they should be collected into dry sterile bottles and kept cool.

When sampling live clinically-affected animals

always collect 2 x 10 ml blood; clotted samples for serology and heparinized samples for virus isolation. If anticoagulant is not available the blood clots can be used for virus isolation, but clots should be separated from the sera before despatch. When paired sera are to be examined, freeze the acute phase samples until the convalescent sera have been collected 2–4 weeks later, and then despatch both to the laboratory.

Submission of samples

All samples must be clearly labelled. They can be transported by hand preferably on wet ice in an insulated container to ensure the minimum of delay but if this is not practicable samples should be posted. Packages must comply with local postal regulations but the following procedure is offered as a guideline.

Pack specimens carefully in a strong insulated container (polystyrene boxes are ideal) using sufficient absorbent material to secure the bottles and soak up liquid in the event of breakage. Add 'freezer gel' bags to maintain low temperature; never use loose ice.

All specimens should be accompanied by a form supplied by the receiving laboratory, which should be completed as fully as possible. The form should be placed in a plastic bag taped to the outside of the container before it is wrapped and addressed. Parcels should be labelled *Pathological Specimens: Fragile, With Care;* the name and address of the sender together with the date of despatch should also be added.

Table A.1 Haematological parameters for normal sheep*

Test	Unit	Range	Mean
Erythrocytes (RBC)	$10^{12/1}$	8.7–14.3	11.5
Hemoglobin	g/dl	10.5–14.3	12.4
MCV	fl	19.3–34.7	27.0
MCHC	g/dl	31.3–50.7	41.0
Hematocrit (PCV)	l/l	0.23–0.37	0.30
platelets	$\times 10^3/mm^3$	250–600	400
Leukocytes (WBC)	$10^{9/1}$	5.2–13.2	9.2
Polymorphonuclears	%	8–40	24
Lymphocytes	%	51–83	67
Eosinophils	%	0–12	4
Basophils	%	0–2.1	0.5
Monocytes	%	0–5.3	2

*After Holman H.H. (1945) *Journal of Comparative Pathology and Therapeutics,* **55,** 229–42.

Table A.2 Biochemistry, serum parameters in the normal sheep

Serum transaminases	Units		Range		
	SGOT/iu/l(30 C)		20–60		
	SGPT/iu/l(30 C)		4–15		
Plasma proteins in g/l	Total	Albumin	IgA	IgG	IgM
(absolute concentrations)	61–79	22–37	0.08–0.84	11–30	0.95–4.70
Nonprotein nitrogenous	Urea nitrogen			Creatinine	
substances	mmol/l			μmol/l	
	2.84–7.12			106–171	
	mmol/l				
Serum calcium levels	2.10–2.80				
Serum phosphorus					
(as inorganic phosphate)	0.90–2.55				
Magnesium	0.78–1.25				
Sodium	146–161				
Potassium	4.0–5.5				
Chloride	98–109				
Serum cholesterol	1.66±0.31				
Plasma glucose	3.6–6.1				

INDEX

Abomasal parasitism 56
Abortion
  border disease 129
  enzootic abortion 119
  salmonella 135
  toxoplasmosis 124
  vibriosis 133
Accessory gland infection 140
Aceto-acetic acid in pregnancy
    toxaemia 148
Acholeplasma oculi 19
  in atypical pneumonia 19
  in ovine keratoconjunctivitis 214
Actinobacillus lignieresi 34
Actinobacillus seminis in arthritis 110
Actinobacillus spp.
  accessory gland infection 140
  mouth diseases 34
  ram epididymitis 140
Actinomycotic dermatitis see mycotic
    dermatitis 200
Acute Laminitis 103
Acute respiratory virus infections 8
  adenoviruses 10
  miscellaneous viruses 11
  parainfluenza virus type 3 9
  reoviruses 11
  respiratory syncytial virus 11
Adenocarcinomas 218
Adenomas 218
Adenoviruses 10
  clinical signs 10
  control 11
  diagnosis 11
  epidemiology 11
  pathology 10
  Pasteurella haemolytica 11
  serotypes 10
  type 1 45
  vaccination 11
Adrenal gland in pregnancy
    toxaemia 149
Adrenal tumours 218
Agalactia see contagious
    agalactia 157
Akabane disease 223
  cause 223
  clinical signs 223
  Culicoides spp. 223
  diagnosis 224
  distribution 223
  transmission and
    epidemiology 223
Allium spp. 232
Alternate grazing systems 251, 252
Alveld see photosensitization 199
Anaplasma ovis 209
Anaplasma mesaeterum 209

Andromedotoxin 233
Anthelmintics 86, 250, 251, 252, 253,
    254
Antitoxin, tetanus 92
Apical pneumonia see atypical
    pneumonia 17
Arthritis 109, 157
  chlamydial polyarthritis 109
    cause 109
    Chlamydia psittaci 109
    clinical signs 109
    contagious agalactia 157
    diagnosis 109
    enzootic abortion 109
    epidemiology and
      transmission 109
    pathology 109
    treatment, control and
      prevention 109
  Erysipelothrix polyarthritis 106
    cause 106
    clinical signs 106
    diagnosis 106
    epidemiology and
      transmission 107
    Erysipelothrix rhusiopathiae 106,
      107
    pathology 107
    prevention and control 107
    treatment 107
    vaccination 107
  neonatal polyarthritis 104
    cause 104
    clinical signs 104
    control 105
    Corynebacterium pyogenes 104
    diagnosis 104
    epidemiology and
      transmission 104
    Erysipelothrix rhusiopathiae 104
    Escherichia coli 104
    Fusobacterium necrophorum 104
    pathology 104
    Streptococci 104
    vaccination 105
  osteoarthritis 109
  post-dipping lameness 108
    cause 108
    clinical signs 108
    control, prevention and
      treatment 108
    diagnosis 108
    epidemiology 108
    Erysipelothrix rhusiopathiae 108
    pathology 108
  Tick pyaemia 105
    cause 105
    clinical signs 105

  control, prevention and
    treatment 106
  Corynebacterium pyogenes 105
  diagnosis 105
  epidemiology 105
  Fusobacterium necrophorum 105
  pathology 105
  Staphylococcus aureus 105
  tick borne fever 105
  unusual infections 110
    Actinobacillus seminis 110
    Corynebacterium ovis 110
    Haemophilus agni 110
    maedi-visna 110
    Mycoplasma capricolum 110
    Pasteurella haemolytica
      biotype T 7
Astrovirus in lamb enteritis 45
Atypical pneumonia 17
  Acholeplasma oculi 19
  clinical signs 19
  control, prevention and
    treatment 22
  diagnosis 22
  epidemiology and transmission 22
  Mycoplasma ovipneumoniae 18, 19,
    20
  Pasteurella haemolytica
    biotype A 18, 19, 20
  Pasteurella multocida 19
  pathology 20
Autofluorescence in
    polioencephalomalacia 85

'B' disease see Border disease 129
Babesia capreoli 213
Babesia major 205
Babesia motasi 213
Bacillary haemoglobinuria 41
  cause 41
  clinical signs 41
  Clostridium oedematiens type D 41
  control, prevention and
    treatment 42
  diagnosis 42
  epidemiology and transmission 42
  pathology 42
  vaccination 42
Bacillus oedematis maligni see
    Clostridium sepicum 39
Bacillus thiaminolyticus in
    polioencephalomalacia 86
Bacterial infections of the central
    nervous system 89
Bacteroides nodosus 98, 100
Balanitis, ulcerative 144
Basic slag poisoning 239
Belland see swayback 82

Bendro see coenurosis 93
Bentback see swayback 82
Bent leg syndrome in
    osteodystrophies 113
Beta hydroxybutyric acid in pregnancy
    toxaemia 148
Big head
    see blackleg 38
    see malignant oedema 42
Big joint see neonatal
    polyarthritis 104
Bile duct tumours 218
Biochemical aspects of pregnancy
    toxaemia 148
Black blowfly (*Phormia terrae-
    novae*) 191
Black disease 40
    cause 40
    clinical signs 40
    Clostridium oedematiens type B 40
    control, prevention and
        treatment 41
    diagnosis 41
    epidemiology and transmission 41
    liver fluke 40, 41, 64
    pathology 40
    treatment 41
    vaccination 41
Black flies (*Simulium*) in Rift Valley
    fever 224
'Black garget' see mastitis, acute
    severe 153
Blackleg 36
    cause 38
    clinical signs 38
    Clostridium chauvoei 38
    control, prevention and
        treatment 39
    diagnosis 39
    epidemiology and transmission 39
    pathology 38
    vaccination 39
Blackquarter see blackleg 38
Blackquarter metritis see blackleg 38
Blood urea in pregnancy
    toxaemia 148
Blowfly myiasis 191
    Calliphora erythrocephala
        (Bluebottle) 191
    cause 191
    clinical signs 191
    control 192
    diagnosis 191
    epidemiology 191
    Lucilia sericata (Greenbottle) 191
    pathology 191
    Phormia terrae-novae (Black
        blowfly) 191
    primary flies 191
    secondary flies 191
    treatment 192
Bluebottle (*Calliphora
    erythrocephala*) 191

Bluetongue 33, *222*
    cause 222
    clinical signs 222
    Culicoides spp. 222
    diagnosis 223
    distribution 222
    epizootic haemorrhagic disease of
        deer 222
    Ibaraki virus 222
    Orbivirus 222
    transmission and epidemilogy 222
Bog asphodel (*Narthrecium
    ossifragum*) 199
Bone atrophy see osteoporosis 111,
    171
Bone diseases see
    osteodystrophies 111
Bone softening see osteomalacia 111,
    165
Border disease 129
    bovine viral diarrhoea/mucosal
        disease virus 129, 132
    'Camel legged' lambs 130, 131
    cause 129
    clinical signs 129
    control 132
    diagnosis 131
    epidemiology and
        transmission 132
    European swine fever virus (hog
        cholera) 129
    pathology 130
    Pestivirus 129
    vaccination 132
Bovine adenovirus type 2 in lamb
    enteritis 45
Bovine viral diarrhoea/mucosal
    disease virus 129, 132
Bracken fern poisoning 218, 234
    acute (haemorrhagic fever) 234
    diagnosis 235
    neoplasia 218, 235
    progressive retinal degeneration
        (bright blindness) 234
    treatment 235
Bradapest see braxy 39
Bradsot see braxy 39
Branhamella ovis (*Neisseria ovis* or
    catarrhalis) 214
Brassica poisoning 231
    Allium spp. 232
    clinical signs 232
    copper deficiency 233
    diagnosis and treatment 232
    glucosinolates 231, 233
    haemolytic anaemia factor 232
    Heinz-Ehrlich bodies 231
    nitrates 231, 233
    pathology 232
    S methylcysteine sulphoxide
        (SMCO) 231, 232, 233
Braxy 39
    cause 39

clinical signs 39
Clostridium septicum 39
control, prevention and
    treatment 40
diagnosis 40
epidemiology and transmission 40
pathology 40
treatment 40
vaccination 40
Breckshuach see braxy 39
Bride see swayback 82
Bright blindness 234
Broken mouth see incisor loss 30, 31
Brucella ovis in ram epididymitis 140
Bunostonum trigonocephalum 58

Calcium 162
    calcitonin 163
    dietary source 163
    hypocalcaemia 163
    lambing sickness 163
    osteodystrophies 113
    parathyroid hormone 163
Calcitonin 163
Calliphora erythrocephala
    (Bluebottle) 191
'Camel legged' lambs see border
    disease 130, 131
Campylobacter fetus 1, 133, 134
    subsp. *coli* 44
    subsp. *intestinalis* 133
    subsp. *jejuni* 44, 133
    in man 135
Campylobacteriosis see vibriosis 133
Canary stain see fleece rot 202
Cappie 113, 165, 236
Carbolic acid poisoning 239
Caries 31
Cats in toxoplasmosis 124, 127, 128
Central nervous system, bacterial
    infections 89
    causes 89
    clinical signs 89
    control, prevention and
        treatment 90
    Corynebacterium pyogenes 89
    diagnosis 90
    epidemiology 90
    pathology and pathogenesis 89
    Staphylococci 89
Cerebrocortical necrosis see
    polioencephalomalacia 85
Chabertia ovina 58
Cheek teeth disease see teeth
    disease 32
Chlamydia 119
    abortion 119
    arthritis 109
    Chlamydia psittaci 109, 119, 214
        atypical pneumonia 19
        group 1 isolates 120
        group 2 isolates 120
        ovine keratoconjuncitivitis 214

Chlamydia trachomatis   119
  elementary body   119
  properties   119
  reticulate body   119
Chlorine   161
Cholangicellular tumours   218
Chondrosarcoma   219
Chorioptes ovis   181, 182
Chronic non-progressive pneumonia
    see atypical pneumonia   17
Chronic respiratory virus
    infections   12
  maedi-visna   15
    clinical signs   15
    control   16
    diagnosis   16
    epidemiology and
      pathogenesis   16
    pathology   15
    retrovirus   15
  pulmonary adenomatosis   12
    cause   12
    herpesvirus   12
    retrovirus   12
Circling disease see listeriosis   79
Cytoecetes ondiri   210
Cytoecetes ovis   210
Cytoecetes phagocytophilia   210
Cling see braxy   39
Clostridial diseases   35
  bacillary haemoglobinuria   41
  black disease   40
  black leg   38
  braxy   39
  enterotoxaemia   35
  lamb dysentery   35
  pulpy kidney   36
  struck   36
  tetanus   91
  vaccination   40
Clostridium chauvoei   38
  alpha toxin   38
  beta toxin   38
  delta toxin   38
  gamma toxin   38
Clostridium chauvoei (feseri)   38, 42
Clostridium chauvoei type A see
    Clostridium septicum   39
Clostridium gigas see *Clostridium*
    oedematiens type B   40
Clostridium haemolyticum see
    Clostridium oedematiens
    type D   41
Clostridium novyi type B see *Clostri-*
    dium oedematiens type B   40
Clostridium novyi type D see *Clostri-*
    dium oedematiens type D   41
Clostridium oedematiens type A   42
Clostridium oedematiens type B   40
  alpha toxin   41
  beta toxin   41
Clostridium oedematiens type D   41
  beta toxin   42

Clostridium perfringens (welchii)   35,
  42
  alpha toxin   35, 36
  antisera   38
  beta toxin   35. 36
  epsilon toxin   35,36, 37, 38
  gamma toxin
  mastitis, acute severe   153
  prototoxin   36, 38
  toxoids   38
  toxin, lethal   36, 37
  types A   35
      B   32, 35, 36
      C   32, 35, 36, 37
      D   32, 35, 36, 38
      E   35
  Vaccination   38
Clostridium septicum   39, 42
  alpha toxin   40
  beta toxin   40
  delta toxin   40
  gamma toxin   40
Clostridium sordelli   42
Clostridium sporogenes   86
Clostridium tetani   91
Clover (subterranean) in ewe
    infertility   143
Cobalt deficiency   168
  in ewe infertility   143
Coccidiosis   49
  cause   49
  clinical signs   50
  control and treatment   52
  diagnosis   51
  Eimeria spp.   49, 51
  pathology   50
Coenurosis   90, 93
  anthelmintic treatment, dogs   96
  cause   94
  clinical signs   94
  Coenurus cerebralis   94
  control and treatment   96
  diagnosis   95
  epidemiology   96
  pathology   94
  Taenia multiceps   94
  Taenia serialis   94
  tapeworm   94
  vaccination   97
Coenurus cerebralis   94
Colesiota conjunctivae   214
Colostrum   256
Congenital axial rotation see foot
    conditions   103
Congenital porphyria (pink
    tooth)   197
Congenital trembles see border
    disease   129
Conjunctivitis see ovine
    keratoconjunctivitis   214
Conjunctivo-keratitis see ovine
    keratoconjunctivitis   214
Contagious agalactia   157

  arthritis   157
  cause   157
  clinical signs   157
  keratoconjunctivitis   157
  Mycoplasma agalactia   157
  Mycoplasma capricolum   157
  Mycoplasma mycoides subsp.
    mycoides   157
Contagious ecthyma see contagious
    pustular dermatitis   185
Contagious foot rot see foot rot   98
Contagious ovine opthalmia see ovine
    keratoconjunctivitis   214
Contagious pustular dermatitis
    (orf)   144, 185
  autogenous vaccine   188
  cause   185
  clinical signs   185
  control   187
  epidemiology and
    transmission   187
  Dermatophilus congolensis   185,
    186, 187
  diagnosis   187
  Fusobacterium (Sphaerophus)
    necrophorum   186
  mastitis   186
  pathology   186
  strawberry foot rot   185
  ulcerative dermatitis   186
  vaccination   188
  venereal disease   186
Cooperia curtecei   58
Copper
  copper calcium EDTA   172
  glycinate   172
  methionate   172
  osteodystrophy   114
  swayback   82
  deficiency
    in Brassica poisoning   233
    cause   171
    clinical signs   170
    diagnosis   171
    molybdenum   171
    osteoporosis   171
    pathology   171
    prevention and treatment   172
    'steely wool'   170
    sulphur   171
    superoxide dismutase   171
    swayback   82, 90, 105, 170, 171
  poisoning   235
    acute   236
    chronic   235, 236
Coproporphyrin in
    photosensitization   197
Corkscrew claw see congenital axial
    rotation   103
Coronavirus-like virus in lamb
    enteritis   45
Corynebacteria
  accessory gland infection   140

ram epididymitis   140
ulcerative balanitis/vulvitis   144
Corynebacterium ovis
arthritis   110
central nervous system
infections   89
Corynebacterium pyogenes   34, 105
central nervous system
infections   89
neonatal polyarthritis   104
orchitis   140
tick pyaemia   105
Cowdria ruminantium   209
Coxiella burnetti   121, 209
Creatine phosphokinase in selenium
deficiency   174
Creutzfeldt–Jakob disease   71
Cripples see tick pyaemia   105
Cruban   113
Cryptorchidism   139
Cryptosporidium in lamb enteritis
45
Cryptosporium leptostromiforme in
lupinosis   238
Culicoides spp. (midge)
Akabane disease   223
bluetongue   222
Cutaneous myiasis see blowfly
myiasis   191
Cystocaulus ocreatus   25
Cytoecetes phagocytophilia   209, 210,
211

Dafties see daft lamb disease   87
Daft lamb disease   87
autosomal recessive gene   88
breeds   88
cause   87
clinical signs   87
control   88
diagnosis   88
epidemiology   88
pathology   87
Damalinea ovis   204
Dental malocclusion   113
Dentigerous cysts   32
Dermatophilosis see mycotic
dermatitis   200
Dermatophilus congolensis   185, 186,
187
fleece rot   203
properties   200
Dermatitis   200
fleece rot   202
mycotic dermatitis   200
ringworm   203
Developmental defects in ewe
infertility   142
Diagnosis of virus diseases   266
Dicrocoelium dendriticum
control   67
life cycle   67
pathogenesis   67

Dictyocaulus filaria   23
Dietary effects in ewe infertility   142
Dietary protein in
osteodystrophies   114
Dimidium bromide   198
Disease investigation   263
Double scalp   165, 236
clinical signs   113
Drug resistance   254

Eimeria spp.   49, 51
life cycle   49
morphological characteristics   51
Elementary body   119
Emphysematous gangrene see
blackleg   38
Enteritis   49
coccidiosis   49
Johne's disease   52
lamb dysentery   44
lamb enteritis   43
parasitic gastroenteritis   56
salmonellosis   136
watery mouth   47
Enterotoxaemia   35
cause   35
Clostridium perfringens (*welchii*)
35
control, prevention and
treatment   38
diagnosis   37
epidemiology and transmission   37
tests   37
Entropion   214
Environmental effects on ewe
infertility   142
Enzootic abortion   119, 138
cause   119
Chlamydia psittaci   119
clinical signs   120
control and prevention   122
diagnosis   121
epidemiology and
transmission   121
pathology   120
treatment   122
vaccination   119, 123
Enzootic ataxia see swayback   82
Enzootic bovine leucosis and
tumours   218
Enzootic haemorrhagic disease of deer
virus   222
Enzootic pneumonia   3
Epididymal hypoplasia or
aplasia   139
Epididymitis   140
Escherichia coli
central nervous system
infections   89
enterotoxigenic   44
K99 antigen   44
lamb enteritis   44
neonatal polyarthritis   104

Erysipelas see *Erysipelothrix*
polyarthritis   106
Erysipelothrix insidiosa see
Erysipelothrix
rhusiopathiae   106
Erysipelothrix polyarthritis   106
Erysipelothrix rhusiopathiae
(*Erysipelothrix insidiosa*)   104,
106, 107, 108
foot conditions   103
neonatal polyarthritis   104
post dipping lameness   108
European swine fever virus (hog
cholera)   129
Evil see swayback   82
Ewe infertility see infertility   142
External parasite control   246, 247
Eye scab see staphylococcal
dermatitis   195

Facial dermatitis see staphylococcal
dermatitis   195
Facial eczema
photosensitization   197, 198
staphylococcal dermatitis   195
Fagopyrin   197
Fasciola gigantica   66
control   66
life cycle   66
pathogenesis   66
Fasciola hepatica   62
Fascioliasis   *62*, 253
Fertilization failure in ewe
infertility   143
'Fibromas' see foot conditions   102
Fibrosarcoma (jaw)   218
Flavivirus   76
Fleece rot   202
Flock management   245
hill flocks   245
external parasite control   246
foot care   246
internal parasite control   245
nutrition   245
perinatal care   246
vaccination   245
lowland flocks   248
housing   248
internal parasite control   248
vaccination   248
upland flocks   246
external parasite control   247
foot care   247
internal parasite control   247
nutrition   247
perinatal care   248
vaccination   247
Fluorosis   31, 114
Flushing ewes   142
Focal symmetrical
encephalomalacia   35, 36
Follicular conjunctivitis see ovine
keratoconjunctivitis   44

Foot care 246, 247
Foot conditions
  acute laminitis 103
  congenital axial rotation (corkscrew
    claw) 103
  contagious pustular dermatitis
    (orf) 102
  Erysipelothrix rhusiopathiae
    infection 103
  'fibromas' 102
  foreign bodies 101
  granulomas 102
  strawberry foot rot 102
  white line disease 101
Foot-and-mouth disease 33, 220
  cause 220
  clinical signs 220
  diagnosis 221
  distribution 221
  serotypes 221
  transmission and
    epidemiology 221
Foot rot 98
  Bacteroides nodosus 98
  Bacteroides nodosus vaccine 100
  benign (scald) 99
  cause 98
  clinical signs 99
  control 100
  diagnosis 99
  Fusobacterium necrophorum 98
  scald 99
  treatment 100
  virulent 99
Forage mite (*Trombicula
    autumnalis*) 204
Fusobacterium necrophorum 98, 100,
    105, 186
  Mouth disease 34
  neonatal polyarthritis 104
  tick pyaemia 105
Fusobacterium sphaerophorus see
    Fusobacterium necrophorum
Fuzzy lambs see border
    disease 129

Gastrointestinal helminth
    control 250
  combined anthelmintics 254
  drug resistance 254
  fascioliasis 253
  nematodiriasis 250
  parasitic gastroenteritis 251
  taeniasis 253
Ganjam virus complex 209
Geeldikkop see
    photosensitization 197
Gefa-gwan see swayback 82
Gid see coenurosis 93
Giddy dunt see coenurosis 93
Glucosinolates in brassica
    poisoning 231, 233
Glutamic oxaloacetate transaminase

in selenium deficiency 174
Glutathione peroxidase in selenium
    deficiency 173
Goats, scrapie 72, 73
Goggle turn see coenurosis 93
Goitrogens 175
Granulomas see foot conditions 102
Grass staggers 165
'Grass tetany' 165
Grazing management in parasitic
    gastroenteritis 253
Greenbottle (*Lucillia sericata*) 191
Grouse, red in louping-ill 78

Haemolytic anaemia factor 232
Haemonchus spp. 56
Haemophilus agni in arthritis 110
Haemophysalis punctata 204
  Babesia major 205
  distribution 205
  Theileria mutans 205
  tick borne-protozoa 213
Haemorrhagic enteritis 36
  clinical signs 36
  pathology 37
Haemorrhagic fever 234
Hairy shaker disease see border
    disease 129
Headfly myiasis 189
  cause 189
  clinical signs 190
  control and treatment 190
  diagnosis 190
  epidemiology 190
  Hydrotea irritans 189
Heather blindness see ovine
    keratoconjunctivitis 214
Heinz–Ehrlich bodies 231
Helianthrone 197
Helpless lambs see daft lamb
    disease 87
Hepatocellular tumours 218, 219
hepatogenous photosensitization 197
Hernia, ventral 151
  cause 151
  clinical signs 151
  prevention 152
  treatment 152
Herpesvirus in pulmonary
    adenomatosis 12
Hill flock management 245
Hog cholera virus 129
Hoof rot see foot rot 198
Hoose see parasitic bronchitis 23
Husk see parasitic bronchitis 23
Hyalomma spp. 213
Hydrotea irritans 189
Hypericin in photosensitization 197
Hypericum perforatum (St. John's
    Wort) 197
Hypericum spp. 197
Hypervitaminosis D in
    osteodystrophies 114

Hypocalcaemia
  clinical signs 163
  pathogenesis 163
  prevention 164
  treatment 163
Hypoglycaemia in pregnancy
    toxaemia 149
Hypomagnesaemia 165
Hypothermia in lambs
  causes 256
  colostrum 256
  diagnosis 258
  ewe nutrition 256
  heat loss 255
  heat production 255
  Moredun lamb thermometer 258
  Moredun lamb warming
    box 259
  pathology 257
  treatment 258

Ibaraki virus 222
Icterogenin 198
Impotence in ram infertility 141
Incisor loss (broken mouth) 29
Inderbitzen technique 24
Infectious necrotic hepatitis see black
    disease 40
Infectious ovine opthalmia see ovine
    keratoconjunctivitis 244
Infertility
  causes, ewe
    developmental defects 142
    dietary effects 142
    environmental effects 142
    fertilization failure 143
    infectious causes 143
    managemental procedures 142
    neoplasia 143
    ovarian inactivity 142
    pathological conditions 143
    selenium deficiency 143
  causes, ram
    accessory gland infection 140
    conditions of penis and
      prepuce 141
    cryptorchidism 139
    epididymal hypoplasia or
      aplasia 139
    epididymitis 140
    impotence 141
    inflammatory causes 140
    neoplasia 141
    orchitis 140
    seasonal effects 141
    semen properties 141
    spermatocoele 139
    testicular atrophy 140
    testicular hypoplasia 139
    tubular degeneration 140
Inherited cerebellar cortical atrophy
    see daft lamb disease 87
Inorganic poisons 231

Internal parasite control 245, 247, 248
Intestinal adenocarcinoma 218, 219
Investigation of disease 263
Iodine deficiency 175
  cause 175
  clinical signs 175
  diagnosis 175
  goitrogens 175
  nitrate 175
  prevention 175
  tetraiodothyronine 175
  thiocyanate 175
  thiouracil 175
  treatment 176
  triodothyronine 175
Iron deficiency 176
Ixodes ricinus 77, 209
  babesiasis 205
  control 79, 205
  epidemiology 204
  life cycle 204
  louping-ill 77
  tick borne fever 210
  tick pyaemia 105
  'tick worry' 205
  see also tick-associated infections 204

Jaagsiekte see pulmonary adenomatosis 12
Jinkback see swayback 82
Johne's disease 52
  cause 52
  clinical signs 53
  control 55
  diagnosis 54
  epidemiology and transmission 54
  Mycobacterium Johnei 52
  pathology 53
Joint evil see neonatal polyarthritis 104
Joint ill see neonatal polyarthritis 104
  central nervous system infections 89

Kale anaemia 231
Kale feeding in ewe infertility 143
Kebbing see enzootic abortion 119
Keds 204
Keratitis see ovine keratoconjunctivitis
Keratoconjunctivitis see ovine keratoconjunctivitis 214
  in contagious agalactia 157
Ketosis in pregnancy toxaemia 148
Kuru 71

'La Bouhite' see maedi-visna 15
Lactation tetany 165
Lamb dysentery 35, 37
  clinical signs 35

Clostridium perfringens (*welchii*) 35
  pathology 35
Lamb enteritis 43
  causes 44
  Campylobacter spp. 44
  central nervous system infections 89
  Cryptosporidium 45
  diagnosis 46
  Escherichia coli 44
  Rotavirus 45
  Salmonellae 44
  treatment and control 47
  viruses 45
Lambing sickness 163
Lamb hypothermia 255
Laminitis, acute 103
Lantama camara 198
Lantadene in photosensitization 198
'La Tremblante' see scrapie 71
Lead in osteodystrophy 114
Lead poisoning 236
  acute 236
  chronic 236
  diagnosis 237
  treatment 237
Lentivirinae 15
Lice 204
Linognathus ovillus 204
Linognathus pedalis 204
'Lip and leg' ulceration 144
Lippia rehmanni in photosensitization 198
Listeria monocytogenes 79, 214
Listeriosis 79
  abortion 80
    cause 79
    clinical signs 80
    control, prevention and treatment 81
    diagnosis 81
    epidemiology and transmission 81
    Listeria monocytogenes 79
    pathology and pathogenesis 80
    properties 79
  meningoencephalitis 80
    clinical signs 80
    diagnosis 81
    pathology and pathogenesis 80
    and moles (*Talpa europaea*) 81
    silage 81
  septicaemia 80
    clinical signs 80
    diagnosis 81
    pathology and pathogenesis 80
  silage 81
Liver enzymes in photosensitization 199
Liver fluke 62
  acute 63
  chronic 63
  black disease 40, 41, 64

  cause 62
  clinical signs 62
  control 253
  diagnosis 64
  epidemiology 64
  Fasciola hepatica 62
  forecasting systems 65
  Lymnaea truncatula 62
  pathology 63
  strategic dosing 66
  subacute 63
  treatment, control and prevention 65
Liver rot see liver fluke 62
Liver tumours 218
Lobar (enzootic) pneumonia see atypical pneumonia 17
Lockjaw see tetanus 91
Louping-ill 76
  cause 76
  clinical signs 76
  control 78
  diagnosis 77
  epidemiology 77
  Flavivirus 76
  grouse, red 78
  Ixodes ricinus 77
  pathology 76
  vaccination 78
Lowland flock management 248
Lucilia sericata (Greenbottle) 191
Lumpy wool see mycotic dermatitis 200
Lupinosis 237
Lupin poisoning 238
Lymnaea auricularia 66
Lymnaea truncatula 62
  life cycle 62
Lymphosarcoma see tumours 217, 218

Macroelement deficiencies see mineral metabolism 161
Mad lambs see daft lamb disease 87
Maedi-visna 15
  in arthritis 110
  cause 15
Magnesium 165
  clinical signs 165
  dietary source 165
  function 165
  hypomagnesaemia 165
  grass staggers 165
  grass tetany 165
  lactation tetany 165
  magnesium tetany 165
  metabolism 166
  prevention 166
  treatment 167
Magnesium tetany 165
Malignant oedema 42

causes 42
clinical signs 42
diagnosis 42
epidemiology, transmission,
    control, prevention and
    treatment 42
pathology 42
Malocclusion, dental 113
Mammary impetigo 194
cause 194
clinical signs 194
diagnosis 194
epidemiology and
    transmission 194
pathology 194
staphylococci 194
treatment 195
Manganese deficiency 176
Managemental procedures and ewe
    infertility 142
Marchi reaction 83
Mastitis 153
acute, severe 153
    cause 153
    clinical signs 154
    Clostridium perfringens
        (*welchii*) 153
    control 156
    diagnosis 154
    epidemiology 154
    Pasteurella haemolytica 153
    Pasteurella haemolytica (type
        A) 154
    pathology 7, 154
    predisposing factors 153
    Staphylococcus aureus 153
    treatment and prevention 154
    vaccine trials 156
contagious pustular
    dermatitis 186
mild chronic 156
    cause 156
    clinical signs 156
    control 157
    diagnosis 156
    incidence 157
    Pasteurella haemolytica biotype
        A 7
Mating schedules 262
Matrix osteoporosis see
    osteoporosis 111
Melanomas 218
Melophagus ovinus 204
Meningitis, *Pasteurella* 7
Microcystic flos-aquae 198
Midges (*Culicoides* spp.) in Rift Valley
    fever 224
Mineral metabolism 161
calcium and phosphorus 162, 164
general 167
macroelement deficiencies 161
magnesium 165
potassium 161

sodium and chlorine 161
sulphur 162
vitamin D 162
Mineral osteoporosis see
    osteomalacia 111
Miscellaneous parasitic
    pneumonias 25
Mites 204
Molar teeth disease see teeth
    disease 32
Moles (*Talpa europaea*) in
    listeriosis 87
Molybdenum in copper
    deficiency 171
Molybdenum in osteodystrophy 114
Monezia expansa 58
Montana progressive pneumonia see
    maedi-visna 15
Moraxella bovis in ovine
    keratoconjunctivitis 214
Mosquitoes in Rift Valley fever 224
Mouth diseases 29
    actinobacillosis 34
    bacterial infections 34
    bluetongue 33
    contagious pustular dermatitis
        (orf) 34
    Corynebacterium pyogenes 34
    foot-and-mouth disease 33
    Fusobacterium necrophorum
        infection 34
    stomatitis 33
Moredun lamb thermometer 258
Moredun lamb warming box 259
Mosquitoes in Rift Valley fever 224
Muellerius capillaris 25
Mycobacterium johnei 52
properties 52
pigmented strains 54
Mycotic dermatitis 200
cause 200
clinical signs 200
control, prevention and
    treatment 202
Dermatophilus congolensis 200
diagnosis 201
epidemiology and
    transmission 202
lumpy wool 200
pathology 201
Pseudomonas spp. 201
Strawberry foot rot 201
Mycoplasma agalactia in contagious
    agalactia 157
Mycoplasma arginini in atypical
    pneumonia 19
in ovine keratoconjunctivitis 214
Mycoplasma capricolum 110
in arthritis 110
in contagious agalactia 157
Mycoplasma conjunctivae in ovine
    keratoconjunctivitis 214
Mycoplasma mycoides subsp. mycoides

in contagious agalactia 157
Mycoplasma ovipneumoniae 18
in atypical pneumonia 18, 19, 20
in ovine keratoconjunctivitis 214

Nairobi sheep disease 227
Nairobi sheep disease virus 209
Narthrecium ossifragum (bog
    asphodel) 199
Nasal adenopapilloma (nasal
    adenocarcinoma) 219
Navel ill see neonatal
    polyarthritis 104
see CNS infections 189
Necrotic ulcerative dermatitis see
    staphylococcal dermatitis 195
Neisseria ovis or *catarrhalis* see
    Branhamella ovis 214
control 250
Nematodiriasis 250
Nematodirus spp. 58
Neonatal polyarthritis see also
    arthritis 104
Neoplasia (bracken poisoning) 236
in ewe infertility 143
in ram infertility 141
Neurofibromas 218
Neostrongylus linearis 25
Nitrates in brassica poisoning 231,
    233
in iodine deficiency 175
Nutrition of flock 246, 247
Nutritional myopathy see Se
    deficiency 173

Oat cells 4, 5, 6
Occlusion (teeth) 33
Oesophagostomum venulosum 58
Oestrus cycle synchronization 261
normal oestrus cycle 261
manipulation of oestrus cycle 261
progestagens 261, 262
prostaglandins 261, 262
pregnant mares serum
    gonadotrophin 261
mating schedules 262
Oral squamous carcinoma 218
Orchitis 140
Orf see contagious pustular
    dermatitis 185
Osteoarthritis 109
Osteodystrophies
bent leg syndrome 113
calcium 113
cappie and double scalp 113
causes 113
clinical signs 113
copper 114
cruban 113
dental malocclusion 113
diagnosis and control 114
dietary protein 114
epidemiology 113

fluorosis   114
hypervitaminosis D   114
lead   114
molybdenum   114
osteomalacia (mineral
          osteoporosis)   111, 165
osteoporosis (matrix
          osteoporosis)   111, 171
pathogenesis   113
phosphorous   113
stunted growth   112
trace elements   114
treatment   115
Trichostrongylus vitrinus   114
vitamin D   113
Osteomalacia   111, 165
Osteoporosis in copper
          deficiency   171
rickets   111
Ostertagia circumcincta   56
Ostertagia trifurcata   56
Ostertagiasis type I   56
Ostertagiasis type II   56
Ovine adenovirus type 1   45
Ovarian inactivity in ewe
          infertility   142
Ovine viral abortion   119
Ovine chlamydial polyarthritis see
          chlamydial polyarthritis   109
Ovine enzootic abortion   119
Ovine keratoconjunctivitis
     Acholeplasma oculi   214
     Branhamella ovis (*Neisseria ovis* or
          catarrhalis)   214
     causal agents   214
     Chlamydia psittaci   214
     clinical signs   214
     Colesiota conjunctivae   214
     control, prevention and
          treatment   216
     diagnosis   216
     epidemiology and transmission   216
     Listeria monocytogenes   214
     Moraxella bovis   214
     Mycoplasma arginini   214
     Mycoplasma conjunctivae   214
     Mycoplasma ovipneumoniae   214
     pathology   215
     Rickettsia conjunctivae   214

Patelata see swayback   82
Paradontal disease see incisor
          loss   29
Parainfluenza virus type 3 (PI3)   9
     clinical signs   9
     control   10
     diagnosis   9
     epidemiology   9
     glycoproteins   9
     Pasteurella haemolytica   10
     Pasteurella haemolytica biotype
          A   5
     pathology   9

tick borne fever   211
vaccination   10
virus vaccine   5
Parasitic bronchitis   23
     cause   23
     clinical signs   23
     diagnosis   24
     Dictyocaulus filaria   23
     epidemiology and transmission   25
     Inderbitzen technique   24
     life cycle and pathogenesis   23
     treatment and control   25
     vaccination   25
Parasitic gastroenteritis   56
     abomasal parasitism   56
     abomasal parasitism – effects   60
     alternate grazing   252
     anthelmintics   251
     control   59, 251
     damage   59
     epidemiology   58
     grazing management   253
     Haemonchus spp.   56
     host nutrition   61
     management and
          anthelmintics   252
     miscellaneous infections   58
          Bunostonum trigonocephalum   58
          Chabertia ovina   58
          Cooperia curticei   58
          Monezia expansa   58
          Oesophagostomum venulosum   58
          Strongyloides papillosus   58
     Nematodirus spp.   58
     Ostertagia spp.   56
     small intestinal parasitism   57
     small intestinal parasitism –
          effects   60
     Trichostrongylus spp.   57
     type I ostertagiasis   56
     type II ostertagiasis   56
Parasitic pneumonias,
          miscellaneous   25
Parathyroid hormone   163
Paratyphoid see salmonellosis   135
Paratuberculosis see Johne's
          disease   52
Pasteurella
     accessory gland infection   140
     meningitis   7
     ram epididymitis   140
     serotypes   3
     in tick borne fever   24
     Pasteurella haemolytica
          adenoviruses   11
          central nervous system
               infections   89
          mastitis   153
          parainfluenza virus type 3   10
          Pasteurellosis   3
          pulmonary adenomatosis   13
     Pasteurella haemolytica biotype
          A   3, 4

atypical pneumonia   18, 19, 20
characteristic properties   3
clinical signs   4
control, prevention and
          treatment   5
diagnosis   4
epidemiology   5
mastitis   7, 154
oat cells   4
pathology   4
parainfluenza type 3 virus   5
pulmonary adenomatosis   5
vaccine   6
Pasteurella haemolytica biotype
          T   3, 4, 6
     arthritis   7
     characteristic properties   3
     clinical signs   6
     control   7
     diagnosis   7
     epidemiology   7
     oat cells   6
     pathology   6
     vaccination   7
Pasteurella multocida   3
     atypical pneumonia   19
     central nervous system
          infections   89
     Pasteurellosis   3
Pasteurellosis   3
     cause   3
     Pasteurella haemolytica   3
     Pasteurella multocida   3
     pneumonic   3
     systemic   3
Pedero see foot rot   98
Penis and prepuce in ram
          infertility   141
Periorbital eczema see staphylococcal
          dermatitis   195
Perinatal care   246, 248
Periodontal disease see incisor
          loss   29
Peste des petits ruminants   225
     cause   225
     clinical signs   225
     distribution   226
     transmission and
          epidemiology   226
     diagnosis   226
Pestivirus   129
Phenolic poisons   239
     sources   240
     clinical signs   240
     pathology   241
     diagnosis   241
     control and treatment   242
Phenothiazine   198
     in photosensitization   198
     sulphoxide   198
Phormia terrae-novae (black
          Blowfly)   191
Phosmopsis rossiana   238

Phosmopsis leptostromiformis   238
Phosphorus   164
  cappie   165
  deficiency signs   165
  dietary requirements   165
  dietary source   164
  double scalp   165
  function   164
  osteodystrophies   113
  osteomalacia   165
  rickets   165
Photosensitization   197
  bog asphodel (*Narthrecium
      ossifragum*)   199
  cause   197
  congenital porphyria (pink
      tooth)   197
  coproporphyrin   197
  diagnosis and investigation   199
  dimidium bromide   198
  epidemiology   198
  facial eczema   198
  fagopyrin   197
  helianthrone   197
  hepatogenous   197
  hypericin   197
  Hypericum spp.   197
  icterogenin   198
  Lantana camara   198
  lantadene   198
  Lippia rehmanni   198
  liver enzymes   199
  Microcystic flos-aquae   198
  phenothiazine   198
  phenothiazine sulphoxide   198
  phylloerythrin   198
  Polygonum fagopyrum   197
  primary   197
  Pythomyces chartarum   198
  secondary   197
  sporedesmin   198
  Tribulus terrestis   198
  uroporphyrin   197
  'yellowses'   198
Phylloerythrin   198
Pietin see foot rot   98
Pink eye see ovine
    keratoconjunctivitis   214
Pink tooth see congenital
    porphyria   197
Plant poisons   231
Plasma creatinine in pregnancy
    toxaemia   148
Plasma glucorticoid in pregnancy
    toxaemia   148
Plochteach see
    photosensitization   197
Plooks see staphylococcal
    dermatitis   193
Pneumonic pasteurellosis   3
Poisons   231
  basic slag poisoning   239
  bracken fern poisoning   234

brassica poisoning   231
carbolic acid   239
copper poisoning   235
inorganic   231
kale anaemia   231
lead poisoning   236
lupin poisoning   237
lupinosis   238
phenolic compounds   239
plant   231
redwater   231
rhododendron poisoning   233
yew poisoning   233
Polioencephalomalacia   85
  and anthelmintic treatment   86
  autofluorescence   85
  Bacillus thiaminolyticus   86
  cause   85
  clinical signs   85
  Clostridium sporogenes   86
  control, prevention and
      treatment   86
  diagnosis   85
  epidemiology   86
  forages   86
  pathology   85
  vitamin B complex   86
Polygonum fagopyrum   197
Potassium   161
Post-dipping lameness see
    arthritis   108
Pregnancy toxamia   147
  aceto-acetic acid   148
  adrenal response   149
  beta hydroxybutyric acid   148
  biochemical aspects   148
  blood urea   148
  clinical pathology   148
  clinical signs   147
  experimental disease   148
  hypoglycaemia   149
  ketosis   148
  pathology   147
  prevention and treatment   149
  plasma creatinine   148
  plasma glucocorticoid   148
Pregnant mares serum gonadotrophin
    in oestrus
    synchronization   261
Primary photosensitization   197
Progestagens   261, 262
Progressive retinal degeneration   234
Prolapse, vagina   151
Proliferative exudative pneumonia see
    atypical pneumonia   17
Prostaglandins   261, 262
Protein in osteodystrophies   114
Protostrongylus rufescens   25, 26
Prototoxin see clostridial diseases   36,
    38
Protozoa
  Babesia capreoli   213
  Babesia motasi   213

  Babesia major   205
  Theileria mutans   205
  Theileria ovis   213
  tick borne   213
  Toxoplasma   124
Pruritic ectoparasite conditions   203
  clinical signs   204
  control   204
  diagnosis   204
  epidemiology and
    transmission   204
  keds   204
  lice   204
  mites   204
  ticks   204
Pseudomonas pyocyanea in central
    nervous system infections   89
Pseudomonas spp. in mycotic
    dermatitis   201
Psittacosis-lymphogranuloma
    microorganisms see
    Chlamydia   119
Psoroptes communis var
    bovis   181
    cuniculi   182
    ovis   181, 182
Psoroptic mange   181
  cause   181
  clinical signs and
    epidemiology   181
  control   183
  diagnosis   182
  economic effects   182
  history of control   183
  Psoroptes communis var
    bovis   181
  Psoroptes communis var
    cuniculi   181
  Psoroptes communis var *ovis*   181
  resistant strains   184
Pteridium aquilinium poisoning   234
Pulmonary adenomatosis   5, 12
  clinical signs   13
  control and treatment   15
  diagnosis   14
  epidemiology   14
  pathology   13
  Pasteurella haemolytica   13
Pulmonary carcinoma see pulmonary
    adenomatosis   12
Pulpy kidney disease   6, 36, 38
  clinical signs   36
  Clostridium perfringens (*welchii*)   36
  pathology   36
Pyodermas   193
  mammary impetigo   194
  staphylococcal dermatitis   195
  staphylococcal folliculitis   193
Pythomyces chartarum   198

Q fever   209
Quarter evil see blackleg   38
Quarter ill see blackleg   38

Rabies 225
  cause 225
  clinical signs 225
  diagnosis 225
  distribution 225
  transmission and
    epidemiology 225
Rain rot see mycotic dermatitis 200
Ram infertility see infertility 139
Redwater 231
Redwater disease see bacillary
  haemoglobinuria 41
Renguerra see swayback 82
Reoviruses 11
  type 1 in lamb enteritis 45
Respiratory syncytial virus 11
Reticulate body 119
Retrovirus 12, 15
  maedi-visna 15
  pulmonary adenomatosis 12
Rhipicephalus bursa in tick borne
  fever 213
Rhipicephalus haemophysaloides in
  tick borne fever 213
Rhododendron poisoning 233
  andromedotoxin 233
Rickets 111, 165
Rickettsia conjunctivae in ovine
  keratoconjunctivitis 214
Rickettsial conjunctivitis see ovine
  keratoconjunctivitis 214
Rickettsias
  Anaplasma mesaetorum 209
  Anaplasma ovis 209
  Cowdira ruminantum 209
  Coxiella burnetti 209
  Cytocetes phagocytophilia 209
  tick borne 209
Rida see scrapie 71
Rift Valley fever 224
  black flies (*Simulium* spp.) 224
  cause 224
  clinical signs 224
  diagnosis 224
  distribution, transmission and
    epidemiology 224
  midges (*Culicoides* spp.) 224
  mosquitoes 224
Rinderpest 225
  cause 225
  clinical signs 225
  diagnosis 226
  distribution 225
  transmission and
    epidemiology 226
Ringworm 203
  cause 203
  clinical signs 203
  diagnosis 203
  treatment 203
  Tricophyton verrucosum 203
Rotavirus in lamb enteritis 45
Ruminal squamous carcinoma 218

Salmonella 135
  abortion 135
    cause 135
    clinical signs 136
    control, prevention and
      treatment 137
    diagnosis 136
    epidemiology and
      transmission 137
    pathology 136
    Salmonella abortus ovis 136
    Salmonella montevideo 136
    vaccine 138
Salmonella abortus ovis 136
Salmonella dublin 136
  in lamb enteritis 44
Salmonella montevideo 136
Salmonella orienburg in lamb
  enteritis 44
Salmonella typhimurium 136
  in lamb enteritis 44
Salmonellosis 135
  cause 135
  clinical signs 136
  control, prevention and
    treatment 137
  diagnosis 136
  epidemiology and
    transmission 137
  pathology 136
Salt poisoning 162
Sarcoptes scabei 181, 182
Saut see photosensitization 197
Scabby mouth see contagious pustular
  dermatitis 185
Scald 99
Schooley see neonatal
  polyarthritis 104
Scrapie 71
  agents 74
    properties 71, 73
    CHI641 73, 74
    class I, II, III 73
    ME7 73
    SSBP/1 74, 73
  autosomal recessive gene 74
  cause 71
  clinical signs 71
  control 74
  diagnosis 72
  epidemiology and transmission 72
  goats 72, 73
  pathology 72
  'sinc' gene 73
  'sip' gene 73
Sertoli cell tumours 141
Seasonal effects on ram
  infertility 141
Secondary photosensitization 197
Selenium deficiency 173
  cause 173
  clinical signs 173
  creatine phosphokinase 174

  diagnosis 173
  ewe infertility 143
  glutamic oxaloacetate
    transaminase 174
  gluthathione peroxidase 173
  infertility 173
  nutritional myopathy (white muscle
    disease) 173
  pathology 173
  prevention 174
  treatment 174
Semen properties and ram
  infertility 141
Seminomas 141
Shear mouth 33
Sheep pox 226
  cause 226
  clinical signs 226
  diagnosis 227
  distribution 227
  transmission and
    epidemiology 227
Sheep scab see psoroptic mange 181
Sheep scabies see psoroptic
  mange 181
'Silly' lambs see daft lamb
  disease 187
'Sinc' gene 73
'Sip' gene 73
Skin tumours 218
S-methylcysteine sulphoxide (SMCO)
  in brassica poisoning 231,
  232, 233
Small intestinal parasitism 57
Snow blindness see ovine
  keratoconjunctivitis 214
Sodium and chlorine 161
  clinical signs 161
  function 161
  salt poisoning 162
Spermatocoele 139
Sphaeroides necrophorus see
  Fusobacterium necrophorum
  98
Sponges, progestagen 261, 262
Sporedesmin 198
Squamous cell carcinoma 219
Staggers see coenurosis 93
Staphylococcal dermatitis 195
  cause 195
  clinical signs 195
  control and treatment 196
  diagnosis 196
  epidemiology and
    transmission 196
  pathology 195
  staphylococcus aureus 195
Staphylococcal folliculitis 193
  cause 193
  clinical signs 193
  diagnosis 193
  epidemiology and
    transmission 193

pathology   193
staphylococci   193
treatment   194
Staphylococci
accessory gland infection   140
central nervous system
   infections   89
mammary impetigo   194
ram epididymitis   140
staphylococcal dermatitis   195
staphylococcal folliculitis   193
Staphylococcus albus in
   pyodermas   196
Staphylococcus aureus   195
dermatitis   195
mastitis   153
tick pyaemia   105
Steely wool   170
Stiff lamb disease   106,
   109
St. John's Wort in
   photosensitization   197
Stomatitis   33
Strawberry foot rot   102, 185
Dermatophilus congolensis
   185
mycotic dermatitis   200
Streptococci
central nervous system
   infections   89
neonatal polyarthritis   104
ram epididymitis   140
Strike see blowfly myiasis   191
Strongyloides papillosus   58
Struck   36
clinical signs   36
pathology   36
Stunted growth   112
Sturdy water brain see coenurosis
   93
Sulphur   162
copper deficiency   162, 171
deficiency signs   162
dietary sources   162
Superoxide dismutase in copper
   deficiency   171
Swayback   82, 90, 105, 170, 171
cause   82
clinical signs   82
control, prevention and
   treatment   84
copper deficiency   82
diagnosis   83
epidemiology   83
Marchi reaction   83
pathology and pathogenesis
   82
Roberts type   82
Swingback see swayback   82
Swingleback see swayback   82
Symptomatic anthrax see
   blackleg   38
Systemic pasteurellosis   3

Taeniasis control   253
Taenia multiceps   94
life cycle   94
Taenia serialis   94
Tapeworm in coenurosis   94
Taxine   234
Taxus baccata (yew) poisoning   233
'Teasing' ewes   142
Teeth diseases   29
anterior swelling of the
   mandible   32
broken mouth   31
caries   31
cheek teeth disease   32
   cause   33
   clinical signs   32
   pathogenesis   33
   treatment   33
discolouration   31
excessive incisor wear   31
fluorosis   31
holes, pitting and discolouration of
   incisor teeth   31
incisor loss (broken mouth)   30
   cause   30
   clinical signs   30
   pathology   30
molar teeth disease   32
occlusion   33
shear mouth   33
wavy mouth   33
Testicular atrophy in ram
   infertility   140
Testicular hypoplasia   139
Tetanus   91
antitoxin   92
cause   91
clinical signs   92
Clostridium tetani   91
control, prevention and
   treatment   92
diagnosis   92
epidemiology and transmission
   92
toxoid   92
toxin   91
treatment   93
vaccination   92
Tetraiodothyronine   175
Theileria mutans   205
Theileria ovis   213
Thiocyanate   175
Thiouracil   175
Tick   204
Tick associated infections   210
Ganjam virus complex   209
louping ill   76
Nairobi sheep disease virus   209
Tick borne fever   105, 210
   cause   210
   central nervous system
      infections   89, 90
   clinical signs   211

control, prevention and
   treatment   213
Cytoecetes phagocytophilia   210,
   211
definition   210
diagnosis   212
epidemiology   212
haematological changes   211
Hyalomma spp.   213
parainfluenza virus type 3   211
Pasteurella   24
pathology   212
relapses   211
Rhipicephalus bursa   213
*Rhipicephalus
   haemophysaloides*   213
sequelae   211
and tick pyaemia   105
tick borne protozoa   213
Babesia capreoli   213
Babesia major   205
Babesia motasi   213
Haemophysalis punctata   213
Theileria ovis   213
tick pyaemia   105
see arthritis   105
central nervous system
   infections   89
in tick borne fever   211
tick borne rickettsias   209
Anaplasma mesaeterum   209
Anaplasma ovis   209
*Cowdria ruminatum
   (ruminantium)*   209
Coxiella burnetti (burnetii)   209
Cytoecetes phagocytophilia   209
tick borne virus encephalitis
   complex   209
Toxin lethal see clostridial
   diseases   36, 37
Toxoplasma gondii   124
antigens   125
isolation   126
life cycle   124
resistance   125
Toxoplasmosis   124
cause   124
cats   124, 127, 128
clinical signs   125
control, prevention and
   treatment   128
diagnosis   126
epidemiology and
   transmission   127
pathology   125
serology   127
Toxoplasma gondii   124
treatment   128
Trace element deficiencies   168
cobalt deficiency   168
copper deficiency   170
iodine deficiency   175
iron deficiency   176

manganese deficiency 176
selenium deficiency 173
zinc deficiency 176
Trace elements in
    osteodystrophies 114
Traberkrankheit see scrapie 71
Transmissible mink
    encephalopathy 71
Trembling see louping-ill 76
Tribulus terrestis 198
Trichostrongylus spp. 56, 57
Trichostrongylus vitrinus in
    osteodystrophies 114
Tricophyton verrucosum 203
Triodothyronine 175
Trombicula autumnalis (forage
    mite) 204
Trombidiform spp. 204
Tubular degeneration in ram
    infertility 140
Tumours 217
    adenocarcinomas 218
    adenomas 218
    adult sheep 218
    bracken 218, 235
    'causal' factors 218
    cholangicellular 218, 219
    chondrosarcoma 219
    and enzootic bovine leucosis 218
    fibrosarcoma (jaw) 218
    hepatocellular 218, 219
    intestinal adenocarcinoma 218, 219
    lymphosarcoma 217, 218
    melanomas 218
    nasal adenocarcinoma 219
    nasal adenopapilloma 219
    neurofibromas 218
    oral squamous carcinoma 218

ruminal squamous carcinoma 218
skin tumours 218
squamous cell carcinoma 219
young sheep 217
Type I ostertagiasis 56
Type II ostertagiasis 56

Udder impetigo see mammary
    impetigo 194
Ulcerative balanoposthitis 144
    cause 144
    clinical signs 144
    contagious pustular dermatitis
        (orf) 144, 146
    epidemiology 146
Ulcerative balanitis, vulvitis 144
    'lip' and 'leg' ulceration 144
Ulcerative dermatitis 144, 186
Upland flock management 246
Uroporphyrin 197

Vaccination of flock 245, 247, 248
Vaginal prolapse 151
    cause 151
    clinical signs 151
    prevention 152
    treatment 152
Venereal disease 186
Ventral hernia 151
Vibriosis 133
    cause 133
    Campylobacter fetus subsp.
        intestinalis 133
    Campylobacter fetus subsp.
        jejuni 133
    Campylobacter fetus in man 135
    clinical signs 133
    control and prevention 134

diagnosis 134
epidemiology and
    transmission 134
pathology 133
treatment 135
vaccination 135
Vibrion septique see *Clostridium
    septicum* 39
Virus disease diagnosis 266
    laboratory techniques 266
    sample selection 266
    specimen collection 268
    submission of samples 268
Vitamin B complex in
    polioencephalomalacia 86
Vitamin D 162
    D$_2$ 162
    D$_3$ 162
    25-hydroxyvitamin D 162
    24-dihydroxyvitamin D 162
    deficiency signs 162
    hypervitaminosis in
        osteodystrophies 114
    osteodystrophies 113
Vulvitis, ulcerative 144

Wavy mouth 33
White line disease 101
White muscle disease (nutritional
    myopathy) 173

Yellowses see photosensitization
    196, 197, 198
Yew poisoning 233
    taxine 234

Zinc deficiency 176
Zwoegerziekte see maedi-visna 15